Solid State Quantum Chemistry

Solid State Quantum Chemistry

the chemical bond and energy bands in tetrahedral semiconductors

ALEXANDER A. LEVIN

Senior Scientist
Kurnakov Institute of General and Inorganic Chemistry
U.S.S.R. Academy of Sciences, Moscow

Translated from the Russian by
Research Servicing Associates

McGRAW-HILL INTERNATIONAL BOOK COMPANY

New York St. Louis San Francisco Auckland Bogotá
Düsseldorf Johannesburg London Madrid Mexico
Montreal New Delhi Panama Paris São Paulo
Singapore Sydney Tokyo Toronto

SOLID STATE QUANTUM CHEMISTRY: The Chemical Bond and Energy Bands in Tetrahedral Semiconductors

First published in the Russian language under the title *Vvedeniye v kvantovuyu khimiyu tverdogo tela* in 1974 by Khimiya Press, Moscow.

1 2 3 4 5 6 7 8 9 0 M A M A 7 8 3 2 1 0 9 8 7

Translated by Research Servicing Associates. This book was set in Press Roman by Hemisphere Publishing Corporation. The editors were Harriet P. Hudson and Mary A. Phillips; the production supervisor was Rebekah McKinney; and the compositor was Susan Jackson. The printer and binder was The Maple Press Company.

Library of Congress Cataloging in Publication Data

Levin, Aleksandr Aronovich.
Solid state quantum chemistry.

Translation of Vvedenie v kvantovuiu khimiiu tverdogo tela.
Includes bibliographical references and index.
1. Chemical bonds. 2. Semiconductors. I. Title.
QD461.L6413 1976 541'.28 76-41179
ISBN 0-07-037435-X

Contents

Preface

This book is addressed to chemists and physicists, both experts and students, interested in the theory of the chemical bond and primarily to those persons involved with problems dealing with the electron structure of solids. This volume serves as an introduction to the theory of the chemical bond in crystals, discussed for the simplest example of "tetrahedral" semiconductors with the structure of diamond and zinc blende.

This choice of material was determined by two considerations. First, purely covalent crystals with the diamond structure and the similar partially ionic $A^{III}B^{V}$ and $A^{II}B^{VI}$ semiconductors are the simplest class of solids and also are of considerable practical importance. Second, the theory of the chemical bond for other classes of crystals has not yet been developed to an adequate degree, at least in the sense in which it is treated in this book. In the study of chemical bonds interest is concentrated not so much on the electron structure of the crystal as such, but on the dependence of the bonds on the properties of the atoms that form the crystal and on the nature of the atomic interaction. The problem is handled in this way because the central problem of solid state chemistry and physical chemistry has become an investigation of the relationship between the properties of solids and their composition and atomic structure.

With the exception of some experts in solid state chemistry and physical chemistry, most chemists, including experts in quantum chemistry, still feel that the electron structure of crystals has no direct relationship to the usual theory of the chemical bond, which has traditionally been restricted to the structure of molecules. The author, therefore, has tended to emphasize the commonality of the problems and methods of the theory of the chemical bond in solids and quantum molecular chemistry.

The presentation in the book is based on band theory (which is identical to the molecular orbital method for molecules), and the primary emphasis is placed on the dependence of the band structure on given characteristics of the bond. The tendency to reduce the theory of the chemical bond in crystals to the problem of charge distribution, the degree of ionicity, etc., is often observed. Such an approach ultimately weakens the theory since the overwhelming majority of the interesting and important properties of solids are related to their band structure.

The author has attempted to present the material in a more accessible manner, avoiding mathematical complexities. Therefore, in particular, any rigorous discussion of group theory had to be omitted, even though it was impossible to completely avoid the symmetry classification of the electron states.

The first two chapters are introductory in nature. Their purpose is to facilitate the study for those who have had no previous preparation in quantum chemistry or solid state theory. Readers familiar with the theory of the chemical bond as taught in universities need only review the first chapter. Readers also familiar with band theory may prefer turning to the third chapter immediately.

The theory of the chemical bond in solids has still not been adequately developed. In addition, except for one brief text by the author [A. A. Levin, "Kvantovaya khimiya kovalentnykh kristallov" (*The Quantum Chemistry of Covalent Crystals*), Moscow, Znaniye Press, 1970], there are no books specifically devoted to the chemical bond in crystals, and consequently there has not been a successful trial of a "canonical" presentation of the theory. It is clear that this book reflects to a certain degree the author's interests and points of view.

The author is grateful to his teachers, Academician Ya. K. Syrkin and Professor M. Ye. Dyatkina. The author also thanks Professor V. L. Bonch-Bruyevich, whose encouraging support played its role in the writing of this book, and Ye. M. Shustorovich, who read the book before its publication and whose friendly advice contributed to its improvement.

Solid State Quantum Chemistry

Introduction

In any system the chemical bond is created by the valence electrons of those atoms constituting it. When these atoms are separated by an appropriate distance, the state of the valence electrons is altered from their state in the free atoms. Therefore, speaking somewhat schematically, there are two possible, different approaches to studying the chemical bond in crystals, just as in molecules. On the one hand, one can ignore the "atomic" nature of the crystal and can consider it simply as a system of many electrons, moving in a certain external field. In such a formulation, which generally leads to *ab initio* calculations, the problem is worked out with the maximum possible precision in the calculation of the energy levels and wave functions of the electrons of the system. Analytical efforts, devoted to finding the electron (band) structure by means of one of the computational procedures that have been developed to date [the orthogonalized plane wave (OPW) method, the associated plane wave (APW) method, the Green function method, etc.], have a similar goal.

On the other hand, a crystal can be considered as a system consisting of atoms (although with a somewhat altered electron structure), the properties and the interaction behavior of which determine the electron structure of the crystal. This approach—the one most often being considered when one speaks of the "chemical bond"—is necessarily dictated by the experimental data, which indicate the presence of numerous correlations between the physical properties and the chemical composition of crystals and also between the properties of crystals and molecules with bonds of similar nature.

A more thorough examination of the test data indicates that the following four problems should be considered as fundamental to the chemical bond theory.

1. It is known that the electron structure of a crystal is determined not only by its crystalline structure or lattice symmetry, but also to a considerable

extent by the individuality of the atoms constituting the crystal. If, for example, one considers such crystals as silicon and germanium, which are isostructural and have identical symmetry, it is easy to see that they have significantly different electron structures (for example, in terms of the location of the minimum in the conduction band or in terms of the order of the *s* and *p* levels $\Gamma^c_{2'}$ and Γ^c_{15}). At the same time, from an "atomistic" viewpoint the only difference between these crystals is the difference in the atoms forming them. This fact leads to a problem that can be formulated as follows:

an investigation of the relationship between the electron structure of a crystal and the kind of atoms and the nature of the interaction between the atoms.

2. As the experimental data show, the electron structure varies in a regular manner in series of chemically and crystallographically similar substances. In this situation other characteristics, such as the heat of atomization and the atomic ionization potential, also vary in parallel. The second problem follows from a consideration of such correlations:

an investigation of the relationship between the change in the electron structure and the regular variation of other physical and chemical properties.

3. The attempt to establish a point of contact between the theory of the chemical bond in crystals and molecular quantum chemistry leads to the formulation of the third problem.

It is known that the atoms of one element (for example, carbon) on the one hand form crystals (diamond, graphite) and on the other hand enter into the composition of certain molecules (hydrocarbons). In this situation the interatomic distances or the binding energies in the crystals and molecules are often nearly identical.* This attests to the close relationship between the electron structures of both types of systems. The third problem follows from this:

an investigation of the relationship between the electron structure of crystals and the electron structure of molecules.

4. Finally, the fourth problem requires no special commentary:

a quantitative calculation or semiquantitative estimate of the electron structure and the transitions in crystals on the basis of data on the properties of the atoms and on the nature of their interaction.

What are suitable methods to use for studying the relationship between the chemical composition, atomic structure, and the electron structure of a crystal? Before the discussion of this question, we intentionally used the general term "electron structure"; one should, in fact, speak of the "band structure" since band theory (one-electron approximation) is also the desired

*For example, the interatomic C–C distance in diamond agrees within 1% with the C–C distance in paraffins, and the C–C bond-breaking energies are identical within about 98% in diamond and in paraffins.

method* although the arguments in its favor may be different for physicists and chemists. For physicists the situation is generally obvious since the energy band theory at present is the only theory that permits interpreting and calculating the optical and electrophysical properties of crystals by a universal means.

To the chemist the situation can be shown to be more complex since, starting in the 1950s, many investigators have used certain ideas of the valence band (VB) method (see Sec. 1.5.3) rather than the one-electron approximation in solid state chemistry. It should be borne in mind, however, that the capabilities of the VB method are limited, in essence, only by the qualitative description of the chemical bond. It is also known that the need to consider a large number of valence structures leads to serious difficulties when the VB method is used to describe the energy spectrum of even comparatively small molecules. These difficulties increase upon the transition to a solid. This makes the VB method of little use for studying the electron structure of crystals. As a historical precedent, one can cite here the evolution of molecular quantum chemistry, the rapid postwar development of which was caused precisely by the changeover from the VB method to the one-electron approximation in the form of the molecular orbital (MO) method, the concept of which is in full agreement with the basic concept of band theory.

The formulation of these four problems prompts certain further considerations of the method that must be used to study the relationship between the band and atomic structure of a crystal. Methods such as the OPW and APW methods are obviously not suitable for describing the band structure in "atomic" terms (since the basic functions in the form of plane or diverging waves do not directly reflect the electron structure of the atoms). Conversely, the linear combination of atomic orbitals (LCAO) method–the traditional method, incidentally, for molecular quantum chemistry–is very well suited for studying the relationship of the band structure to the kind of atoms and to their interaction with one another. The drawbacks of the LCAO method can be corrected by a semiempirical formulation of the theory, as is well known from molecular quantum chemistry. Here it is best to recall the semiempirical methods used to study organic molecules since the $A^N B^{8-N}$ tetrahedral semiconductors are similar to them in the sense that they have a common structure and a more or less uniform charge distribution. It must be emphasized, however, that regardless of this the semiempirical version by itself is perhaps the most adequate for those problems of chemical bond theory that have been listed above.

Finally, let us point out that for the $A^N B^{8-N}$ class of crystals to be considered (by means of a prior changeover from an atomic function basis to the so-called equivalent orbital basis) the entire theory can be formulated in

*As long as one is speaking of nontransition elements or compounds of nontransition elements.

"analytic" form; explicit formulas can be obtained that describe the dependence of the band structure and band-to-band transitions on the atomic levels and on the parameters characterizing the interaction of the atomic functions of valence-bound atoms. The possibility of such an analytic form of the theory is, of course, a fortuitous circumstance since this makes it possible to investigate in general form the relationship between the atomic and electron structure of crystals.

1

One-Electron Approximation in the Theory of the Chemical Bond

The theory of the chemical bond, as became clear at the end of the 1920s, reduces to an investigation of electrons in molecules and solids by means of quantum mechanics, which came into being shortly before this. Quantum mechanics,* in principle, makes it possible to calculate the properties of any molecule or crystal from the basic equation of this theory—the Schroedinger equation. It is only necessary to take into consideration the attraction of each electron to all the nuclei and the mutual repulsion of each pair of electrons.

However, the rigorous solution of this "many-electron" problem was found to be impossible because of mathematical difficulties. Therefore different approximations are used in practical applications of the quantum mechanics of many-electron systems. One of these—the so-called "one-electron approximation"—now predominates in molecular quantum chemistry and also in the theory of the chemical bond in crystals, to which this book is devoted.

In this and subsequent chapters we will first discuss (or review) briefly the general principles of the one-electron model, as well as its use for describing the chemical bond. As we will see, the use of the one-electron approximation for atoms, molecules, and solids leads, respectively, to the well-known shell theory of the atom, the molecular orbital (MO) method in molecular quantum chemistry, and the band theory of solids.

*References [1, 2] can be recommended for a detailed study of quantum mechanics. For a more concise presentation of the principles of quantum mechanics, see also the textbooks on quantum chemistry cited in the references.

1.1 ONE-ELECTRON DESCRIPTION OF MANY-ELECTRON SYSTEMS

1.1.1 Physical Principles of the One-Electron Model

The concept of the one-electron approximation began with the prequantum mechanics (Bohr) theory of complex atoms, which can be formulated as follows.

The motion of the electrons in a many-electron system is extremely complex in nature since all the electrons interact with one another; consequently, the behavior of any of them is determined by the motion of all the rest of the electrons.

The one-electron model is based on the assumption that the action on a given electron of all the nuclei and all the rest of the electrons of the system can be approximately replaced by the action of some average "effective" field, in which the potential energy of an electron–the so-called "effective one-electron potential"

$$V = V(\mathbf{r}) = V(x, y, z) \tag{1.1}$$

–depends only on the coordinates (x, y, z) of this electron. In this way the investigation of different many-electron systems is reduced to the much simpler investigation of the motion of one electron in fields with different potentials.

We shall not be interested in just any state of the electron in the field (1.1), but only the so-called "stationary" states. In the Bohr model of the atom these states correspond to the stable electron orbits. In fact an electron, like any microparticle, does not move along a trajectory (the wave nature of the electron). Therefore in quantum mechanics the Bohr orbits for the stationary electron states are replaced by one-electron "wave functions"

$$\psi(\mathbf{r}) = \psi(x, y, z) \tag{1.2}$$

defined in the entire three-dimensional space.

In the theory of atoms and in quantum chemistry one-electron functions of the form (1.2) (not necessarily for stationary states) are called "orbitals"–a term that simultaneously expresses the similarity and difference between the Bohr orbits and the functions (1.2).

The stationary states correspond to certain energy levels. Thus the sequences of one-electron orbitals for the stationary electron states in the field (1.1)

$$\psi_1, \psi_2, \ldots, \psi_n, \ldots \tag{1.3}$$

correspond to the sequence of one-electron levels

$$\varepsilon_1, \varepsilon_2, \ldots, \varepsilon_n, \ldots \tag{1.4}$$

or the one-electron "energy spectrum" of the system.

Such a correspondence does not assume that two different functions necessarily correspond to different levels. It may be that several of the functions (1.3) correspond to one and the same energy level. Such a level is called degenerate, and the number of different functions, corresponding to this level, is called the level degeneracy multiplicity (single levels are said to be nondegenerate).

All of this may appear extremely trivial to the reader; however, this is sufficient to understand essentially what it is that nearly all quantum chemists (except those who prefer the valence bond method) and those dealing with the electron structure of crystals are concerned with. They compute the orbitals (1.3) and levels (1.4) for different systems since in the one-electron approximation the sets (1.3) and (1.4) cover everything that can be said about the electron structure and the electron properties of any system. (Sometimes, however, in band structure cases they are generally interested only in finding the levels.)

A system of N electrons is described in the one-electron model simply as an ensemble of particles, each of which is in a certain orbital (1.3); thus the ground (lowest energy) and any of the excited states of the system are represented as shown in Fig. 1.1. The striving for minimum energy forces the electrons to occupy the lowest possible levels; however, in view of the Pauli principle no more than two electrons must be in each orbital (in this case they have opposite spins). Then the optical properties of the system are determined by one-electron transitions from one state (ψ_i) to another (ψ_j), and the energy necessary to remove an electron from some orbital ψ_i (the so-called orbital ionization potential) will be equal to the energy corresponding to the one-electron level ε_i, taken with the opposite sign.*

Whereas the scheme of levels (1.4) gives information about the optical properties of the system (and in the case of a crystal, information about such electrical properties as the electrical conductivity), the form of the orbitals (1.3) makes it possible to draw conclusions concerning the electron density distribution in the system.

This fact is associated with the physical meaning of the wave function. The square of the wave function at some point of space, and more precisely the square of the modulus (the wave function can be complex), is proportional to

*This last statement (within the Hartree-Fock scheme, see Sec. 1.3) constitutes the gist of Koopmans' theorem and appears totally obvious since the energy scale is chosen such that the energy of an electron at infinity is equal to zero and is negative for an electron in a bound state. Let us emphasize, however, that such obvious considerations must be handled with a certain caution since the energy of the transition $\varepsilon_i \to \varepsilon_j$ in the Hartree-Fock scheme is not necessarily equal to the difference in the one-electron energies; this is true for crystals but not for small molecules.

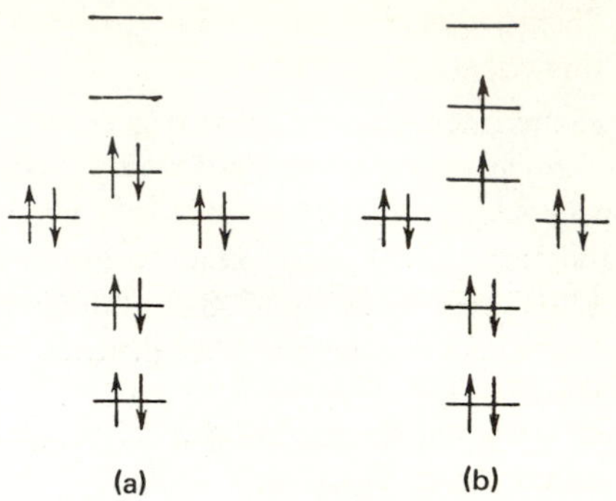

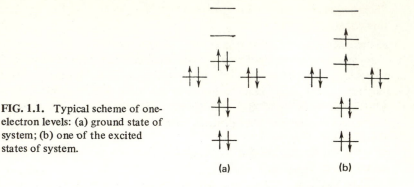

FIG. 1.1. Typical scheme of one-electron levels: (a) ground state of system; (b) one of the excited states of system.

the probability of finding an electron at this point of space. In order that the value of $|\psi|^2$ not only be proportional but also equal to the probability, the wave function ψ must be "normalized," i.e., multiplied by a numerical coefficient such that the equality

$$\int |\psi|^2 \, dv = 1 \tag{1.5}$$

is satisfied (the integral is taken over all space). Then, if the orbitals (1.3) have already been normalized, as is usually done, the value of $|\psi|^2$ for each orbital will give the probability distribution for an electron in the corresponding orbital. This distribution is usually represented in the form of an "electron cloud," the density of which is the corresponding probability. We will now assume, as is assumed in atomic physics and quantum chemistry, the charge of an electron is a unit of charge. In this case the electron density, corresponding to any orbital (1.4), is obviously equal to $|\psi|^2$ when one electron is in an orbital and $2|\psi|^2$ when two electrons are in it. Finally, the resulting electron density for the entire system is defined by the expression

$$\rho(\mathbf{r}) = 2 \sum_{\substack{\text{doubly occupied} \\ \text{orbitals}}} |\psi_i|^2 + \sum_{\substack{\text{singly occupied} \\ \text{orbitals}}} |\psi_j|^2 \tag{1.6}$$

For a system in which all ground state orbitals are doubly occupied, i.e., for a "system with filled shells," the second summation in (1.6) is absent.

1.1.2 One-Electron Schroedinger Equation and One-Electron Hamiltonian

How are the one-electron spectrum (1.3) and orbitals (1.4) found? In the Bohr theory of the atom the quantization rule, postulated by Bohr, is used to find the energy levels and electron orbits. In quantum mechanics the one-electron levels and orbitals are determined from the one-electron

"Schroedinger equation for stationary states." In the atomic system of units*
this equation has the form

$$-\frac{1}{2}\Delta\psi + V\psi = \varepsilon\psi \tag{1.7}$$

Here, $\Delta\psi$ denotes the sum of the second derivatives of the function ψ with respect to the coordinates: $\Delta\psi = \partial^2\psi/\partial x^2 + \partial^2\psi/\partial y^2 + \partial^2\psi/\partial z^2$, and V is the effective one-electron potential (1.1); for any multiatomic system it can be assumed to be known, in principle, as soon as the number and type of atoms are given, as well as their spatial arrangement.

Thus, in the one-electron approximation any problem of the theory of atoms, molecules, or crystals is reduced to that of solving Eq. (1.7) for different specific situations.

Let us only point out the well-known terminology and some of the mathematical properties of the Schroedinger (1.7) and wave functions (1.3), which are used almost as frequently in quantum mechanical calculations as the associative or commutative laws are in arithmetic.

The Schroedinger equation (1.7) is usually written in the abbreviated form

$$\hat{H}\psi = \varepsilon\psi \tag{1.8}$$

Here $\hat{H}\psi$ is the sum of the second derivatives of the function ψ plus the product of ψ by the potential V, so that $\hat{H}$ is referred to as the (one-electron) Hamiltonian operator of the Hamiltonian

$$\hat{H} = -\frac{1}{2}\Delta + V \tag{1.9}$$

which can "operate" on the function ψ, and $\hat{H}\psi$ is referred to as the result of the operation of the Hamiltonian on the wave function.

The fact that the wave functions of stationary states satisfy the Schroedinger equation (1.7) or (1.8) can be expressed in operator terms by saying that the operation of the Hamiltonian on these functions reduces to multiplying them by the value of the energy of the corresponding stationary state.

A function, for which the operation of an operator on it is reduced to a multiplication of it by a number, is called an eigenfunction of this operator, and the corresponding number is called an eigenvalue of the operator. Therefore the wave functions (1.3) are spoken of as the "eigenfunctions" of the Hamiltonian (1.9), and the energies (1.4) are said to be the "eigenvalues" of this Hamiltonian.

*In this system the electron charge e, electron mass m, and Planck's constant $\hbar = h/2\pi$ are adopted as the units.

It is seen from Eq. (1.8) that the eigenvalue of the Hamiltonian is easily found from its eigenfunction. Multiplying both sides of (1.8) by ψ^* and integrating over all space, we have

$$\varepsilon = \frac{\int \psi^* \hat{H} \psi \, dv}{\int \psi^* \psi \, dv} \tag{1.10}$$

(The asterisk here and below denotes the complex conjugate.)

If it is considered that $\hat{H}$ is defined by Eq. (1.9), then it is easy to establish the following two properties for the Hamiltonian:

1. "Linearity," which means that the equality

$$\hat{H}\{c_1\chi_1 + c_2\chi_2 + \cdots\} = c_1\hat{H}\chi_1 + c_2\hat{H}\chi_2 + \cdots \tag{1.11}$$

is satisfied for the arbitrary functions $\chi_1(\mathbf{r}), \chi_2(\mathbf{r}), \ldots$ (not necessarily the solutions of the Schroedinger equation) and the arbitrary numerical coefficients $c_1, c_2, \ldots$. This property is used over and over in quantum chemistry calculations since the wave functions of the electrons in molecules and crystals are nearly always represented in the form of sums of other, simpler functions;

2. "Self-consistency," which means that the relation

$$\int \chi_1^* \hat{H} \chi_2 \, dv = \left\{ \int \chi_2^* \hat{H} \chi_1 \, dv \right\}^* \tag{1.12}$$

is true for any two functions χ_1 and χ_2, or, in another notation

$$\langle \chi_1 | \hat{H} | \chi_2 \rangle = \langle \chi_2 | \hat{H} | \chi_1 \rangle^*$$

where

$$\langle \chi_1 | \hat{H} | \chi_2 \rangle = \int \chi_1^* \hat{H} \chi_2 \, dv \tag{1.13}$$

Thus, if the complex conjugate is ignored, it makes no difference whether $\hat{H}$ inside the integral sign is applied to the first (χ_1) or to the second (χ_2) function.

It follows from the linearity of $\hat{H}$ that the product of the eigenfunction ψ by any number will also be an eigenfunction with the same eigenvalue. Therefore each solution of the Schroedinger equation is defined with an accuracy to an arbitrary numerical coefficient. This means that this solution can always be chosen such that it is normalized. For normalized eigenfunctions all the relations are simplified and, in particular, Eq. (1.10) assumes the form

$$\varepsilon = \int \psi^* \hat{H} \psi \, dv = \langle \psi | \hat{H} | \psi \rangle \tag{1.14}$$

The self-consistency of $\hat{H}$ also leads to an important property of the eigenfunctions: Any two eigenfunctions ψ_1 and ψ_2, corresponding to the different eigenvalues ε_1 and ε_2, are "orthogonal," i.e., the equality

$$\int \psi_1^* \psi_2 \, dv = \langle \psi_1 | \psi_2 \rangle = 0 \tag{1.15}$$

is true for them. Although the proof of this equality is very simple, we will not give it here; the reader is referred to the textbooks on quantum mechanics.

1.1.3 Hartree-Fock Method

A more rigorous approach to the one-electron model is possible, the separate results of which will be useful below. It is given by the "self-consistent field (SCF) method" (see [1, 2] and, for more detail, [3, 4]) or the "Hartree-Fock method,"* and we shall discuss it briefly here, adhering to the well-known paper by Roothaan [7].

In the SCF method, to describe the stationary states of a system of N electrons in an external field one starts from the many-electron Schroedinger equation

$$\left\{-\frac{1}{2}\Delta_1-\frac{1}{2}\Delta_2-\cdots-\frac{1}{2}\Delta_N+V(1)+V(2)+\cdots+V(N)+\frac{1}{2}\sum_{i\neq j}\frac{1}{r_{ij}}\right\}\Psi=E\psi \tag{1.16}$$

which must satisfy the many-electron wave function ψ, depending on all N electrons

$$\Psi=\Psi(1, 2, \ldots, N) \tag{1.17}$$

Here E is the energy of the system, V is the potential energy of an electron in the external field, and the terms $1/r_{ij}$ describe the electron interaction.

The one-electron model arises if it is assumed that all the electrons in the system possess individual one-electron wave functions.†

This assumption is equivalent to the following specific form of the function Ψ. Besides the orbital, let us introduce the concept of "spin orbital" as the wave function of one electron, taking spin into consideration. In this case each orbital ψ corresponds to two spin orbitals—$\psi\alpha$ and $\psi\beta$. Here α is the so-called spin function for an electron with spin "up," and β is the spin function for an electron with spin "down" (it is assumed that the spin state of an electron is independent of the coordinate part of the wave function ψ), and all the electrons occupy different spin orbitals. Then the one-electron approximation is equivalent to the assumption that the many-electron function is written in the form of a determinant composed of the spin orbitals for all the electrons—the so-called Slater determinant.

*For an explanation of the SCF method (applied to atoms) by one of its originators, see Hartree [5]. Fock [6] is responsible for the final formulation of the method.

†This is actually an assumption. In classical mechanics each of the interacting particles must have its own trajectory. In quantum mechanics, however, the individual one-particle functions exist only for noninteracting particles; for interacting particles in the strict sense one can speak only of the wave function of the system as a whole.

Assume, for example, we have a system with filled shells in the ground state (Fig. 1.1a). In this case N will always be even ($N = 2n$) and the system state will be described by n orbitals

$$\psi_1, \psi_2, \ldots, \psi_n \tag{1.18}$$

which correspond to $2n$ spin orbitals:

$$\psi_1\alpha, \psi_1\beta, \psi_2\alpha, \psi_2\beta, \ldots, \psi_n\alpha, \psi_n\beta \tag{1.19}$$

Then Ψ is written in the form of the Slater determinant:

$$\Psi = \frac{1}{\sqrt{N!}} \text{Det} \begin{Vmatrix} \psi_1(1)\alpha(1) & \psi_1(1)\beta(1) \cdots \psi_n(1)\alpha(1) & \psi_n(1)\beta(1) \\ \psi_1(2)\alpha(2) & \psi_1(2)\beta(2) \cdots \psi_n(2)\alpha(2) & \psi_n(2)\beta(2) \\ \cdots & \cdots & \cdots \\ \cdots & \cdots & \cdots \\ \psi_1(N)\alpha(N) & \psi_1(N)\beta(N) \cdots \psi_n(N)\alpha(N) & \psi_n(N)\beta(N) \end{Vmatrix} \tag{1.20}$$

Writing Ψ in determinant form still does not guarantee that the orbitals (1.18) will be the solutions of the one-electron Schroedinger equation (1.7). We impart this form to the equation if we require that the determinant Ψ best approximates the true solution of the many-electron equation (1.16) for the ground state of the system, i.e., that the total energy of all the electrons of the system is a minimum. In this case the orbitals (1.18) can be found from the following Hartree–Fock equations:

$$-\frac{1}{2}\Delta\psi_i(1) + V\psi_i(1) + \sum_{j=1}^{n} 2\int \frac{|\psi_j(2)|^2 \psi_i(1)}{r_{12}} dv_2 - \left\{\int \frac{\psi_j^*(2)\,\psi_i(2)}{r_{12}} dv_2\right\} \psi_j(1) = \varepsilon_i \psi_i(1) \tag{1.21}$$

The equations (1.21) are also often written in operator form [7]:

$$\hat{F}\psi_i = \varepsilon_i \psi_i \tag{1.22}$$

Here the Hartree–Fock operator $\hat{F}$ is the sum of two operators

$$\hat{F} = \hat{h} + \hat{G}$$

where

$$\hat{h} = -\frac{1}{2}\Delta + V; \qquad \hat{G} = \sum_{j=1}^{n} 2\hat{J}_j - \hat{K}_j \tag{1.23}$$

and the operation of the "Coulomb" $\hat{J}_j$ and "exchange" $\hat{K}_j$ operators on the arbitrary function ψ is defined by the relations

$$\hat{J}_j\psi(1) = \int \frac{|\psi_j(2)|^2 \psi(1)}{r_{12}} dv_2$$
$$\hat{K}_j\psi(1) = \left\{\int \frac{\psi_j^*(2)\psi(2)}{r_{12}} dv_2\right\} \psi_j(1) \qquad (1.24)$$

It is seen from Eq. (1.21) that the Hartree-Fock equation for each one-electron orbital is similar to the Schroedinger equation (1.7) with the only difference being that the form of the one-electron effective potential is elaborated here. In the Hartree-Fock scheme the latter has the form $V + \hat{G}$, where V is the potential of an electron in the external field (of a nucleus or nuclei, for example), and $\hat{G}$ is equal to the sum of the Coulomb potential, taking into consideration the repulsion of the other electrons, and the exchange potential, which has no classical interpretation and is caused by the anti-symmetrization of the one-electron functions in the expressions for ψ. The exchange potential is nonlocal in nature; in calculations for atoms and crystals and also in the Slater-Johnson method for molecules it is often replaced by a local potential, for which the Slater [8] approximation $-(3/2)(2\rho/\pi)^{1/3}$ (where ρ is the electron density) or various modifications of the Slater potential are usually used.

As follows from Eq. (1.21) the Coulomb and exchange potentials in the equations for each function ψ_i depend on the solutions ψ_j of all the other equations with $j \neq i$. Therefore the Hartree-Fock equations form a system of interrelated equations that is solved by the successive approximation method (self-consistency).

The eigenvalues ε_i on the right side of Eqs. (1.21) and (1.22) are called the one-electron energies. They play the role of the eigenvalues of the effective one-electron Hamiltonian (1.7) in the "simple" (non-self-consistent) formulation of one-electron theory (see Secs. 1.1.1 and 1.1.2). Within the framework of the Hartree-Fock scheme it can be shown that the energy to remove an electron, described by the function ψ_i, from the system ("the orbital ionization potential") is equal to $-\varepsilon_i$ (Koopman's theorem [4]; see also [7]). Let us note, however, that in the Hartree-Fock approximation the energy E of the system of electrons is not equal to the sum of the one-electron energies.

1.2 MODEL OF ATOMIC SHELLS

1.2.1 Atomic Orbitals

The simplest example of the one-electron model is the atomic shell model, with which everyone is so familiar, that is often used as an exact description of the structure of an atom.

In accordance with the general theory, in the shell model each electron of the atom is considered to be moving in the effective field (1.1) created by the attraction of the nucleus and the repulsion of the other electrons. This field is assumed to be spherically symmetrical: $V(\mathbf{r}) = V(r)$. Then the stationary states of the electrons in the atom correspond to the set of one-electron $1s$, $2s$, $2p$, $3s, \ldots$ levels, and each of these levels corresponds to the appropriate one-electron functions: $1s$, $2s$, $3s$ atomic orbitals (AOs). Let us recall that these notations correspond to the classification of the atomic levels and AOs in terms of the values of the quantum numbers–principal $n = 1, 2, 3, \ldots$ and orbital $l = 0, 1, 2, \ldots$; the value $l = 0$ for any n corresponds to s orbitals, the value $l = 1$ corresponds to the p orbitals, etc. Besides n and l there is the magnetic quantum number m, which assumes values of $-l \leqslant m \leqslant +l$ for each l. Therefore all the s levels are nondegenerate, and each corresponds to a single s orbital having spherical symmetry (Fig. 1.2a). Conversely, all the rest of the levels are degenerate and, for example, each p level corresponds to three p orbitals, differing in spatial orientation and oriented along the three coordinate axes (Fig. 1.2b).

As follows from the general theory (see Sec. 1.3), the atomic orbitals, related to the different energy levels, are orthogonal. Thus the orthogonality of the s and p functions is directly evident from Fig. 1.2 since the product, let us say, $s \cdot p_z$ assumes positive values for $z > 0$ and negative values, but with the same modulus, for $z < 0$, so that the integral of the product of the functions goes to zero. (The orthogonality of the functions with different n and identical l, for example, 1s and 2s AOs, is not evident from the figure, of course; the proof for it must take into account the radial dependence of the AOs.)

The different AOs, belonging to each degenerate level, are also chosen to be mutually orthogonal. In particular, the orthogonality of the p functions is clearly evident from Fig. 1.2. In addition, for convenience all the AOs are normalized, so that we will assume they are normalized in all the calculations.

In accordance with the general scheme of the one-electron model, it is assumed in the atomic shell model that the electrons successively occupy the one-electron states in the order of increasing level energy and in accordance with the Pauli principle. Thus in the shell model the state of the many-electron atom is described by the electron configuration of this atom, where the occupied one-electron levels and the number of electrons in them (the level population) are indicated.

Let us recall that the AOs for degenerate levels can be written in two forms–real and complex. For example, the p functions in Fig. 1.2 are real AOs. In a spherical coordinate system (r, ϑ, φ) they are written in the form:

$$\begin{aligned} p_x &= R_{n,l}(r) \sin\vartheta \cos\varphi \\ p_y &= R_{n,l}(r) \sin\vartheta \sin\varphi \\ p_z &= R_{n,l}(r) \cos\vartheta \end{aligned} \tag{1.25}$$

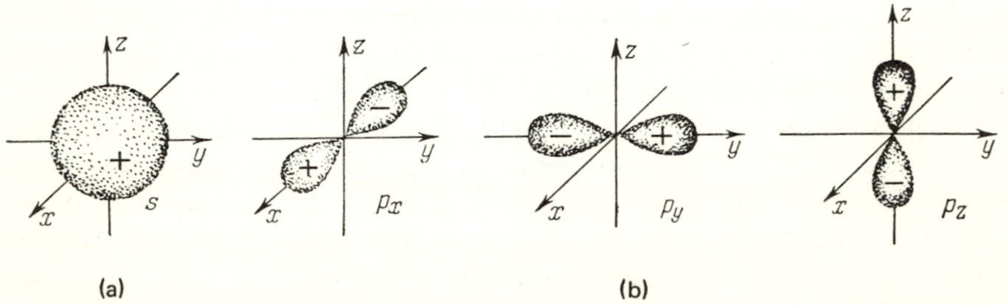

FIG. 1.2. Atomic s orbitals (a) and p orbitals (b).

where $R_{n,l}$ is the radical part of the AO, depending on the principal and orbital quantum numbers. Real functions, generally speaking, do not correspond to defined values of the magnetic quantum number m. Thus the functions p_x and p_y simultaneously correspond to $m = +1$ and $m = -1$. Complex AOs are linear combinations of real and vice versa. Thus, the p AOs in complex form are:

$$p_{-1}=\left(\frac{1}{\sqrt{2}}\right)\{p_x-ip_y\}\sim R_{n,l}\sin\vartheta\exp(-i\varphi)$$

$$p_{+1}=\left(\frac{1}{\sqrt{2}}\right)\{p_x+ip_y\}\sim R_{n,l}\sin\vartheta\exp(+i\varphi) \qquad (1.26)$$

$$p_0=p_z=R_{n,l}\cos\vartheta$$

The complex AOs correspond to certain m values and differ from the real in their transformation properties with rotations about the z axis.

1.2.2 Finding the Atomic Levels and Atomic Functions

As already mentioned, the energy of each atomic level depends on the state of all electrons. The same must be said with respect to the form of each AO, or, more precisely, its radial part $R_{n,l}$; because of the spherical symmetry of the potential, the angular part has a standard form for all atoms. Thus finding the one-electron levels of atoms and ions as well as the radial parts of the corresponding AOs is a complicated procedure (except for the trivial case of a hydrogen atom and hydrogenlike ions–in this case the levels and the radial parts of the AOs are expressed by explicit formulas).

Corresponding calculations are performed by the SCF method. The levels and AOs calculated by this method are given in tables in [9, 10]. For simplicity, in practice, functions are also used that have radial parts specified in analytic form: in the form of one exponent $ae^{-\alpha r}$ or the sum of several such exponents $a_1\exp(-\alpha_1 r) + a_2\exp(-\alpha_2 r) + \ldots$, where the coefficients and exponents are chosen such that the best possible approximation to the radial parts of the self-consistent functions is achieved [11, 12]. The Slater radial functions

$$R(r)=Ar^{(n^*-1)}\exp(-\alpha r) \qquad (1.27)$$

in which the parameters n^* and α are calculated from simple rules (see, for example, [3]) are used quite often.

Another possibility exists besides these "nonempirical" calculations for finding the one-electron levels of an atom. Since the energy of the one-electron level determines the energy to remove an electron from an atom, the one-electron levels can be determined from experimental data. (Let us emphasize that here we are speaking not only of the energy to remove an electron from an upper level–the usual ionization potential–but also the energy to remove an electron from any level: the so-called orbital ionization potentials.)

Such an empirical method is often used in quantum chemistry and is most suitable for our purposes. Since the problem involves, in particular, the fact that the actual energy band structure is related to the actual atomic levels, the most natural way to solve this problem is to take the atomic level energies from experimental data.

We will give only certain illustrations of the "empirical" procedure of finding the atomic levels since the orbital ionization potentials cannot always be taken directly from atomic spectra tables [13].

The orbital ionization potential of an atom is equal to the energy difference of this atom and the ion whose electron configuration differs from that of the atom by the absence of one electron at the appropriate level. In the general case, however, the electron configuration of the atom (ion) does not completely determine its energy. Each configuration corresponds to several different levels–terms* (separated from one another by an appreciable distance of about 1 eV).

The terms no longer characterize the one-electron states of the atom, but rather the state of the atom as a whole, and the very fact of their existence attests to the approximate nature of the one-electron model. For not too heavy atoms the terms are classified in terms of the values of the total orbital angular momentum L and the total spin S of the electrons (the so-called Russell–Saunders scheme). Thus, for group IV elements with an ns^2np^2 valence electron configuration three terms are possible: ${}^1S$, ${}^3P$, and ${}^1D$, corresponding to values of $L = 0$, $S = 0$, $L = 1$, $S = 1$, $L = 2$, $S = 0$, respectively. (The letters $S, P, D, F, \ldots$ denote states with $L = 0, 1, 2, 3, \ldots$.) Here, the numbers standing at the upper left of the letter symbol are equal to the value of $2S + 1$ and correspond to the so-called "term multiplicity." Each term ${}^{2S+1}L$ consists of $2S + 1$ (for $L \geqslant S$) or $2L + 1$ (for $L < S$) closely spaced components–the fine structure levels (at separations of the order of 0.1 eV). These components correspond to different methods of the relative orientation of the total spin and total orbital momenta of the system. In addition, each level of the fine structure is degenerate–$(2L + 1)$-fold (for $L \geqslant S$) or $(2S + 1)$-fold (for $L < S$)–so that the total number of levels in the term is equal to $(2S + 1)(2L + 1)$ with their multiplicity taken into consideration. Thus, for example, for group IV atoms the ${}^1S$ term contains one fine structure level and is nondegenerate; the ${}^3P$ term consists of three fine structure levels, each of which is triply degenerate; and the ${}^1D$ term contains one fivefold degenerate fine structure level.

However, the presence of the atomic terms is not "included" within the framework of the MO theory or the band theory of solids. In these theories the one-electron orbitals of the system are constructed from atomic orbitals, so that the MO energies or the "Bloch function" energies in solid state theory must be expressed in terms of the one-electron atomic levels. Therefore here it is necessary to determine what these levels, i.e., the energies of an atom (ion) with a given electron configuration, must mean.

This can be done since the separation between terms of one configuration is, as a rule, much less than the average separation between terms for different configurations. Then, according to Slater [14], it can be assumed approximately that the energy of an atom (ion) with a given configuration is equal to its average energy, where the averaging is done over all terms within the configuration limits, and each fine structure level is taken into account as many times as its multiplicity. Thus by the energy of a carbon atom with

*The theory of terms or, as it is called, "atomic multiplets" is explained in considerable detail in [3, 4] and in less detail in [1, 16].

the configuration $1s^2 2s^2 2p^2$ we mean the average energy of this atom in the ^1S, ^3P, and ^1D states, i.e., the quantity

$$\langle E\,(\mathrm{C}s^2p^2)\rangle = \frac{1}{15}\{E\,(^1S)+9E\,(^3P)+5E\,(^1D)\} \tag{1.28}$$

where, in turn, by $E(^{2S+1}L)$ we mean the average energy of the atom (ion) in the ^{2S+1}L state:

$$E\,(^{2S+1}L)=\begin{cases}\dfrac{1}{2S+1}(E_1+E_2+\cdots+E_{2S+1}), & L\geqslant S\\[2mm] \dfrac{1}{2L+1}(E_1+E_2+\cdots+E_{2L+1}), & L<S\end{cases} \tag{1.29}$$

and E_i is the energy of an atom in the state corresponding to the ith fine structure level.

In this same way one can determine the energy of an atom (ion) in the excited state. Thus, for a C^{1+} carbon atom in the excited $1s^2 2s2p^2$ state we have the terms ^2S, ^2P, ^4P, and ^2D. Correspondingly, by the energy of the C^{1+} ion in this state we mean the average quantity

$$\langle E\,(\mathrm{C}^{1+}2s2p^2)\rangle = \frac{1}{30}\{2E\,(^2S)+6E\,(^2P)+12E\,(^4P)+10E\,(^2D)\} \tag{1.30}$$

The question of finding the energy and orbital ionization potentials for the atoms and ions in which we are interested will be considered in more detail in Sec. 2.4.1.

1.3 LCAO METHOD

1.3.1 Multiatomic Systems in One-Electron Theory and the Concept of the LCAO Method

In the previous section we spoke of the application of the one-electron model in atomic theory. The modern theory of the chemical bond—the molecular orbital (MO) method* and the band theory of solids—is an application of this same model to multiatomic systems. Such an application usually presumes two additional assumptions.

1. Adiabatic approximation.† Strictly speaking, in a multiatomic system the relative motion of the nuclei must be taken into consideration along with the electron motion. Nuclei, however, are ~10^3-10^5 times heavier than the electrons. Therefore they move much slower, so that in studies of the motion of the electrons the nucleus can be considered stationary in most cases. This is

*See references [15-17] for a detailed description of the MO method; the MO method was first suggested by Hund (1927) and developed by Millikan (1929). The principles of band theory were set forth at about this same time, primarily by Bloch (1928) and Brillouin (1930).

†See, for example, [17] for a more detailed justification of the adiabatic approximation. See [18] for a discussion of deviations from adiabaticity.

called the adiabatic approximation or the Born–Oppenheimer approximation.

2. Valence approximation. It is assumed (and verified by experiment) that not all of the electrons of the atoms in a molecule or crystal play an important role in the bond, but only the valence electrons.

Thus in the adiabatic and valence approximations the main problem in the theory of the chemical bond is finding the one-electron levels and the one-electron orbitals of the system by solving the Schroedinger equation (1.7). Here the effective one-electron potential (1.1), acting on each (valence) electron, is assumed to be the sum of the potentials of the atomic cores plus the resulting potential of all the rest of the valence electrons of the system. By the solution of Eq. (1.7) we mean, of course, an approximate solution since the potential (1.1) for a molecule or crystal is even more complicated than for an atom because of the presence of many attracting centers–nuclei or atomic cores.

This multicenter nature of the problem suggests, however, a certain method of solving Eq. (1.7). It is natural to assume that in the vicinity of any given atom the potential V of the entire system is more or less close to the potential of this atom. Then it is logical to assume that each solution of the Schroedinger equation for the entire system in the vicinity of a given atom is also close to the solution of the equation for this atom, i.e., to some of its AOs. In principle, of course, one can suggest many methods of obtaining one-electron functions of a similar form. However, a simple and convenient, from a mathematical viewpoint, form of this function is a linear combination of the AOs of all the atoms entering into the system. It is known that the atomic functions decrease very rapidly (exponentially) with distance from the nucleus. Therefore in the vicinity of each atom the contribution of the AOs of "foreign" atoms is small and any function of this form reduces primarily to the AO of the atom "itself" only.

This method of solving the Schroedinger equation for a multiatomic system by expanding the eigenfunctions of the effective one-electron Hamiltonian in a sum of AOs is called the linear combination of atomic orbitals (LCAO) method. In the case of molecular systems it is called the MO LCAO method; in the case of crystals it is also often called the tight-binding method. This LCAO method is an especially suitable method of studying chemical bond problems, at least in the sense in which they were formulated in the introduction. Actually, the multiatomic system is considered here to consist of atoms (and not simply of electrons and nuclei). This point of view is directly reflected in the mathematical tools of the theory in which the solutions of the Schroedinger equation are also considered as consisting of atomic functions.

From the mathematical point of view, however, the LCAO method is only a special case of a class of methods, using the expansion of the eigenfunctions in terms of simpler "basic" functions. In the study of crystals, in addition to the LCAO expansion, other expansions are also used continually (see Secs. 2.4.1–2.4.6 for more details), and recently similar methods have been used to study molecules (see, for example, [19]).

1.3.2 Eigenfunctions and the Secular Equation

Now let us turn to a more specific examination of the LCAO method. Let us assume we have a system consisting of an arbitrary number (two, three and it can be 10^{23}) atoms, each of which can have any number of AOs. Let us assume, moreover, that

$$\chi_1, \chi_2, \ldots, \chi_M \tag{1.31}$$

will be the atomic orbitals of all these atoms, renumbered in an arbitrary sequence from 1 to M. Then in the LCAO method any solution of the Schroedinger equation for our system is written in the form

$$\psi = c_1\chi_1 + c_2\chi_2 + \cdots + c_M\chi_M \tag{1.32}$$

where $c_1, c_2, \ldots, c_M$ are unknown coefficients, the specific values of which define a given solution of the Schroedinger equation (1.7).

In order to find these coefficients, let us substitute the expansion (1.32) in Eq. (1.7) or, what is the same thing, in Eq. (1.8). Then, using the linearity of the operator $\hat{H}$, we have

$$\hat{H}\psi = c_1\hat{H}\chi_1 + c_2\hat{H}\chi_2 + \cdots + c_M\hat{H}\chi_M = \varepsilon\,(c_1\chi_1 + c_2\chi_2 + \cdots + c_M\chi_M) \tag{1.33}$$

from which, taking all terms to the left side we obtain

$$c_1(\hat{H} - \varepsilon)\chi_1 + c_2(\hat{H} - \varepsilon)\chi_2 + \cdots + c_M(\hat{H} - \varepsilon)\chi_M = 0 \tag{1.34}$$

We shall successively multiply the relation (1.34) by the AOs (1.31), integrating each resulting product over all space [if the AOs are taken in complex form, then (1.34) is multiplied by functions that are the complex conjugates of the AOs (1.31)].

For simplicity we shall assume here that the AOs of adjacent atoms overlap only very slightly, so that the corresponding overlap integrals

$$S_{ij} = \langle \chi_i \mid \chi_j \rangle = \int \chi_i^* \chi_j \, dv \tag{1.35}$$

are small* and these AOs are approximately orthogonal. (The different AOs

*This approximation was continuously used in the "simple" non-self-consistent MO method (Hückel method [15, 16]. Later it was introduced, under the name "zero differential overlap approximation," within the framework of the semiempirical SCF method for π-electron systems [20], and then it was extended to all systems [21]. See [22] for a justification of this approximation for small values of the overlap integrals. Let us also point out that regardless of this justification for the semiempirical form of the theory the question of taking into consideration the overlap integrals is not particularly important since the empirical selection of parameters compensates the absence of these integrals.

of one and the same atom are always orthogonal in the strict sense.) Then the relation (1.34) leads to a system of linear homogeneous algebraic equations:

$$
\begin{aligned}
(H_{11}-\varepsilon)\, c_1+H_{12}c_2+\cdots+H_{1M}c_M&=0\\
H_{21}c_1+(H_{22}-\varepsilon)\, c_2+\cdots+H_{2M}c_M&=0\\
\cdots\cdots\cdots\cdots\cdots\cdots\cdots&\\
H_{M}c_1+H_{M_2}c_2+\cdots+(H_{MM}-\varepsilon)\, c_M&=0
\end{aligned}
\tag{1.36}
$$

where the H_{ij} denote the matrix elements of the effective electron Hamiltonian $\hat{H}$ of the system in the AO basis:

$$H_{ij}=\langle \chi_i \,|\, \hat{H} \,|\, \chi_j \rangle = \int \chi_i^* \hat{H} \chi_j \, dv; \quad H_{ij}=H_{ji}^* \tag{1.37}$$

They are expressed in terms of the known Hamiltonian $\hat{H}$ of the system and in terms of the AOs (1.31) that are also known. [The equality $H_{ij} = H_{ji}^*$ is a special case of Eq. (1.12) and follows from the self-consistency of the Hamiltonian.]

Thus the unknown coefficients c_1, $c_2, \ldots$ in the expansion of the eigenfunction (1.32) will be the solutions of the system of equations with known coefficients and can be determined if the unknown eigenvalues ε are found beforehand. This is also easy to do. The solvability condition of the system of homogeneous linear algebraic equations (1.36) is that its determinant become zero:

$$\text{Det}\begin{Vmatrix} H_{11}-\varepsilon & H_{12} & \cdots & H_{1M} \\ H_{21} & H_{22}-\varepsilon & \cdots & H_{2M} \\ \cdots & \cdots & \cdots & \cdots \\ H_{M1} & H_{M2} & \cdots & H_{MM}-\varepsilon \end{Vmatrix}=0 \tag{1.38}$$

(The determinant $\text{Det}\|H_{ij} - \varepsilon\vartheta_{ij}\|$ is called the secular determinant.)

The corresponding secular equation (1.38) is obviously an Mth degree algebraic equation. It has M roots: ε_1, $\varepsilon_2, \ldots, \varepsilon_M$, which give the M possible one-electron levels of our system. Substituting each of these roots in turn into the system (1.36) and determining from it the unknown coefficients c_1, $c_2, \ldots, c_M$, we obtain each eigenfunction corresponding to this level, so that finally the secular equation (1.38) in combination with the system (1.36) gives the set of desired one-electron levels together with the corresponding one-electron functions.

Let us only point out that (as follows from the procedure of solving a system of linear homogeneous algebraic equations) the solutions of the Schroedinger equation obtained will not be normalized and it is necessary to normalize them by multiplying by an appropriate normalizing factor, equal to the quantity

$$A=\{|c_1|^2+|c_2|^2+\cdots+|c_M|^2\}^{-1/2}$$

for each eigenfunction (1.32).

It is clear that the number M of basic AOs (1.31) cannot be less than one-half the number of electrons $N/2$ in the system, and in all the cases of interest to us $M > N/2$.* Therefore the total number of levels, obtained from the secular equation (1.38), is always greater than the number of electron pairs that occupy these levels, starting with the lowest. The remaining empty levels correspond to the one-electron excited states of the system, and the filled correspond to its ground state.

Let us make only one comment concerning the physical meaning of the roots of the secular equation (1.38), corresponding to the excited states of the system.

The somewhat intuitive concept of the effective one-electron Hamiltonian assumes a more precise meaning in the Hartree–Fock method, which is called the Roothaan method in the LCAO approximation [7]. As stated in Sec. 1.1.3, in the Hartree–Fock method the operator $\hat{F}$ plays the role of the operator $\hat{H}$. Let us assume the expansion, in terms of AOs, of the one-electron eigenfunctions of the ground state of the system, entering into the Slater determinant, has the form

$$\psi_i=\sum_{j=1}^{M} c_{ji}\chi_j \tag{1.39}$$

and let us assume the potentials of the atomic nuclei (or cores) will be V_A. Then in the LCAO approximation the matrix elements of the operator $\hat{F}$ in the AO basis

$$F_{ij}=h_{ij}+G_{ij} \tag{1.40}$$

have the form

$$h_{ij}=\langle \chi_i|-\frac{1}{2}\Delta+\sum_A V_A|\chi_j\rangle \tag{1.41}$$

$$G_{ij}=\sum_{k,\,l=1}^{M} P_{kl}\left\{\left(ij\left|\frac{1}{r_{12}}\right|kl\right)-\frac{1}{2}\left(il\left|\frac{1}{r_{12}}\right|jk\right)\right\} \tag{1.42}$$

where

$$\left(ij\left|\frac{1}{r_{12}}\right|kl\right)=\int\frac{\chi_i^*(1)\,\chi_j(1)\,\chi_k^*(2)\,\chi_l(2)}{r_{12}}\,dv_1\,dv_2 \tag{1.43}$$

*The equality $M = N/2$ corresponds to the case when all the AOs, included in the basis (1.31), are filled with electrons in the free atoms. This case corresponds to atoms with filled shells (inert gases), which generally do not interact with one another chemically.

and P_{kl} is the "population matrix," the elements of which are expressed in terms of the coefficients for the AOs in the eigenfunctions (1.39) in the following manner:

$$P_{kl} = 2 \sum_{i=1}^{n} c_{ki}^{*} c_{li} \tag{1.44}$$

In this case the one-electron levels ε_i are determined as before by the roots of the secular equation

$$\text{Det} \, \| F_{ij} - \varepsilon \delta_{ij} \| = 0 \tag{1.45}$$

However, the difference in the one-electron energies here is not equal to the actual energy expenditure when one electron makes a transition from a doubly filled level ε_i to an unoccupied level ε_j. This difference is expressed in the general case by the formula

$$\Delta E = \varepsilon_j - \varepsilon_i - \{(J_{ij} - K_{ij}) \pm K_{ij}\} \tag{1.46}$$

The plus sign here refers to a singlet excited state; the minus sign refers to the triplet; and the quantities J_{ij} and K_{ij} (Coulomb and exchange integrals) are determined by the formulas

$$J_{ij} = \int \frac{|\psi_i(1)|^2 \, |\psi_j(2)|^2}{r_{12}} \, dv_1 \, dv_2 \tag{1.47}$$

$$K_{ij} = \int \frac{\psi_i^*(1) \, \psi_j(1) \, \psi_i^*(2) \, \psi_j(2)}{r_{12}} \, dv_1 \, dv_2 \tag{1.48}$$

where ψ_i, ψ_j are the eigenfunctions of the operator $\hat{F}$. Nevertheless, the term in the braces in (1.46) must be considered only for molecules; it goes to zero for crystals, so that for the latter the excitation energy is equal to the difference of the one-electron energies (see also Sec. 2.3.2).

1.3.3 Matrix Elements in LCAO Method

Let us now study in more detail the matrix elements H_{11}, $H_{12}, \ldots$ in the secular determinant (1.38). Since the secular equation (1.38), with the exception of the unknown ε, contains only the matrix elements H_{ij}, $ij = 1, 2, \ldots, M$ in the final analysis they completely define the possible energy levels and the wave functions of the system.

The matrix elements H_{ij} are divided into two types. The first type of matrix elements lies on the principal diagonal of the secular determinant, joining the upper left corner with the lower right. These "diagonal" matrix elements have the form H_{ii} and are defined by the equations (1.37) for $i = j$. They are called the Coulomb integrals and are usually denoted by α_i. Thus, the ith Coulomb integral is equal to the average value of the Hamiltonian of the system in the ith atomic orbital

$$\alpha_i = H_{ii} = \langle \chi_i | \hat{H} | \chi_i \rangle \tag{1.49}$$

In order to clarify the physical meaning of the Coulomb integral, let us separate the potential V of the system into the sum of the potentials of the individual atoms. Then the α_i can be rewritten in the form

$$\alpha_i = \langle \chi_i | -\frac{1}{2}\Delta + V_0 | \chi_i \rangle + \langle \chi_i | \sum_A{}' V_A | \chi_i \rangle \tag{1.50}$$

where V_0 is the potential of that atom to which the orbital χ_i belongs, and $\Sigma'_A V_A$ is the sum of the potentials of all the rest of the atoms of the molecule or crystal.

In this case, as Eq. (1.50) shows, a second kind of situation can arise. For purely covalent systems or systems with a very small effective charge on the atoms the attraction action of the atomic cores of "foreign" atoms is neutralized by the repulsion of the electrons of these atoms. In such a case the potentials of all the atoms have a short-range interaction, the second term in Eq. (1.50) is small, and consequently, the Coulomb integral α_i has approximately the same value it would have for an isolated atom:

$$\alpha_i = \langle \chi_i | -\frac{1}{2}\Delta + V_0 | \chi_i \rangle = \varepsilon_{aT} \tag{1.51}$$

where ε_{at} is the energy of the atomic level to which the function χ_i belongs.

Conversely, in systems with a marked degree of ionicity "foreign atoms will not be neutral, and their potential is a long-range interaction Coulomb potential. In the case of interest to us–in a crystal–these ions are located at lattice sites in an ordered manner and their joint effect on an electron of a given atom can be taken into consideration by replacing, as is usually done, the potential $\Sigma'_A V_A$ by the Madelung potential of the lattice.

Moreover, the first term in Eq. (1.50) can also be interpreted as the energy of an electron in an isolated atom (ion) in the case of a partially ionic system, although, of course, it is necessary to take into consideration here that the energy levels of the ion differ from the energy levels of the neutral atom (see Sec. 5.3.1).

Thus, the diagonal matrix elements in the LCAO method describe the dependence of the energy spectrum of the system on the energy levels of the individual atoms or ions with the total field of the lattice also taken into consideration when necessary.

The nondiagonal matrix elements H_{ij} with $i \neq j$ have a different physical meaning. These matrix elements, in turn, can also be divided into two classes, with the first including elements of the form $H_{ij} = \langle \chi_i | \hat{H} | \chi_j \rangle$, where χ_i and χ_j are different atomic functions of the same atom, say the s and p_x orbitals or the p_x and p_y orbitals. Let us write these matrix elements in the form

$$\langle \chi_i | \hat{H} | \chi_j \rangle = \langle \chi_i | -\frac{1}{2}\Delta + V_0 | \chi_j \rangle + \langle \chi_i | \sum_A{}' V_A | \chi_j \rangle \tag{1.52}$$

similar to (1.50). Then the first term in (1.52) is equal to zero because of the orthogonality of the different AOs of one atom. The second term does not go to zero in the general case (although it will be small). However, for crystals with a tetrahedral or octahedral symmetry of the potential (and we will consider only such crystals) the second term is also equal to zero at the lattice sites, so that these elements H_{ij} are of absolutely no interest.

The nondiagonal matrix elements $H_{ij} = \langle \chi_i | \hat{H} | \chi_j \rangle$, where χ_i and χ_j are the AOs of different atoms, constitute a second class. These matrix elements are denoted by β_{ij}:

$$\beta_{ij} = H_{ij} = \langle \chi_i | \hat{H} | \chi_j \rangle \tag{1.53}$$

They are called resonance integrals or the interaction integrals of the atomic orbitals (AOs χ_i and χ_j) and describe the interaction of the AOs of different atom pairs in a molecule or crystal.

If this interaction is "excluded" and all the resonance integrals are artifically set equal to zero, then the secular determinant (1.38) assumes the form $\mathrm{Det}\|H_{ii}\delta_{ij} - \varepsilon\delta_{ij}\| = 0$. In this case the roots of Eq. (1.38) will be equal to the energy levels of the individual atoms (ions). Thus the resonance integrals in the LCAO method cause a difference in the energy spectrum of the system from the simple sum of atomic levels.

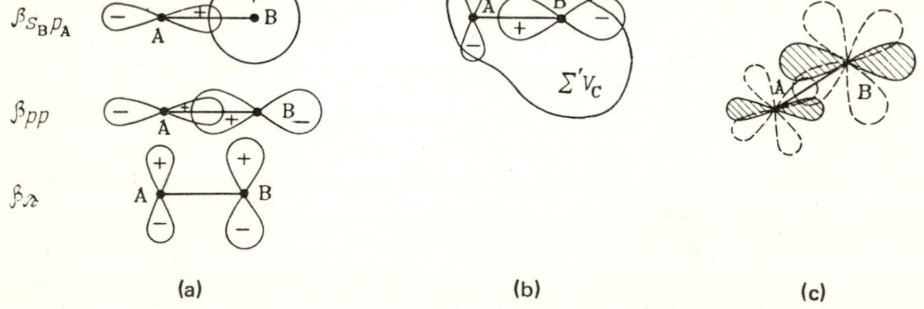

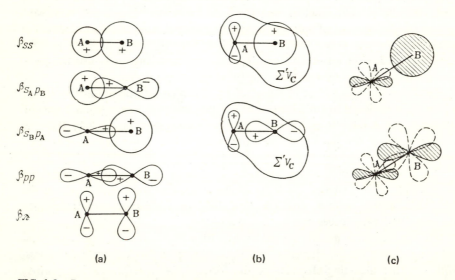

FIG. 1.3. Resonance integrals for system consisting of atoms with s and p orbitals.

It is not hard to see that for each pair of atoms A, B there is always a finite, and often small, number of different interaction integrals. Thus if both atoms have only s orbitals, then for them there exists only one resonance integral, which we denote by β_{ss}. If the atoms A and B have both s and p orbitals, then five resonance integrals are possible: β_{ss}, β_{pp}, $\beta_{s_A p_B}$, $\beta_{s_B p_A}$, and β_{π} (Fig. 1.3a). In fact, let us again resort to the expansion of the system potential V in the sum $V = V_A + V_B + \Sigma'_C V_C$, where V_A and V_B are the potentials of a given pair of atoms A, B and $\Sigma'_C V_C$ is the potential of all the other atoms. Then it is easy to see that for the resonance integrals of the type depicted in Fig. 1.3b that are possible at first glance the term $\langle\chi_i|-\frac{1}{2}\Delta + V_A + V_B|\chi_j\rangle$ is exactly equal to zero, and the term $\langle\chi_i|\Sigma'_C V_C|\chi_j\rangle$ is very small (for the crystals considered below they also become exactly zero). At the same time the resonance integrals of the type depicted in Fig. 1.3c can always be reduced to the five integrals considered by expanding each of the atomic p functions in a sum of other p functions, oriented parallel and perpendicular to the line A–B.

Let us also point out that the type of interaction of the atomic functions, described by the integrals β_{ss}, $\beta_{s_A p_B}$, $\beta_{s_B p_A}$, and β_{pp}, is called the σ interaction functions themselves are called $s\sigma$–$s\sigma$, $s\sigma$–$p\sigma$, and $p\sigma$–$p\sigma$ integrals, while the interaction described by the integral β_{π} is called a π interaction, and the integral β_{π} is the $p\pi$–$p\pi$ integral.

There are basically two different approaches to finding the Coulomb and resonance integrals—nonempirical and empirical (and also a number of hybrid methods in which elements of both approaches are combined).

In the nonempirical method all the integrals are found by purely analytical means using some atomic functions, while in the empirical method they are taken from experimental data. The Coulomb integrals are taken from atomic spectra data (often with corrections that take acocunt of the potential of the remainder of the system), and the resonance integrals are usually determined from the electron spectra data of other systems with a similar atom interaction behavior. Below we shall use a semiempirical method since it permits the most direct establishment of the dependence of the electron structure of a multiatomic system on the properties of the atoms and on the nature of the bond.

Let us make several more comments with regard to the empirical and nonempirical methods in quantum chemistry (see [23] for a detailed discussion).

A nonempirical method is used in those cases when the necessary parameters are hard or impossible to find from experimental data (for example, in the calculation of many inorganic and, in particular, complex compounds with their varied and dissimilar structure). In addition, the nonempirical method must be used, of course, when it is desired to calculate the system *ab initio* although such calculations have proved successful only for a small number of fairly simple systems. Conversely, the semiempirical method is usually used for a long series of systems with a fairly similar bond nature. (A classical example is the Hückel method or the Pariser-Parr-Pople method, applied to molecules with conjugated bonds, for which the semiempirical method is extremely effective.)

Similar series include, in particular, series of tetrahedral $A^N B^{8-N}$ crystals and saturated molecules similar to them in chemical structure.

Let us point out that besides the advantages stated above (important for the development of a version of the theory of interest to us) the semiempirical method makes it possible to correct to a considerable degree, in molecular quantum chemistry as well, the inadequacies of the approximation since the empirical values of the parameters take implicit account of a number of factors omitted in the nonempirical approach. For this reason, for example, in the calculation of the band structure of crystals, empirical correction parameters often enter into the nonempirical calculation. Such a hybrid method turns out to be much more precise, whereas the purely nonempirical approach often leads to results considerably different from experimental ones (See Sec. 2.4.5 for more detail.)

1.3.4 Coefficients for Atomic Functions in LCAO Method and the Effective Charge on Atoms

Besides the matrix elements of the Hamiltonian, the coefficients in the expansion of the eigenfunctions in terms of AOs are also important for studying the nature of the bond: They determine the so-called AO populations, which in turn determine the values of the effective charges on the atoms.

Let us assume (1.32) is one of these eigenfunctions, where the coefficients $c_1, c_2, \ldots, c_M$ have already been found. Then the quantity

$$|\psi|^2 = |c_1|^2 |\chi_1|^2 + |c_2|^2 |\chi_2|^2 + \cdots + |c_M|^2 |\chi_M|^2$$
$$+ c_1^* c_2 \chi_1^* \chi_2 + \cdots + c_{M-1}^* c_M \chi_{n-1}^* \chi_M \qquad (1.54)$$

gives, as usual, the electron density distribution corresponding to the function (1.32). (For a real function the square of the modulus is equal, of course, to its square.) Integrating (1.54) over all space and again assuming all the atomic functions are orthogonal, we have

$$\int |\psi|^2 dv = |c_1|^2 \int |\chi_1|^2 dv + |c_2|^2 \int |\chi_2|^2 dv + \cdots + |c_M|^2 \int |\chi_M|^2 dv \qquad (1.55)$$

According to Eq. (1.55) the total charge of an electron $\int|\psi|^2 dv$, in the state ψ (1.32) and delocalized over the entire system, is composed of the charges $\int|\chi_i|^2 dv$ on the individual AOs. These charges are taken with the weights $|c_i|^2$ and, consequently, the square of the modulus of each of the coefficients c_i defines the "partial" population of the ith AO in a given eigenfunction.

In order to determine the total population of this AO, it is necessary to take into consideration that a given AO can simultaneously enter into several different eigenfunctions and also to take into consideration that either one or two electrons can exist in each state ψ. If this is kept in mind in the subsequent case (a system with filled shells), then the total population v_i of the atomic orbital χ_i is expressed in the form of a sum:

$$\nu_i = 2 \sum_{\substack{\text{over all eigenfunctions of the ground state} \\ \text{in which } \chi_i \text{ enters}}} |c_i|^2 \tag{1.56}$$

Finally, in the general case each atom of the system has not one but several AOs. Therefore the concept of the effective number of electrons in an atom enters into the LCAO method. The effective number of electrons n_A in a certain atom A is defined as the sum of the populations of all the atomic orbitals of the atom. Thus, it is calculated from the formula

$$n_A = \sum_{\substack{\text{over all AOs} \\ \text{of atom A}}} \nu_i = \sum_{\substack{\text{over all AOs} \\ \text{of atom A}}} 2 \sum_{\substack{\text{over all eigenfunctions} \\ \text{in which } \chi_i \text{ enters}}} |c_i|^2 \tag{1.57}$$

In turn, the effective number of electrons defines the effective charge Z_A of an atom, which in the valence approximation is equal to the charge z_A of the atomic core minus the effective number of valence electrons n_A:

$$Z_A = z_A - n_A \tag{1.58}$$

It is obvious that for homoatomic systems, such as the H_2 molecule or a diamond-type crystal, the charge of the core is equal to the effective number of valence electrons and the effective charges of all the atoms are equal to zero.

In the general case, however, in the formation of heteroatomic molecules or crystals the electrons are redistributed so that the atoms cease to be neutral. Then for each atom A, generally speaking, $n_A \neq z_A$ and, consequently, the effective charge $Z_A \neq 0$.

Let us emphasize that in the LCAO method the charges of the ions in a molecule or crystal, unlike the charges of free monatomic ions, must not always be integers. It is seen from Eqs. (1.57) and (1.58) that the effective charges Z_A of the atoms are continuous functions of the coefficients c_i. Thus the LCAO method makes it possible to describe any gradations between purely covalent (or, stated better, electrically neutral; see Sec. 5.1) and purely ionic states of the system when the effective charges of the atoms are equal to their formal charges (valence).

1.3.5 Two-Center Problem in LCAO Method

In conclusion let us consider a special case that is isolated from the general theory by the fact that it is the connecting link between the one-electron model and the classical structural theory in chemistry. In essence, the problem is to find the levels and eigenfunctions—the MOs—for a diatomic system AB with one valence orbital for each atom; it is assumed here that the atoms A, B can be the same or different.

As we will see below, the description of multiatomic systems, being "saturated" in the sense that their structural formula can be uniquely represented in the form of atoms joined by ordinary two-center valence bonds, is reduced to this problem in many respects. (In the latter case, however, it is better to speak of the "quasi-two-center" problem since the two-center orbitals will not be the eigenfunctions of the Hamiltonian of the entire system.)

Let us first consider the homoatomic system A_2, for example, the hydrogen molecule H_2. The eigenfunctions (MOs) for it in the LCAO method are written in the form

$$\psi = c_1\chi_A + c_2\chi_B \tag{1.59}$$

the system of equations (1.36) has the form

$$\begin{cases} (H_{11}-\varepsilon)\,c_1 + H_{12}c_2 = 0 \\ H_{21}c_1 + (H_{22}-\varepsilon)\,c_2 = 0 \end{cases} \quad \text{or} \quad \begin{cases} (\alpha-\varepsilon)\,c_1 + \beta c_2 = 0 \\ \beta c_1 + (\alpha-\varepsilon)\,c_2 = 0 \end{cases} \tag{1.60}$$

and the secular equation (1.38) has the form

$$\mathrm{Det}\begin{Vmatrix} H_{11}-\varepsilon & H_{12} \\ H_{21} & H_{22}-\varepsilon \end{Vmatrix} = 0 \quad \text{or} \quad \mathrm{Det}\begin{Vmatrix} \alpha-\varepsilon & \beta \\ \beta & \alpha-\varepsilon \end{Vmatrix} = 0 \tag{1.61}$$

Here, it has been taken into consideration that the Coulomb integrals H_{11} and H_{22} are identical because the atoms A and B are identical:

$$H_{11} = H_{22} = \alpha, \quad H_{11} = \langle \chi_A | \hat{H} | \chi_A \rangle, \quad H_{22} = \langle \chi_B | \hat{H} | \chi_B \rangle \tag{1.62}$$

and β denotes the resonance integral; in this case it is obviously

$$\beta = \langle \chi_A | \hat{H} | \chi_B \rangle \tag{1.63}$$

The resulting secular equation will be a quadratic equation $(\alpha - \varepsilon)^2 - \beta^2 = 0$. It has the two roots

$$\varepsilon^{(+)} = \alpha + \beta; \quad \varepsilon^{(-)} = \alpha - \beta \tag{1.64}$$

It therefore gives the values of the one-electron energies for the two possible eigenfunctions that can be formed from χ_A and χ_B. To find these functions, according to the standard procedure for solving systems of linear homogeneous equations, it is first necessary to make the substitution $\varepsilon = \varepsilon^{(+)}$ in Eq. (1.60) and then the substitution $\varepsilon = \varepsilon^{(-)}$. The substitution $\varepsilon = \varepsilon^{(+)}$ gives $c_2 =$

c_1, so that the first MO $\psi^{(+)}$ has the form const$(\chi_A + \chi_B)$ or, when normalized to unity

$$\psi^{(+)} = \frac{1}{\sqrt{2}}\{\chi_A + \chi_B\} \tag{1.65}$$

Similarly, for $\varepsilon = \varepsilon^{(-)}$ we have $c_2 = -c_1$; hence

$$\psi^{(-)} = \frac{1}{\sqrt{2}}\{\chi_B - \chi_A\} \tag{1.66}$$

Let us point out that $\psi^{(+)}$ and $\psi^{(-)}$ are orthogonal, just as they should be according to the general theory.

Thus in the two-center problem the two AOs produce two MOs and two atomic levels–two one-electron levels (1.64) with a separation of $2|\beta|$ (Fig. 1.4). The lower of these levels is $\varepsilon^{(+)}$ (the resonance integral β is negative). It is below both atomic levels, so that the electron in the $\varepsilon^{(+)}$ level binds both centers. Therefore the $\varepsilon^{(+)}$ level and the orbital $\psi^{(+)}$ are called bonding. On the other hand, the energy level $\varepsilon^{(-)}$ is greater than the energy of any of the atomic levels. An electron in the $\varepsilon^{(-)}$ level destabilizes the system; accordingly the level $\varepsilon^{(-)}$ and orbital $\psi^{(-)}$ are called antibonding. In the ground state the electrons of the molecule are obviously in the $\varepsilon^{(+)}$ level, so that the $\varepsilon^{(-)}$ level corresponds to the excited state of the system (it is assumed here that each of the atoms A, B has one electron in each of its valence AOs). The effective numbers of valence electrons in the atoms A and B are determined in a trivial manner from Eq. (1.57). For a homoatomic molecule A_2 with one valence electron per atom they are equal to one, as one should expect: $n_A = n_B = 1$, and the effective charges of the atoms are zero: $Z_A = Z_B = 0$.

For a heteroatomic system AB the discussion proceeds in the same way as for the homoatomic, except that the coefficients c_1 and c_2 in the bonding and antibonding MOs should by no means be equal to each other in absolute magnitude. Since the equality

$$c_1^2 + c_2^2 = 1 \tag{1.67}$$

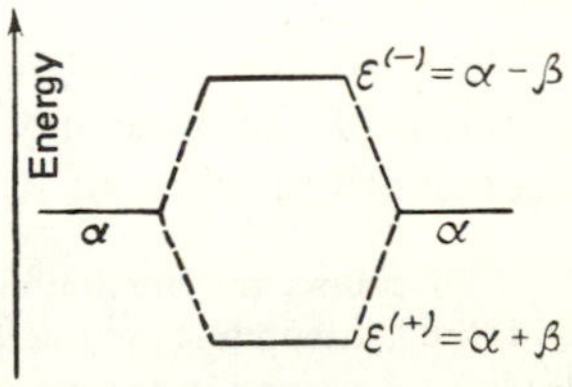

FIG. 1.4. Energy level diagram for homoatomic two-center system with one AO for each atom.

is true for the normalized orbital, the two coefficients c_1, c_2 are not independent and can be expressed as functions of the one parameter λ. It is convenient to introduce this parameter after assuming

$$c_1 = \frac{\lambda}{\sqrt{1+\lambda^2}}; \quad c_2 = \frac{1}{\sqrt{1+\lambda^2}} \tag{1.68}$$

for the bonding orbital, so that this orbital is written in the form

$$\psi^{(+)} = \frac{1}{\sqrt{1+\lambda^2}} \{\chi_B + \lambda\chi_A\} \tag{1.69}$$

Then the antibonding orbital is written in the form*

$$\psi^{(-)} = \frac{1}{\sqrt{1+\lambda^2}} \{\lambda\chi_B - \chi_A\} \tag{1.70}$$

since this is the only combination of χ_A and χ_B orthogonal to (1.69).

As seen from Eqs. (1.68)-(1.70), the parameter λ can vary within the interval $0 \leqslant \lambda < \infty$. Then for $\lambda = 1$, Eqs. (1.69) and (1.70) become (1.65) and (1.66) for the MOs in a purely covalent molecule, while for $\lambda = 0$ or for $\lambda = \infty$ they become

$$\begin{aligned} &\psi^{(+)} = \chi_B; \quad \psi^{(-)} = \chi_A \quad (A^+B^-); \\ &\psi^{(+)} = \chi_A; \quad \psi^{(-)} = \chi_B \quad (A^-B^+) \end{aligned} \tag{1.71}$$

This corresponds to the limiting case of the purely ionic structures A^+B^- or A^-B^+. In this case the bonding MO is identical to the AO of one of the atoms, and the antibonding MO is identical to the AO of the other. This, however, is obviously *a priori* since in a purely ionic molecule the two bonding electrons belong entirely to one of its atoms.

The range of variation of λ can, of course, be restricted if it is stipulated that B will always be the more electronegative atom. Then, obviously, λ will always fall within the interval $0 \leqslant \lambda \leqslant 1$.

Thus, the value of the parameter λ determines the degree of covalency of the bond in the molecule and therefore below λ will be called the "covalency parameter."

We shall now obtain the expression for the effective charge Z on the atoms in terms of the covalency λ. Let us stipulate that Z denotes the charge on the electropositive atom A: $Z = Z_A$.

*Of course, the bonding orbital could be written in the form $(1+\lambda^2)^{-1/2}\{\chi_A + \lambda\chi_B\}$ and the antibonding in the form $(1+\lambda^2)^{-1/2}\{\lambda\chi_A - \chi_B\}$. Such a rearrangement only leads to a change in the numerical values of the parameter λ. Thus, for the LiH molecule $0 < \lambda < 1$ in the old notation, whereas in the new notation $1 < \lambda < \infty$.

Since the coefficient for χ_A is equal to $\lambda/(1 + \lambda^2)^{1/2}$ in Eq. (1.69) for the filled MO, the population of the AO χ_A is equal to $2\lambda^2/(1 + \lambda^2)$. Then from Eq. (1.58) the effective charge on the atom A will be equal to the difference between the charge of the core (+1) and the population $2\lambda^2/(1 + \lambda^2)$, i.e.,

$$Z = Z_A = 1 - \frac{2\lambda^2}{1+\lambda^2} = \frac{1-\lambda^2}{1+\lambda^2} \tag{1.72}$$

In a similar manner it is easy to find the charge on the electronegative atom B. The population of the orbital χ_B in this case is equal to $2/(1 + \lambda^2)^{1/2}$ and

$$Z_B = 1 - \frac{2}{1+\lambda^2} = -\frac{1-\lambda^2}{1+\lambda^2} \tag{1.73}$$

so that $Z_B = -Z_A = -Z$, as should be expected, of course.

1.4 SYMMETRY IN ONE-ELECTRON THEORY

1.4.1 Classification of States in Terms of Irreducible Representations of a Symmetry Group

One of the guiding principles for using the one-electron approximation is the relationship between the one-electron states and the symmetry properties of the system,* and we shall discuss it briefly without giving a mathematical proof.

Here it is easy to grasp the basic concept for the example of the classification of atomic levels. As stated, in the one-electron theory each electron of an atom is described by the Schroedinger equation (1.7) with a centrally symmetrical potential $V = V(r)$. In view of this spherical symmetry there are no isolated directions in the atom. Therefore the observed properties of an atom cannot depend on whether any specific Cartesian coordinate system (with origin at the nucleus) or any other system, obtained from the first one by some rotation in space or a reflection in any plane passing through the nucleus, is used to describe it.

Although this is trivial at first glance, it leads to important consequences. Let us consider, for example, three p AOs for the same principal quantum number n. By a simple relabeling of the axes it is easy, for example, for the function p_x in the "old" coordinates to become the function p_y or p_z in the "new" coordinates. Then it follows from the equivalency of all the coordinate systems that the functions p_x, p_y, p_z must correspond to the same energy, i.e., they must belong to the triply degenerate level $\varepsilon_{n,p}$. Conversely, there is no coordinate transformation that can convert the np function into, for example, the ns function. Therefore from a symmetry viewpoint the level $\varepsilon_{n,s}$

*See [24] for a detailed discussion, [1] for a briefer discussion, and [15-17] and [25], for example, for the application to quantum chemistry.

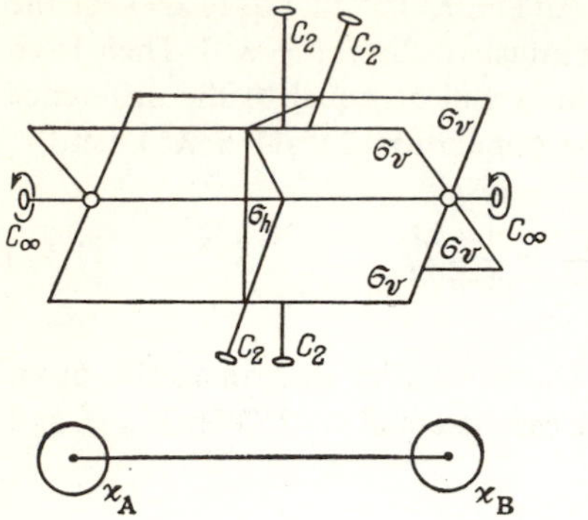

FIG. 1.5. Symmetry of homo-atomic diatomic molecule.

in a centrally symmetrical field is not necessarily identical to the $\varepsilon_{n,p}$ level. It is not, in fact, identical to the latter (except for the special case of the Coulomb potential $V = 1/r$ for the hydrogen atom and hydrogenlike ions). In this same manner it is not hard to see that by means of rotations and reflections of the coordinate system the *nd* functions will always be converted into other *nd* functions (or linear combinations of *nd* functions), but they will never be converted into *ns* or *np* orbitals so that, generally speaking, the *nd* level must be different from the *ns* and *np* levels.

An examination of the hydrogen-type molecule A_2 gives another example of this same kind (Fig. 1.5). It is also "symmetrical" in the sense that there are many coordinate systems in which this molecule looks the same. This same symmetry property can also be described in other expressions by stating that there are a number of symmetry transformations or "operations," upon the completion of which the molecule, considered in a given coordinate system, occupies a position that cannot be distinguished from the previous one. (Such a point of view is even more obvious when one is studying molecules and crystals.)

The ensemble of symmetry transformations for the molecule A_2 (it is called the "symmetry group" and is denoted by $D_{\infty h}$ for molecules of this type) includes all the transformations that transform the system into itself, to wit: any rotations about the axis (C_∞) of the molecule, 180° rotations about lines (C_2) passing through the center of the molecule perpendicularly to the molecular axis, reflections in the planes (σ_v) passing through this axis, and a reflection in the plane σ_h perpendicular to the molecular axis..

Let us now examine the behavior of the eigenfunctions during the symmetry operations. It is clear that the AOs χ_A and χ_B remain in place during certain transformations (for example, during rotations about the molecular axis), and are interchanged during other transformations. However, the eigenfunctions $\psi^{(+)}$ and $\psi^{(-)}$ are transformed into themselves during all symmetry operations (the function $\psi^{(-)}$ may change sign), but the symmetrical function $\psi^{(+)}$ is never converted into the antisymmetrical $\psi^{(-)}$ and vice versa.

The conclusion, which follows from these two examples, is as follows. With respect to transformations from the symmetry group the one-electron eigenfunctions can always be divided into several individual sets, such that the functions of different sets are not "mixed," but functions from the same set are not amenable to further division. Then all the functions of one set belong to the same one-electron level, which is therefore degenerate with a multiplicity equal to the number of functions in the set (remember the examples!).

Although what has been said is generally sufficient to grasp the essence of this subject, let us repeat the same material once more, using the special (universally used) terminology, but without entering into a deep mathematical description.

The ensemble of all the symmetry operations of a system is called, as stated above, the symmetry group G of the system, and the transformation of the wave functions during the symmetry operations forms a "representation" of the symmetry group. Then the functions, belonging to each of the sets mentioned above, are said to belong (or correspond) to the irreducible representation; the number of functions in a set is called the dimensionality of this representation. Thus the *s*, *p*, and *d* AOs belong, respectively, to the one-, three-, and five-dimensional irreducible representations of the atom symmetry group (it is identical to the O_3 symmetry group of a sphere). Similarly, the functions $\psi^{(+)}$ and $\psi^{(-)}$ belong to the two one-dimensional irreducible representations of the $D_{\infty h}$ symmetry group of the A_2 molecule.

Using the terminology and the examples discussed above, one can formulate the following general rules.

1. A certain irreducible representation of the symmetry group of a system belongs to each energy level for an electron in an arbitrary system. In this situation the eigenfunctions of the Hamiltonian of the system, corresponding to this level, belong to this representation of the symmetry group.

2. The multiplicity of the degeneracy of each level is equal to the dimensionality of the corresponding irreducible representation. In particular, the levels corresponding to the one-dimensional representation are nondegenerate (these are the *s* levels in a centrally symmetrical field).

3. Although different levels generally correspond to different irreducible representations, the inverse statement is not true and the same representation can correspond to different levels.

Thus the representation corresponding to the *s*, *p*, *d*, etc., levels is different; however, the same irreducible representation of the O_3 group corresponds to different *s* levels (1*s*, 2*s*, . . .). Similarly, identical representations correspond to the 2*p*, 3*p*, . . . levels, etc. Thus in the ensemble ("basis") of eigenfunctions we have chosen, the same irreducible representation can be realized several times, and a specific energy level corresponds to each such realization.

1.4.2 Use of Symmetry in LCAO Method

The relationship between the one-electron states and the irreducible representations is important from two viewpoints. On the one hand it provides the guiding principle for the classification of the one-electron levels and functions. Using symmetry, one can state, without solving the secular equations, how many different levels the system has and what the multiplicities of these levels are. Thus for the A_2 molecule discussed above it is easy to see that two combinations can be formed from its two AOs χ_A and χ_B: symmetrical and antisymmetrical, belonging to two irreducible representations. These representations, denoted by A_{1g} and A_{1u}, are one-dimensional, so that in the basis of the orbitals χ_A, χ_B the molecule A_2 has two nondegenerate levels.

The number and multiplicity of the levels for more complicated systems are determined in a similar manner. Let us take, for example, a tetrahedral molecule AH_4 like methane (Fig. 1.6a). Here A can be any group IV element (this example is necessary later). In such a system it is natural to take the valence *ns* and *np* AOs of the central atom and the 1*s* AOs of the ligands as the basis. In the presence of the tetrahedron symmetry transformations (T_d group) the *s* orbital of the central atom is invariant, whereas the *p* AOs are transformed into linear combinations of these same AOs. Thus the AOs of the central atom belong to two irreducible representations: one-dimensional A_1 and three-dimensional T_2

$$\begin{array}{lll} s & \textbf{representation} & A_1 \\ p_x,\ p_y,\ p_z & \textbf{representation} & T_2 \end{array} \tag{1.74}$$

Similarly, four linear combinations, belonging to these same representations, can also be formulated from the four orbitals of the ligands. In fact, the combination

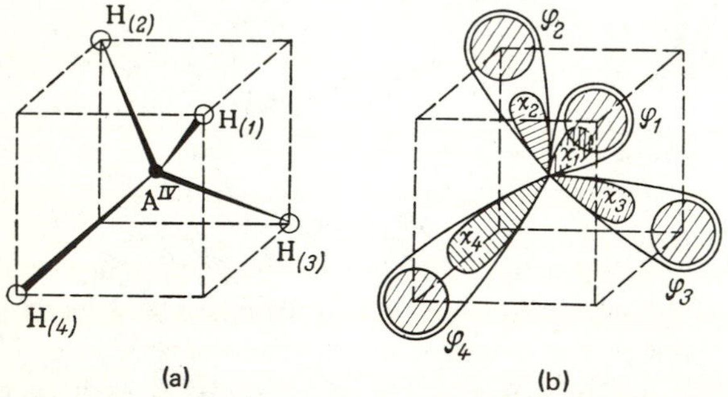

FIG. 1.6. Tetrahedral methane-type molecule (a) and the equivalent orbitals of this molecule (b).

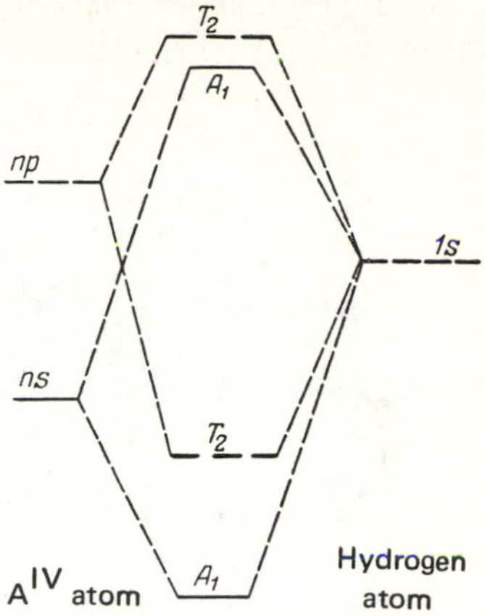

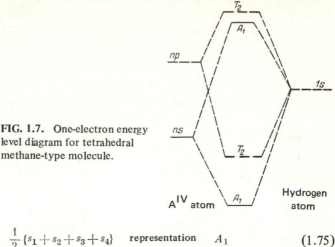

FIG. 1.7. One-electron energy level diagram for tetrahedral methane-type molecule.

$$\frac{1}{2}\{s_1+s_2+s_3+s_4\} \quad \text{representation} \quad A_1 \tag{1.75}$$

remains invariant, like the s orbital of the central atom, to group T_d operations.

On the other hand it is not hard to prove that the three combinations

$$\left.\begin{array}{l}\frac{1}{2}\{s_1-s_2-s_3+s_4\}\\ \frac{1}{2}\{s_1-s_2+s_3-s_4\}\\ \frac{1}{2}\{s_1+s_2-s_3-s_4\}\end{array}\right] \quad \text{representation} \quad T_2 \tag{1.76}$$

are transformed just like the p functions of the central atom.*

Thus, in the basis of the s and p functions of the central atom and the s functions of the ligands the representation A_1 is realized twice and the representation T_2 is realized twice, so that in this basis the energy level scheme of the AH_4 molecule consists of two singly and two triply degenerate levels, as shown in Fig. 1.7.† (Of course, the relative arrangement of the levels and the separations between them cannot be obtained from symmetry considerations only.)

*The validity of this is evident, even without proof, directly from Fig. 1.6a. In fact, each of the combinations (1.76) is similar in its form to one of the p functions: it assumes positive values along the positive direction of the corresponding coordinate axis and equal in modulus, but negative values along the negative direction of this axis.

†This scheme is in total agreement with experimental data. In particular, the presence of two different ground state levels for the AH_4 molecule causes two different values of the first ionization potential for the valence electrons. For example, these values are 13.0 and 23.1 eV for the CH_4 molecule.

Accordingly, in the AH_4 molecule there are two MOs, belonging to the A_1 representation, and six MOs, belonging to the T_2 representation:

$$\psi_1^{(\pm)} = \frac{1}{\sqrt{2}}\left\{c_1 s \pm \frac{1}{2} c_2 (s_1 + s_2 + s_3 + s_4)\right\} \qquad \text{representation } A_1$$

$$\left.\begin{aligned} \psi_2^{(\pm)} &= \frac{1}{\sqrt{2}}\{c_1' p_x \pm c_2' (s_1 - s_2 - s_3 + s_4)\} \\ \psi_3^{(\pm)} &= \frac{1}{\sqrt{2}}\{c_1' p_y \pm c_2' (s_1 - s_2 + s_3 - s_4)\} \\ \psi_4^{(\pm)} &= \frac{1}{\sqrt{2}}\{c_1' p_z \pm c_2' (s_1 + s_2 - s_3 - s_4)\} \end{aligned}\right] \qquad \text{representation } T_2 \tag{1.77}$$

Here, as usual, the superscripts (+) and (−) refer to the filled (bonding) and empty (antibonding) functions.

In conclusion let us touch upon a second, less fundamental aspect of symmetry theory (although for systems with $\sim 10^{23}$ atoms it transcends the boundaries of being purely mechanical). This side of the problem consists of the fact that taking the symmetry into consideration facilitates the actual finding of the levels of the wave functions by lowering the degree of the corresponding secular equations.

Assume the system, which was the subject in Sec. 1.3.2, has a certain symmetry and we are able to distribute the AOs (1.31) among irreducible representations by replacing, where necessary, the AOs by their appropriate linear combinations ("symmetrized AOs"). Then the secular equation (1.38) of degree M can be replaced by several lower degree secular equations ("factorization of the secular equation"). Then each of these equations corresponds to one irreducible representation, such that the number of different equations is equal to the number of different representations, and the degree of each is equal to the multiplicity with which a given irreducible representation is realized.*

Thus for the AH_4 molecule, ignoring symmetry, an eight-degree secular equation must be solved. By taking the tetrahedral symmetry of the system into consideration, one can limit the problem to the solution of two second-order secular equations: one for the s AOs of the central atom and the symmetrized combination of the $1s$ AOs of the ligands and one for any of the p AOs of the central atom and the corresponding symmetrized combination of $1s$ AOs of the ligands.

*This is simple to prove. The eigenfunctions are classified in terms of the irreducible representations. Therefore it certainly is sensible to seek the functions, corresponding to some representation, in the form of combinations of only those symmetrized AOs that belong to this same representation.

1.5 ONE-ELECTRON MODEL, LOCALIZED ORBITALS AND CLASSICAL STRUCTURAL CHEMISTRY

1.5.1 Delocalized and Localized Orbitals. Collective Properties*

One of the most characteristic properties of the eigenfunctions in multiatomic systems is their delocalized form. An electron in a state, describable by an eigenfunction, is not in the general case localized at an atom or small group of atoms, but is delocalized over the entire system. This fact is the natural consequence of the general procedure of finding the eigenfunctions since from the very outset they are sought in the form of linear combinations of all the AOs of the system.

From another point of view the delocalization of the eigenfunctions follows from the fact that they belong to irreducible representations of the symmetry group of the system. In fact, for a sufficiently symmetrical system nearly every (or even every) atom is equivalent in terms of symmetry to several other atoms, located in different portions of the system. Therefore linear combinations of the AOs of all such atoms are at once assigned to each irreducible representation. This is clearly evident from the example considered above of methane-type tetrahedral molecules, for which both the AOs of the central atom as well as the AOs of the ligands enter into each of the representations A_1 and T_2.

At first glance, however, such a completely delocalized nature of the eigenfunctions is sharply divergent from the familiar concepts of chemical structure theory. According to this theory in its well-known "Lewis" interpretation the bond in saturated compounds (such as methane CH_4, other paraffins C_nH_{2n+2}, their analogs $A_n^{IV}H_{2n+2}$, as well as A^NB^{8-N} tetrahedral crystals) is accomplished by electron pairs, localized at the two-center A^{IV}—H, A^{IV}—A^{IV}, or A^N—B^{8-N} bonds.

Such a concept of isolated bonds actually agrees well with several experimental facts in certain molecules and crystals. It is known, for example, that the C—C and C—H distances in all the saturated hydrocarbons C_nH_{2n+2} molecule is equal, within an accuracy of 2%, to the sum of the heats for breaking all of its C—H and C—C bonds (see, for example, [15, 26]), so that the concept of localized two-center bonds certainly agrees with some real aspects of the behavior of electrons in saturated systems.

Thus it is natural to ask the question: How does this concept fit in with the concept of completely delocalized eigenfunctions? An answer to this question was given in papers by Coulson [28], Lennard-Jones, Pople, and Hall

*A detailed discussion of many of the problems dealing with the contents of this section can be found, for example, in [15, 26, 27].

[29]: Localization of the bonds not only does not contradict the one-electron theory, but also is a direct consequence of the one-electron model.

In order to understand how this result is obtained, let us reemphasize that (a) the concept of localized bonds is not ascribed to all systems by any means, but only to saturated and, furthermore, only in their ground state; (b) it describes only those properties that depend on all the electrons of the system at once, for example, properties such as the total binding energy or the charge distribution (following Dewar [15, 26], we shall call these properties "collective"). Then, as can be proved, for each saturated system all the eigenfunctions of the ground state $\psi_i^{(+)}$ can be expressed in terms of those one-electron functions $\varphi_i^{(+)}$, that, first, will be mutually orthogonal and, second, will be localized at the two-center bonds that enter into the classical structure theory (localized orbitals, LOs). It turns out in this situation that in the calculation of any collective property with the LOs one can proceed in the same manner as with the eigenfunctions of the Hamiltonian of the system, so that in studying collective properties it makes no difference what functions are used: localized or delocalized.

The formulated statement is often expressed as follows. The transformation for changing from one orthogonal basis to another (with the same number of functions) is called a "unitary transformation," and the bases that can be interrelated by this transformation are said to be "unitarily equivalent"* (the multidimensional analog of coordinate systems in analytical geometry that are rotated with respect to one another). Then it is stated that in the case of saturated systems the unitarily equivalent basis of functions $\varphi_i^{(+)}$, localized at two-center bonds, is found for the basis of the eigenfunctions $\psi_i^{(+)}$, and the collective quantities automatically turn out to be constant (invariant) with respect to the changeover to the first basis.†

For simplicity let us demonstrate this for the example of the saturated AH_4 molecule (see Fig. 1.6a). For this molecule the eigenfunctions–MOs–are given by (1.77), in which for simplicity we set $c_1 = c_2$, $c_1' = c_2'$; this means zero effective charges on the atoms. We now form the following four combinations of the MOs of the ground state:

$$
\begin{aligned}
\varphi_1^{(+)} &= \frac{1}{2}\{\psi_1^{(+)} + \psi_2^{(+)} + \psi_3^{(+)} + \psi_4^{(+)}\} \equiv \frac{1}{\sqrt{2}}\{\chi_1 + s_1\} \\
\varphi_2^{(+)} &= \frac{1}{2}\{\psi_1^{(+)} - \psi_2^{(+)} - \psi_3^{(+)} + \psi_4^{(+)}\} \equiv \frac{1}{\sqrt{2}}\{\chi_2 + s_2\} \\
\varphi_3^{(+)} &= \frac{1}{2}\{\psi_1^{(+)} - \psi_2^{(+)} + \psi_3^{(+)} - \psi_4^{(+)}\} \equiv \frac{1}{\sqrt{2}}\{\chi_3 + s_3\} \\
\varphi_4^{(+)} &= \frac{1}{2}\{\psi_1^{(+)} + \psi_2^{(+)} - \psi_3^{(+)} - \psi_4^{(+)}\} \equiv \frac{1}{\sqrt{2}}\{\chi_4 + s_4\}
\end{aligned}
\qquad (1.78)
$$

*See, for example, [30] for more details about unitary transformations.

†Strictly speaking, for invariance it is sufficient that only the transformation of the basis be unitary, regardless of whether the functions $\varphi_i^{(+)}$ are localized orbitals or not.

where the χ_i denote the following linear combinations of the AOs of the central atom (it is easy to prove that they are all mutually orthogonal):

$$\chi_1 = \frac{1}{2}\{s + p_x + p_y + p_z\}$$

$$\chi_2 = \frac{1}{2}\{s - p_x - p_y + p_z\}$$

$$\chi_3 = \frac{1}{2}\{s - p_x + p_y - p_z\} \tag{1.79}$$

$$\chi_4 = \frac{1}{2}\{s + p_x - p_y - p_z\}$$

These functions χ_i are the well-known hybrid tetrahedral sp^3 orbitals, directed from the central atom to the ligands (see Fig. 1.6b). The designation sp^3 indicates that each of them is the sum of one s and three p AOs, and the term "hybrid" means that the functions χ_i, although purely atomic, belong to different atomic levels; consequently, they are not AOs in the sense in which this term has been used up to now.

It is seen from Eq. (1.78) that each of the functions $\varphi_i^{(+)}$ is the sum of one of the hybrid sp^3 functions of the central atom and the AOs of that ligand to which this sp^3 orbital is directed (see Fig. 1.6b). Therefore the four orbitals $\varphi_i^{(+)}$ are localized at just the four A—H bonds that figure in the classical

structure formula $\begin{array}{c} H \\ \diagdown \\ H \diagup \end{array} A \begin{array}{c} H \\ \diagup \\ \diagdown H \end{array}$, and all four functions $\varphi_i^{(+)}$ are orthogonal (as

usual, we neglect the overlapping of the AOs of the ligands). Now it is easy to understand how the classical structural chemistry fits in with the one-electron model. By means of (1.78) we express the eigenfunctions $\psi_i^{(+)}$ in terms of the LOs $\varphi_i^{(+)}$:

$$\psi_1^{(+)} = \frac{1}{2}\{\varphi_1^{(+)} + \varphi_2^{(+)} + \varphi_3^{(+)} + \varphi_4^{(+)}\}$$

$$\psi_2^{(+)} = \frac{1}{2}\{\varphi_1^{(+)} - \varphi_2^{(+)} - \varphi_3^{(+)} + \varphi_4^{(+)}\}$$

$$\psi_3^{(+)} = \frac{1}{2}\{\varphi_1^{(+)} - \varphi_2^{(+)} + \varphi_3^{(+)} - \varphi_4^{(+)}\} \tag{1.80}$$

$$\psi_4^{(+)} = \frac{1}{2}\{\varphi_1^{(+)} + \varphi_2^{(+)} - \varphi_3^{(+)} - \varphi_4^{(+)}\}$$

Let us write down the standard formulas, expressing the total energy E of all the electrons and the electron density ρ in MO terms:

$$E = 2\sum_i \langle \psi_i^{(+)} | \hat{H} | \psi_i^{(+)} \rangle \tag{1.81}$$

$$\rho = 2\sum_i |\psi_i^{(+)}|^2 \tag{1.82}$$

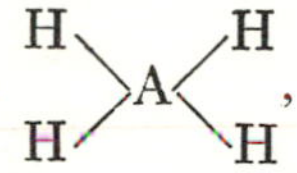

Then, substituting instead of $\psi_i^{(+)}$ their expressions in terms of $\varphi_i^{(+)}$, after trivial manipulations we obtain

$$E = 2 \sum_i \langle \varphi_i^{(+)} | \hat{H} | \varphi_i^{(+)} \rangle \qquad (1.83)$$

$$\rho = 2 \sum_i [\varphi_i^{(+)}]^2 \qquad (1.84)$$

It is seen from Eqs. (1.83) and (1.84) that E and ρ for the AH_4 molecule can be written in the form of a sum of "increments" along the individual bonds, which in a given case also gives a justification of the structure theory. An analogous proof can be given in general form too, so that Eqs. (1.83) and (1.84) are valid for any saturated systems, where $\psi_i^{(+)}$ means the eigenfunctions and $\varphi_i^{(+)}$, the LOs.

1.5.2 Equivalent Orbitals

Let us point out one fact that relates to the form of the LO. For an arbitrary saturated system the LOs, corresponding to different kinds of bonds, are by no means equivalent. Thus the LOs for ethane $\begin{smallmatrix} H \\ H \end{smallmatrix} \!\! > \!\! C\!-\!C \!\! < \!\! \begin{smallmatrix} H \\ H \end{smallmatrix}$, belonging to the C—C bond and the C—H bond, are different. Similarly, for the butane molecule

$$
\begin{array}{cccc}
H & H & H & H \\
| & | & | & | \\
H-C_{(1)}-C_{(2)}-C_{(3)}-C_{(4)}-H \\
| & | & | & | \\
H & H & H & H
\end{array}
$$

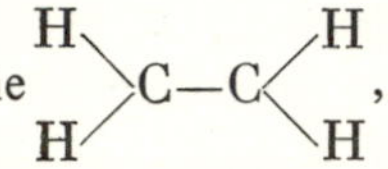

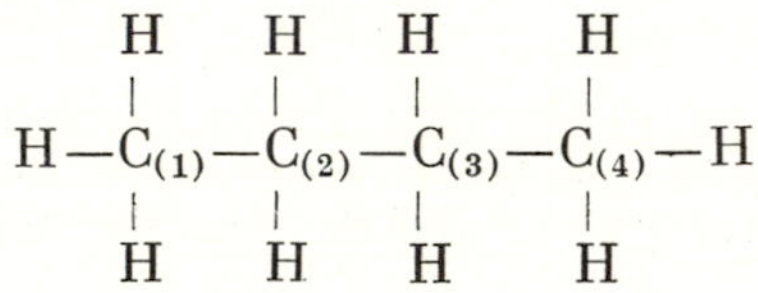

the LOs belonging to the $C_{(1)}\!-\!C_{(2)}$ and $C_{(2)}\!-\!C_{(3)}$ bonds, strictly speaking, are also different since no symmetry operations exist that transform the $C_{(1)}\!-\!C_{(2)}$ bond into the $C_{(2)}\!-\!C_{(3)}$ (although, according to experimental data, these LOs are nearly identical).

However, in the case of (saturated) systems with high symmetry, such as methane or diamond, for any pair of bonds there is an operation that transforms one bond into another. For these systems the LOs can be expressed such that all the LOs will be equivalent, i.e., only their position in space will be different. Such LOs are called "equivalent orbitals" (EOs). Equations (1.83) and (1.84) for the LOs are, of course, also valid for the EOs. For the EOs, however, Eq. (1.83) is replaced by the simpler one

$$E = 2 \times \text{(number of bonds)} \times \langle \varphi^{(+)} | \hat{H} | \varphi^{(+)} \rangle \qquad (1.85)$$

since here all the quantities $\langle \varphi_i^{(+)} | \hat{H} | \varphi_i^{(+)} \rangle$ for the EOs are identical.

1.5.3 Additional Comments with Regard to LOs

In addition to what has been said, let us make three comments with regard to LOs and their relationship to structure theory.

The first of these is related to hybrid orbitals, which in the modern treatment are introduced simply as linear combinations of AOs, arising due to the changeover from eigenfunctions to the LOs. In the original treatment, however, the meaning of hybridization was not this. Historically, the concept of hybridization was introduced by Pauling in connection with the question of the state of the carbon atom in CH_4. He showed that one can select those combinations of $2s$ and $2p$ AOs for which the resulting functions will be directed to the vertices of the tetrahedron. Pauling assumed–and this opinion was universally accepted for quite a long time–that the formation of hybrid orbitals is the determining cause of stereochemistry (the "theory of directional valences"; see, for example, the classical monograph [31]). From the point of view of the physical proof of the theory of directional valences, however, it was never sufficiently convincing, so that this aspect of the original Pauling theory has now been discarded. In any case from the viewpoint of the one-electron model hybridization is the result, rather than the cause, of stereochemistry.

The second comment is concerned with the proof of Eqs. (1.83) and (1.84). As mentioned, this proof is based only on the unitary nature of the transformation of the basis:

$$\psi_i^{(+)} = \sum_\alpha v_{\alpha i} \varphi_\alpha^{(+)} \tag{1.86}$$

In fact, the condition for a unitary transformation has the form

$$\sum_\alpha v^*_{i\alpha} v_{j\alpha} = \delta_{ij} \tag{1.87}$$

Then we have

$$\begin{aligned} \rho &= 2\sum_i |\psi_i^{(+)}|^2 = 2\sum_i \sum_k \sum_l v^*_{ki} v_{li} \varphi_k^{(+)*} \varphi_l^{(+)} \\ &= 2\sum_k \sum_l \left\{\sum_i v^*_{ki} v_{li}\right\} \varphi_k^{(+)*}\varphi_l^{(+)} = 2\sum_k\sum_l \delta_{kl} \varphi_k^{(+)*}\varphi_l^{(+)} = 2\sum_k [\varphi_k^{(+)}]^2 \end{aligned} \tag{1.88}$$

$$\begin{aligned} E &= 2\sum_i \langle \psi_i^{(+)} | \hat{H} | \psi_i^{(+)} \rangle = 2\sum_i\sum_k\sum_l v^*_{ki} v_{li} \langle \varphi_k^{(+)} | \hat{H} | \varphi_l^{(+)} \rangle \\ &= 2\sum_k \langle \varphi_k^{(+)} | \hat{H} | \varphi_k^{(+)} \rangle \end{aligned} \tag{1.89}$$

which was to be proved.

A drawback of the proof presented is the assumption that $E = 2\Sigma \langle \psi_i^{(+)} | \hat{H} | \psi_i^{(+)} \rangle$, whereas this equality is valid only for the non-self-consistent version of the theory. The proof for the Hartree–Fock scheme is based on writing the total wave function of the ground state in the form of the Slater determinant (1.20). This is possible since a saturated system certainly has filled shells.

As first noted by Fock [6], the choice of the one-electron orbitals $\psi_i^{(+)}$ of the ground state in the Slater determinant is ambiguous. It is defined to within an arbitrary unitary transformation since in this situation the determinant is only multiplied by an unimportant factor, equal to one in absolute value. This factor has no effect on the values of collective quantities since in the expression for each of them Ψ^* enters along with Ψ. Thus in describing the ground state the sets $\psi_i^{(+)}$ and $\varphi_i^{(+)}$ must be considered equally justified, and there is no basis for assuming (as sometimes stated [15, 26]) that the electrons in saturated compounds are "in fact" delocalized and only behave as if they were localized. In reality for the ground state both descriptions are equally real; they complement each other in the sense that each eigenfunction belongs to a certain level, but by itself it characterizes neither the binding energy nor the distribution ρ.

On the other hand, each EO characterizes both (if, of course, the LOs are EOs), but does not belong to any energy level.

Finally, let us point out that the interpretation of structure theory presented above (in the Lewis treatment) is not the only one possible since the very transfer of the Lewis concept into the language of quantum mechanics is not totally unambiguous.

The possibility of a quantum-mechanical interpretation of the Lewis electron pair was first shown by Heitler and London (in 1927). In the Heitler–London method a two-electron function, not a one-electron orbital, corresponds to each A—B bond. To obtain this function we first assume that electron 1 is localized at atom A and electron 2 at atom B, after which (in order to take into consideration the indistinguishability of the electrons) we add another solution in which the electrons change places. The resulting two-electron wave function

$$\varphi(1, 2) = \frac{1}{2}\{\chi_A(1)\chi_B(2) + \chi_B(1)\chi_A(2)\}\{\alpha(1)\beta(2) - \beta(1)\alpha(2)\} \quad (1.90)$$

where α, β are the spin functions of the electrons, gives in the Heitler–London approximation another interpretation of the valence aspect and lies at the base of the well-known valence scheme method (VS method) in quantum chemistry.* It is easy to extend the VS method by adding the "ionic terms" $\chi_A(1)\chi_A(2)$, $\chi_B(1)\chi_B(2)$, which take account of the possibility that the two electrons simultaneously reside at atom A or at atom B. In its conventional or extended form the VS method was sometimes used to calculate the binding energy of diamond or diamond-like crystals [32–34], and it is used repeatedly in the works of Krebs, Muzer, Pearson, and Syushe for a qualitative description of the chemical bond in other solids too (see [35] and the bibliography in [36]). It must, however, be always remembered that the VS method has several very serious drawbacks and is considerably inferior to the one-electron approximation in universality and simplicity. The VS method, being associated with the necessary assumption spin pairing in each electron pair, is unsuitable for describing the nature of the bond even in the simplest possible molecule—the molecular hydrogen ion H_2^+—and is sharply at variance with experiment with regard to the magnetic properties of paramagnetic molecules with an even number of electrons (for example, O_2). In the case of systems described by more than one valence scheme, such as benzene and ferrocene molecules or crystals of most structure types, the use of the VS method is extremely complicated. If the very considerable mathematical difficulties involved in the study of excited states are added to this, it becomes understandable why the VS method was discarded long ago in most problems of quantum chemistry. These drawbacks of the VS method are by no means reduced by switching over to solid state theory, and this

*This method was proposed in the early 1930s by Slater and Pauling. For a description of the VS method see, for example, [16, 17].

method in the general case can hardly serve as the basis of a logically integrated and practically useful theory of the chemical bond in solids. For this reason for a uniform treatment we prefer the interpretation of valence aspects in the language of localized one-electron orbitals, which follows directly from the general one-electron model.

1.5.4 Localized Orbitals and One-Electron Properties

Although the collective properties of saturated systems are described with equal success in the basis of eigenfunctions, as well as in the basis of LOs (EOs), no such equivalence exists for properties such as the orbital ionization potentials or the transition energy between the one-electron levels. These properties are determined not by all the electrons at once, but by the states of the individual electrons of the system (called "one-electron" properties by Dewar [15, 26]). For describing the one-electron properties only the eigenfunctions of the Hamiltonian of the system are suitable since only they correspond to definite energy levels. At the same time the LOs, generally speaking, are combinations of the eigenfunctions for the different levels and therefore they do not belong to any certain energy.

This is in complete agreement with the fact that the LO does not belong to any irreducible representation of the symmetry group of the system (but it belongs to a reducible one). This is clearly seen in the example of AH_4 molecule, for which the EOs $\varphi_i^{(+)}$ are combinations of the MO $\psi_1^{(+)}$ of the A_1 representation and the MOs $\psi_2^{(+)}$, $\psi_3^{(+)}$, $\psi_4^{(+)}$ of the T_2 representation.

Nevertheless, the LOs can be used, which is sometimes useful for studying one-electron properties.

First, the "average" one-electron properties can be characterized by means of the LOs. (Then, however, the one-electron properties become collective—this is a problem of terminology.)

It was mentioned above that the LOs or EOs for the ground state are bonding in nature [see, for example, Eq. (1.78)]. Besides the bonding LOs $\varphi_i^{(+)}$, let us now also introduce the antibonding LOs $\varphi_i^{(-)}$, orthogonal to the bonding:

$$\langle \varphi_i^{(-)} | \varphi_j^{(+)} \rangle = 0 \text{ for all } j, \text{ including when } j = i \tag{1.91}$$

Then the LOs $\varphi_i^{(-)}$ will be related by a unitary transformation to the eigenfunctions $\varphi_i^{(-)}$, belonging to the excited states of the system. For example, for the case being considered (the AH_4 molecule) the $\varphi_i^{(-)}$ are written in the form

$$\begin{aligned} \varphi_1^{(-)} &= \frac{1}{\sqrt{2}}\{\chi_1 - s_1\} \\ \varphi_2^{(-)} &= \frac{1}{\sqrt{2}}\{\chi_2 - s_2\} \end{aligned} \tag{1.92}$$

$$\varphi_3^{(-)} = \frac{1}{\sqrt{2}} \{\chi_3 - s_3\}$$
$$\varphi_4^{(-)} = \frac{1}{\sqrt{2}} \{\chi_4 - s_4\} \qquad (1.92)\ (continued)$$

and their relationship to the MOs $\psi_i^{(-)}$ for the excited states is given by the formulas

$$\begin{aligned}
\varphi_1^{(-)} &= \frac{1}{2} \{\psi_1^{(-)} + \psi_2^{(-)} + \psi_3^{(-)} + \psi_4^{(-)}\} & \psi_1^{(-)} &= \frac{1}{2} \{\varphi_1^{(-)} + \varphi_2^{(-)} + \varphi_3^{(-)} + \varphi_4^{(-)}\} \\
\varphi_2^{(-)} &= \frac{1}{2} \{\psi_1^{(-)} - \psi_2^{(-)} - \psi_3^{(-)} + \psi_4^{(-)}\} & \psi_2^{(-)} &= \frac{1}{2} \{\varphi_1^{(-)} - \varphi_2^{(-)} - \varphi_3^{(-)} + \varphi_4^{(-)}\} \\
\varphi_3^{(-)} &= \frac{1}{2} \{\psi_1^{(-)} - \psi_2^{(-)} + \psi_3^{(-)} - \psi_4^{(-)}\} & \psi_3^{(-)} &= \frac{1}{2} \{\varphi_1^{(-)} - \varphi_2^{(-)} + \varphi_3^{(-)} - \varphi_4^{(-)}\} \\
\varphi_4^{(-)} &= \frac{1}{2} \{\psi_1^{(-)} + \psi_2^{(-)} - \psi_3^{(-)} - \psi_4^{(-)}\} & \psi_4^{(-)} &= \frac{1}{2} \{\varphi_1^{(-)} + \varphi_2^{(-)} - \varphi_3^{(-)} - \varphi_4^{(-)}\}
\end{aligned} \qquad (1.93)$$

We will define the average level $\langle\varepsilon^{(+)}\rangle$ for the occupied and the average level $\langle\varepsilon^{(-)}\rangle$ for the excited states as the arithmetic average of the individual levels, and each level ε_i is taken as many times as its multiplicity n_i:

$$\begin{aligned}
\langle \varepsilon^{(+)} \rangle &= \frac{1}{\text{total number of levels}} \sum_i n_i^{(+)} \varepsilon_i^{(+)} \\
\langle \varepsilon^{(-)} \rangle &= \frac{1}{\text{total number of levels}} \sum_i n_i^{(-)} \varepsilon_i^{(-)}
\end{aligned} \qquad (1.94)$$

Let us now write the quantities $\langle\varepsilon^{(+)}$, $\langle\varepsilon^{(-)}\rangle$ in the form

$$\begin{aligned}
\langle \varepsilon^{(-)} \rangle &= \frac{1}{\text{total number of levels}} \sum_i \langle \psi_i^{(+)} | \hat{H} | \psi_i^{(+)} \rangle \\
\langle \varepsilon^{(-)} \rangle &= \frac{1}{\text{total number of levels}} \sum_i \langle \psi_i^{(-)} | \hat{H} | \psi_i^{(-)} \rangle
\end{aligned} \qquad (1.95)$$

and pass from the eigenfunctions $\{\psi_i^{(+)}\}$, $\{\psi_i^{(-)}\}$ to the unitarily equivalent bases of LOs $\{\varphi_i^{(+)}\}$, $\{\varphi_i^{(-)}\}$. Then, in view of the unitary equivalence [compare result of Eq. (1.89)], we have

$$\begin{aligned}
\langle \varepsilon^{(+)} \rangle &= \frac{1}{\text{total number of bonds}} \sum_{\text{over bonds}} \langle \varphi_i^{(+)} | \hat{H} | \varphi_i^{(+)} \rangle \\
\langle \varepsilon^{(-)} \rangle &= \frac{1}{\text{total number of bonds}} \sum_{\text{over bonds}} \langle \varphi_i^{(-)} | \hat{H} | \varphi_i^{(-)} \rangle
\end{aligned} \qquad (1.96)$$

The equations (1.96) assume an especially simple form in the case when all LOs are EOs. Then all the terms on the right sides of Eq. (1.96) are identical and we obtain

$$\begin{aligned} \langle \varepsilon^{(+)} \rangle &= \langle \varphi^{(+)} | \hat{H} | \varphi^{(+)} \rangle \\ \langle \varepsilon^{(-)} \rangle &= \langle \varphi^{(-)} | \hat{H} | \varphi^{(-)} \rangle \end{aligned} \tag{1.97}$$

where $\varphi^{(+)}$ is any bonding EO and $\varphi^{(-)}$ is any antibonding EO.

The equations make it possible to determine the "average" ionization potential $\langle I \rangle$ of the system and its "average" excitation energy $\langle \Delta\varepsilon \rangle$. For arbitrary LOs

$$\langle I \rangle = -\langle \varepsilon^{(+)} \rangle = -\frac{1}{\text{total number of bonds}} \sum_i \langle \varphi_i^{(+)} | \hat{H} | \varphi_i^{(-)} \rangle \tag{1.98}$$

$$\langle \Delta\varepsilon \rangle = \langle \varepsilon^{(-)} \rangle - \langle \varepsilon^{(+)} \rangle = \frac{1}{\text{total number of bonds}} \times \sum_i \{ \langle \varphi_i^{(-)} | \hat{H} | \varphi_i^{(-)} \rangle - \varphi_i^{(+)} | \hat{H} | \varphi_i^{(+)} \rangle \} \tag{1.99}$$

and for the EOs

$$\langle I \rangle = -\langle \varphi^{(+)} | \hat{H} | \varphi^{(+)} \rangle \tag{1.100}$$

$$\langle \Delta\varepsilon \rangle = \langle \varphi^{(-)} | \hat{H} | \varphi^{(-)} \rangle - \langle \varphi^{(+)} | \hat{H} | \varphi^{(+)} \rangle \tag{1.101}$$

The meaning of Eqs. (1.100) and (1.101) is especially simple. They mean that $\langle I \rangle$ is equal to the average value of the Hamiltonian of the system in an arbitrary filled EO, taken with a minus sign, while $\langle \Delta\varepsilon \rangle$ is the difference of the average values in antibonding and in bonding EOs.

Finally, let us emphasize one more fact concerning the question of the relation of the LOs (EOs) to the one-electron properties. Although the individual LOs are not related to definite levels, one can, however, by specifying the complete set of all LOs—both bonding as well as antibonding—find these levels after constructing linear combinations of the LOs such as Eq. (1.93), which are identical to the one-electron eigenfunctions. Such a method of constructing the eigenfunctions is especially convenient in the case when the LOs are EOs, as is the case, for example, for the saturated hydrocarbons or crystals such as diamond and ZnS (see Sec. 2.6.2).

2

Band Theory of Solids

Band theory* is the application of the one-electron model to crystals; it is fundamentally identical to the MO method in molecular quantum chemistry, and in the LCAO approximation it is identical to the MO LCAO method. Thus the study of the electron structure of a crystal reduces within the framework of band theory to the solution of Eq. (1.8), i.e., to finding the eigenvalues and eigenfunctions of the effective one-electron Hamiltonian (1.9), where by V we mean the effective potential of the crystal. The one-electron eigenfunctions thus obtained are called "Bloch functions" (BFs, the analog of MOs in molecules).

As is characteristic of eigenfunctions, each BF is delocalized over the entire system. Therefore the BF in the LCAO method must be represented in the form of linear combinations of the AOs of all the atoms of the crystal. In principle this is done the same as in the general case (see Sec. 1.3.2). However, the large number of atoms in a crystal (of the order of 10^{23}) as well as the periodicity of the structure impart a unique form to the energy spectrum and the eigenfunctions.

The enormous number of AOs in the system leads to the appearance of a corresponding number of levels and eigenfunctions. At the same time the periodicity of the structure of crystals causes the appearance of a universal translational symmetry. Although a crystal can have symmetry elements besides translational ones (see Sec. 2.5.1), it is the translational symmetry that leads to the characteristic classification of the energy levels and eigenfunctions.

In this regard it should be mentioned that taking the symmetry into consideration together with the classification of the states makes it possible to reduce the degree of the secular equations (see Sec. 1.4.2). For molecules with

*A description of band theory can be found in textbooks on solid state theory, see, for example, [37, 38]. References [40–42] and a considerable portion of [39] are devoted to this question specifically; in particular the monographs [39, 40] are recommended for an in-depth study of band theory. For a simple description of band theory see also [43].

a relatively small number of atoms this aspect of group theory is rather "technical." For crystals, however, it becomes essential since ignoring the symmetry in this case would necessitate solving an equation of $\sim 10^{23}$ degree(!). As we shall see, consideration of one translational symmetry makes it possible to deal with a fully acceptable secular equation, the degree of which is equal to the number of atomic functions in the calculation in the unit cell of the crystal.

2.1 ONE-DIMENSIONAL CRYSTAL

2.1.1 Bloch Functions for One-Dimensional Line

The importance of translational symmetry is conveniently exhibited for a model of a crystal in the form of a line of N atoms, equally spaced from one another and having one valence AO* (Fig. 2.1). In this situation, as is usually done in solid state theory, we will assume that the line is a closed ring. Such a "cyclic Born–Karman condition" greatly simplifies the investigation. Actually, in the ring-shaped line all atoms are equivalent, whereas in an open line the atoms at its ends exist under different conditions from the atoms located in the middle.

The hypothetical H_N molecule (of N hydrogen atoms) or the system of $p\pi$ orbitals of a cyclic polyene, two-dimensional and with bonds of equal length (for $N = 6$ benzene C_6H_6 is such a polyene) can represent an example of such a "crystal."

The unit cell of such a line will be any segment of length equal to the shortest interatomic distance a (the lattice period). Correspondingly, the translational symmetry of the line consists of the fact that all atoms of the line coincide with others of these same atoms for any translations that are multiples of the lattice period a, or, what is the same thing, for rotations by any angle that is a multiple of $2\pi/N$ (C_N symmetry group). Then the fundamental role of the translational symmetry follows from the following theorem: the BF eigenfunctions of the one-electron Hamiltonian of the line can be chosen such that they belong to irreducible representations of the group of translations of the line† (the C_N group).

We will now look for linear combinations of the AOs of the line that are transformed into irreducible representations of the C_N group.

It follows from the general theory that (a) all representations of the C_N group are one-dimensional, (b) there are N of them, (c) the wave function

*See [17, 43] for a simple description of this model. See [41] for the case of an arbitrary one-dimensional periodic structure.

†See [17, 24] concerning the relationship between the BFs and the group of translations, or refer to any of the band theory books cited above. References [44, 45] and the review [46] can be recommended in particular for dealing with the question of the application of group theory to solid state theory.

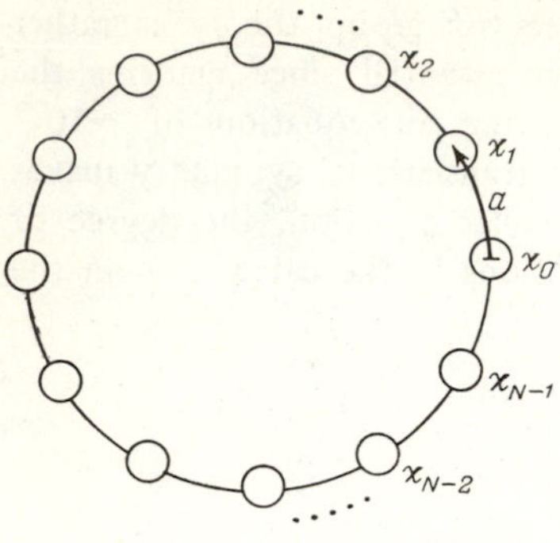

FIG. 2.1. Cyclic line, consisting of atoms with *s* orbitals.

belonging to each *n*th irreducible representation is multiplied by the number exp $[2\pi i(n-1)/N]$ upon rotation by the minimum allowable angle $2\pi/N$.

It is not hard to see that the combinations of the AOs of our line that satisfy all these conditions have the following form (for convenience the N atoms and N BFs are designated by numbers from 0 to $N-1$; $1/N$ is a normalization factor):

$$\psi_0 = \frac{1}{\sqrt{N}}\{\chi_0 + \chi_1 + \ldots + \chi_{N-1}\}$$

$$\psi_1 = \frac{1}{\sqrt{N}}\left\{\chi_0 + e^{2\pi i \frac{1\cdot 1}{N}}\chi_1 + e^{2\pi i \frac{1\cdot 2}{N}}\chi_2 + \ldots + e^{2\pi i \frac{1(N-1)}{N}}\chi_{N-1}\right\} \quad (2.1)$$

$$\cdots\cdots\cdots\cdots\cdots\cdots\cdots\cdots$$

$$\psi_{N-1} = \frac{1}{\sqrt{N}}\left\{\chi_0 + e^{2\pi i \frac{(N-1)\cdot 1}{N}}\chi_1 + e^{2\pi i \frac{(N-1)\cdot 2}{N}}\chi_2 + \ldots + e^{2\pi i \frac{(N-1)(N-1)}{N}}\chi_{N-1}\right\}$$

or in general form

$$\psi_n = \frac{1}{\sqrt{N}}\sum_{m=0}^{N-1} e^{2\pi i \frac{nm}{N}}\chi_m, \quad n = 0, 1, \ldots, N-1 \quad (2.2)$$

In fact, the fulfillment of condition (b) is obvious; it remains to prove that conditions (a) and (c) are satisfied. To do this it is sufficient to show that for a rotation by $2\pi/N$ each of the functions is transformed into itself to within the numerical coefficient exp $[-2\pi in/N]$.

Let us actually rotate the ring in Fig. 2.1 by an angle $2\pi i/N$ in the counterclockwise direction. Then the 0 atom becomes atom 1, atom 1 becomes atom 2, etc., so that the transformation of the AOs is represented by the scheme: $\chi_0 \to \chi_1$, $\chi_1 \to \chi_2, \ldots, \chi_{N-2} \to \chi_{N-1}$, $\chi_{N-1} \to \chi_0$. It is obvious that this cyclic permutation of the AOs generally does not alter the first of the functions (2.1), i.e., for the stated rotation ψ_0 is multiplied by $N-2 = \exp$

$2\pi i0/N$. The second function ψ_1 for the stated cyclic permutation of the atomic orbitals undergoes the following transformation:

$$\psi_1 = \frac{1}{\sqrt{N}}\left\{\chi_0 + e^{2\pi i\frac{1\cdot 1}{N}}\chi_1 + e^{2\pi i\frac{1\cdot 2}{N}}\chi_2 + \ldots + e^{2\pi i\frac{1\,(N-1)}{N}}\chi_{N-1}\right\}$$

$$\longrightarrow \psi_1' = \frac{1}{\sqrt{N}}\left\{\chi_1 + e^{2\pi i\frac{1\cdot 1}{N}}\chi_2 + e^{2\pi i\frac{1\cdot 2}{N}}\chi_3 + \ldots + e^{2\pi i\frac{1\,(N-1)}{N}}\chi_0\right\}$$

Taking the factor $\exp[-2\pi i1/N]$ outside the brackets, we have:

$$\psi_1' = e^{-2\pi i\frac{1}{N}} \cdot \frac{1}{\sqrt{N}}\left\{\chi_0 + e^{2\pi i\frac{1\cdot 1}{N}}\chi_1 + \ldots + e^{2\pi i\frac{1\,(N-1)}{N}}\chi_{N-1}\right\} = e^{-2\pi i\frac{1}{N}}\psi_1$$

so that for a rotation by $2\pi/N$ the function ψ_1 is multiplied by $\exp\,[-2\pi i1/N]$. By applying a similar consideration to the functions $\psi_2, \ldots,$ etc., one can prove that ψ_2 is multiplied by $\exp\,[2\pi i2/N]$, ψ_3 by $\exp\,[-2\pi i3/N]$, etc.

Equations (2.1) and (2.2), obtained for the BFs (or MOs) of an atom line, can be considered to be in final form. Nevertheless, let us rewrite them again in different notations in order to arrive at the standard form used in band theory. All operations required to do this reduce to an identity transformation of the exponent for the imaginary exponent terms in Eqs. (2.1) and (2.2).

(a) To extend the discussion to the case of a three-dimensional lattice, we shall characterize the atoms of the line (the one-dimensional lattice) not by the number m, but by the radius vector

$$\mathbf{R} = m\mathbf{a} \tag{2.3}$$

where $|\mathbf{a}| = a$ is the period of the lattice. Then the atomic orbital of each mth atom is written in the form $\chi_m = \chi(\mathbf{r} - m\mathbf{a}) = \chi(\mathbf{r} - \mathbf{R})$.

(b) In the exponents of the imaginary exponential terms in Eqs. (2.1) and (2.2) we introduce the radius vectors $\mathbf{R}$ of the atoms instead of their numbers m, so that we write the exponent $2\pi inm/N$ in each exponential term in the following manner: $2\pi inm/N = i(2\pi/a)(n/N)ma = i(2\pi/a)(n/N)R$, where R is the length of the vector $\mathbf{R}$.

(c) We now represent the number $(2\pi/a)(n/N)$ by the symbol k:$k = (2\pi/a)(n/N)$; we note that for a given lattice period a and a quantity of atoms N the number k depends only on n. Thus k is uniquely related to the number of the BF and can be used equally as well as the number n for numbering the different BFs.

(d) Let us point out, finally, that the product of two numbers can always be represented in the form of a scalar product of two vectors. Therefore let us write the transformed exponent in the form $(2\pi/a)(n/N)R = kR = \mathbf{k}\cdot\mathbf{R}$, i.e., in the form of a scalar product of the radius vector **R** of the atom by another vector **k** of the form:

$$\mathbf{k}=\frac{n}{N}\,\mathbf{b};\;\; b=\frac{2\pi}{a} \tag{2.4}$$

We finally obtain the standard form of writing the BFs for a one-dimensional lattice. Each BF in this notation is characterized by its own value of the vector **k** and is written in a uniform manner (the summation is done over all sites of the one-dimensional lattice):

$$\psi(\mathbf{k}\,|\,\mathbf{r})=\frac{1}{\sqrt{N}}\sum_{\mathbf{R}} e^{i\mathbf{kR}}\chi(\mathbf{r}-\mathbf{R}) \tag{2.5}$$

so that the BF depends on two vectors–the variable radius vector **r** of an electron and the constant vector **k** that is enumerating a given BF.

Let us also point out how the transformations of the BF during operations from the group of line translations are written in the new notation. As pointed out, for a displacement by a lattice period a each nth BF is multiplied by exp $[-2\pi in/N]$, or, in the new notations, by exp $[-i\mathbf{k}\mathbf{a}]$. Thus

$$\psi(\mathbf{k}\,|\,\mathbf{r}-\mathbf{a})=e^{-i\mathbf{ka}}\psi(\mathbf{k}\,|\,\mathbf{r}) \tag{2.6}$$

Let us mention that Eq. (2.6) can be rewritten in a more general form by considering translations by an arbitrary integer number of periods $\mathbf{R}_0 = p\mathbf{a}$. Then, obviously

$$\psi(\mathbf{k}\,|\,\mathbf{r}-R_0)=e^{-i\mathbf{kR}_0}\psi(\mathbf{k}\,|\,\mathbf{r}) \tag{2.7}$$

Equations (2.6) and (2.7) emphasize the group-theory meaning of the vector **k** that enters into the notation of (2.5) for the BF. They show that each **k** vector does not simply denote a corresponding Bloch function, but also indicates the irreducible representation of the group of translations to which a given BF belongs.

If it is not required that the BFs belong to irreducible representations of the group of translations, then the BFs of the line can be written in real form when the coefficients for the AOs will be cosines or sines. As will be shown below, the vectors **k** for a one-dimensional line can be drawn in both the positive as well as negative directions with respect to the coordinate origin (see Sec. 2.1.2), and the BFs $\psi(\mathbf{k}/\mathbf{r})$ and $\psi(-\mathbf{k}/\mathbf{r})$ correspond to the same energy. Therefore in addition to these functions one can introduce real eigenfunctions of the form

$$
\begin{aligned}
\psi' &= \frac{1}{2}\,\{\,(\mathbf{k}\,|\,\mathbf{r}) + \psi\,(-\mathbf{k}\,|\,\mathbf{r})\} \\
\psi'' &= \frac{1}{2i}\,\{\psi\,(\mathbf{k}\,|\,\mathbf{r}) - \psi\,(-\mathbf{k}\,|\,\mathbf{r})\}
\end{aligned}
\tag{2.8}
$$

The MOs for the π electrons of benzene or other conjugate systems are often written in real form.

2.1.2 k Space and First Brillouin Zone

Let us now discuss the properties of the vectors **k**. This leads to the concepts of **k** space and the first Brillouin zone.

In addition to the atomic line (Fig. 2.1) let us construct another ("reciprocal") lattice that consists of sites located a distance $(2\pi/a) = b$ from each other (Fig. 2.2). Then the period of the reciprocal lattice will be equal to b, so that the periods of the direct and reciprocal lattices are actually proportional.

Let us now divide each of the unit cells of the reciprocal lattice into N parts–the number of atoms in the real lattice–and choose an arbitrary site of the reciprocal lattice as the origin. Then to each of those points that divide the unit cells of the reciprocal lattice into N small segments one can associate a vector **k** with origin at the coordinate origin of the reciprocal lattice and terminating at this point (Fig. 2.2).

The resulting discrete ensemble of **k** vectors is called **k** space, and the importance of this construction is the fact that the Bloch functions of the actual lattice correspond to the vectors of **k** space (here, of course, both **k** space and the actual lattice are one dimensional). It is in fact easy to see that the length of each vector **k** of **k** space is equal to $k = (n/N)b = (2\pi/a)(n/N)$, such that for $n = 0, 1, \ldots, N - 1$ all these vectors are identical to the **k** vectors introduced in the previous section for numbering the BFs.

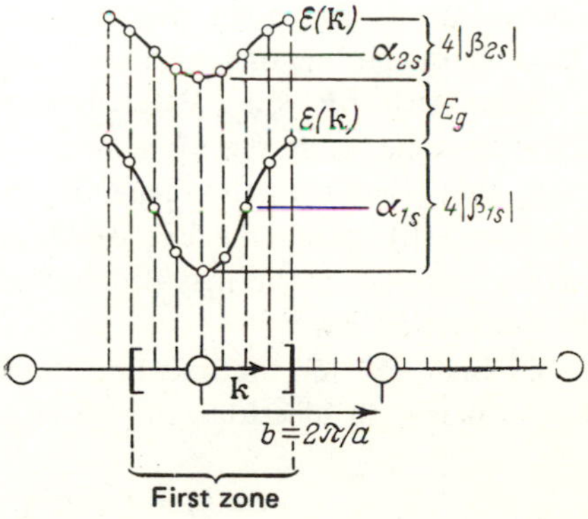

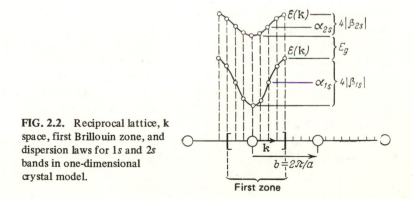

FIG. 2.2. Reciprocal lattice, k space, first Brillouin zone, and dispersion laws for 1s and 2s bands in one-dimensional crystal model.

This correspondence, however, is not a one-to-one correspondence. Whereas one BF of the form (2.5) corresponds to each vector **k** of **k** space, one and the same BF is described by different **k** vectors. In fact, let us take two vectors **k** and **k**′, differing by an integer number of reciprocal lattice periods $p\mathbf{b} = \mathbf{K}$:

$$\mathbf{k}' = \mathbf{k} + \mathbf{K} \tag{2.9}$$

$$\mathbf{K} = p\mathbf{b} \tag{2.10}$$

and let us substitute them into Eq. (2.5) for the BF. Then for the vector **k** the coefficients in (2.5) for the AOs will be equal to exp $(i\mathbf{kR})$, while for the vector **k**′ they will be equal to exp $(i\mathbf{k}'\mathbf{R})$ = exp $(i\mathbf{kR})$ exp $(i\mathbf{KR})$, where the second factor is equal to one in view of the relation (2.10): exp $(i\mathbf{KR})$ = exp $(ip\mathbf{bR})$ = exp $[i(2\pi/a)p \cdot ma]$ = exp $[2\pi i \times \text{an integer}] = 1$.

Thus any vectors, differing from each other by an integral number of periods of the reciprocal lattice, correspond to one and the same BF, so that the **k** vectors from any segment of the reciprocal lattice of length b are sufficient to describe all the BFs of the line. It is convenient to take as this segment the one that is symmetrically located on both sides of the coordinate origin of the reciprocal lattice. This segment, containing N vectors of **k** space, is called the first Brillouin zone.

Let us emphasize that the construction described above has a very clear meaning from a group-theory point of view. Since the C_N group of translations of a line of N atoms has N irreducible (one-dimensional) representations, for numbering the BFs (2.5), it is sufficient to take N **k** vectors from the entire **k** space. It is convenient to choose these vectors such that they fill the segment that is also the first Brillouin zone.

2.1.3 Dispersion Laws and Energy Bands

Let us now turn to an investigation of the electron energy levels in the one-dimensional line of atoms.

Since each BF (2.5), defined by the value of a vector ε from the first Brillouin zone, is an eigenfunction, a one-electron energy level ε corresponds to it and, therefore, also depends on this value **k**: $\varepsilon = \varepsilon(\mathbf{k})$. By taking the function $\varepsilon(\mathbf{k})$ for all the BFs of the line (i.e., for all vectors **k**) we obtain a certain function in the first Brillouin zone. This function $\varepsilon(\mathbf{k})$ is usually called the "dispersion law."

To find $\varepsilon(\mathbf{k})$ for the one-dimensional line of atoms being considered here, it is sufficient to substitute the BF into the formula (1.14):

$$\varepsilon(\mathbf{k}) = \langle \psi(\mathbf{k} \mid \mathbf{r}) \mid \hat{H} \mid \psi(\mathbf{k} \mid \mathbf{r}) \rangle \tag{2.11}$$

Then, taking into consideration the LCAO form of the BF (2.5), we find that $\varepsilon(\mathbf{k})$ is determined by the value of the Coulomb integral:

$$\alpha = \langle \chi_m | \hat{H} | \chi_m \rangle = \langle \chi(\mathbf{r}-\mathbf{R}) | \hat{H} | \chi(\mathbf{r}-\mathbf{R}) \rangle \tag{2.12}$$

and by the values of the resonance integrals describing the interaction of the atoms of the line with one another. If only the interaction of the adjacent, "valence-bound" atoms in the line is taken into consideration, then only one type of resonance integral will enter into the dispersion law:

$$\beta = \langle \chi_m | \hat{H} | \chi_{m+1} \rangle = \langle \chi(\mathbf{r}-\mathbf{k}) | \hat{H} | \chi(\mathbf{r}-\mathbf{k}-\mathbf{a}) \rangle \tag{2.13}$$

In this case the specific form of the dispersion law is completely determined by the integrals (2.12) and (2.13).

Thus it is not hard to derive that for a line composed of atoms with only s functions (the hypothetical H_N molecule) or with only the $p\pi$-type functions (the π system of a cyclic polyene), the dispersion law has the form

$$\varepsilon(\mathbf{k}) = \alpha + 2\beta \cos \mathbf{ka} \tag{2.14}$$

where $\beta = \beta_s$ in the first case and $\beta = \beta_\pi$ in the second (Fig. 2.2).

The dispersion law for a line of atoms with type $p\sigma$ atomic functions has a similar form

$$\varepsilon(\mathbf{k}) = \alpha - 2\beta \cos \mathbf{ka} \tag{2.15}$$

except for the fact that because of the oddness of the atomic functions, the β becomes $-\beta$ in the formula (2.15).

The dispersion law can be represented graphically by plotting the $\mathbf{k}$ values along the abscissa axis and the values of $\varepsilon = \varepsilon(\mathbf{k})$ along the ordinate axis.

In view of the discrete nature of $\mathbf{k}$ space the graph of $\varepsilon(\mathbf{k})$ is not a continuous curve, but is represented by individual points, the number of which is equal to the number of atoms in the system. For large N, however, this discrete nature can be ignored, and the points on the $\varepsilon(\mathbf{k})$ curve nearly merge. For example, for the line depicted in Fig. 2.2, the graph of $\varepsilon(\mathbf{k})$ (2.14) assumes the form of a continuous cosine curve, shifted by the amount α along the ordinate axis. This graph shows one characteristic feature of the energy spectrum of a crystal.

Let us project the points of the $\varepsilon(\mathbf{k})$ curve onto the energy axis. Then their projections fill a continuous interval along the ε axis, i.e., all N levels of our line of atoms are located in one band. This band is called the energy band, the energy zone (not to be confused with the first Brillouin zone!).

In particular, for the dispersion law (2.14) the average level of the band is located at the point $\varepsilon = \alpha$, and the band width amounts to $4|\beta|$. The physical meaning of these results is obvious. In a diatomic molecule, such as H_2, the interaction of the AOs splits the system of levels of the two identical atoms into two molecular levels (see Fig. 1.4). Analogously, in the line of N atoms (each with one valence AO) the interaction of the atoms splits the N-fold degenerate level with energy $\varepsilon = \alpha$ into N levels, which fill the entire band. The amount of splitting in this situation is naturally proportional to the "strength" of the interaction of the AOs, i.e., to the resonance integral $|\beta|$. For small values of $|\beta|$ the splitting is small and the band becomes narrow, approaching a purely atomic level in the limit.

2.1.4 Band Structure

Until now we have been considering the simplest example (one-dimensional) of a crystal in which only one AO is present in a unit cell. In this case the eigenfunctions of the Hamiltonian are wholly determined by the symmetry of the system, and no solution of the secular equations is required to find the eigenfunctions and levels.

The case when several AOs are present in the unit cell in a one-dimensional crystal is more complicated. Such a crystal can be realized in various ways. For example, it can be a line consisting of one kind of atom, but with different bond lengths, so that several atoms will be contained within a unit cell (a polyene with alternating bonds). Another kind of example can be a line consisting of different atoms, such as the hypothetical $(LiH)_N$ molecule of alternating lithium and hydrogen atoms. Finally, one can take a homoatomic line with identical interatomic distances, but consisting of atoms having several valence AOs (for example, a line of hydrogen atoms, where besides the $1s$ orbitals of the H atoms their excited $2s$ orbitals are also taken into consideration).

Formally, the difference between these lines of atoms is inconsequential. According to the general scheme of the LCAO method the only important thing here is the total number of AOs to be considered in a unit cell of the crystal.

Thus let us assume there is a complex line of atoms, consisting of N unit cells, in one cell of which there are m AOs, so that for all the atoms of the line there are a total on mN AOs. Then the eigenfunctions of the Hamiltonian of the line of atoms will be linear combinations of these mN AOs, belonging to irreducible representations of the group of translations and classified in terms of the corresponding values of the vector $\mathbf{k}$.

These combinations can be found in the following manner. In any unit cell let us choose some certain, let us say the jth, AO. Then in view of the equivalence of all the unit cells, that same AO is found in any other cell of the crystal that differs from the originally chosen one only in that it has been translated. Let us now take the N translationally equivalent AOs for all N unit cells of the line. Considered by themselves, these translationally equivalent AOs $\chi_j(\mathbf{r}-\mathbf{R})$ form the simple line, already familiar to us above, whose BFs have the form

$$\psi_j(\mathbf{k}\,|\,\mathbf{r}) = \frac{1}{\sqrt{N}} \sum_{\mathbf{R}} e^{i\mathbf{k}\mathbf{R}} \chi_j(\mathbf{r}-\mathbf{R}) \tag{2.16}$$

We construct these BFs for each of the m AOs χ_j, $j = 1, 2, \ldots, m$ and we will consider the corresponding BFs as the basic BFs of the complex line of atoms. For a given value of the vector $\mathbf{k}$ each of the BFs (2.16) for each j value is transformed into one and the same irreducible representation of the group of translations. Therefore the eigenfunctions of the Hamiltonian of the line must be sought in the form of linear combinations of the basic BFs (2.16) of the type

$$\psi(\mathbf{k}\,|\,\mathbf{r}) = c_1\psi_1(\mathbf{k}\,|\,\mathbf{r}) + c_2\psi_2(\mathbf{k}\,|\,\mathbf{r}) + \ldots + c_m\psi_m(\mathbf{k}\,|\,\mathbf{r}) \tag{2.17}$$

[let us emphasize that one and the same $\mathbf{k}$ value is taken for all the BFs in (2.17)].

To find the coefficients in (2.17) and the corresponding energy levels, according to Sec. 1.3.2 a secular equation of degree m must be solved:

$$\text{Det}\begin{Vmatrix} \langle\psi_1|\hat{H}|\psi_1\rangle - \varepsilon & \langle\psi_1|\hat{H}|\psi_2\rangle & \ldots & \langle\psi_1|\hat{H}|\psi_m\rangle \\ \langle\psi_2|\hat{H}|\psi_1\rangle & \langle\psi_2|\hat{H}|\psi_2\rangle - \varepsilon & \ldots & \langle\psi_2|\hat{H}|\psi_m\rangle \\ \cdots & \cdots & \cdots & \cdots \\ \langle\psi_m|\hat{H}|\psi_1\rangle & \langle\psi_m|\hat{H}|\psi_2\rangle & \ldots & \langle\psi_m|\hat{H}|\psi_m\rangle - \varepsilon \end{Vmatrix} = 0 \tag{2.18}$$

The roots, $\varepsilon_1, \varepsilon_2, \ldots, \varepsilon_m$ of Eq. (2.18) depend on $\mathbf{k}$ (since the matrix elements of the secular determinant depend on $\mathbf{k}$). Thus the secular equation (2.18) defines the m functions

$$\varepsilon = \varepsilon_1(\mathbf{k}),\ \varepsilon = \varepsilon_2(\mathbf{k}),\ \ldots,\ \varepsilon = \varepsilon_m(\mathbf{k}) \tag{2.19}$$

in the first Brillouin zone, i.e., m dispersion laws or, as it is said, m branches of the dispersion law $\varepsilon(\mathbf{k})$.

The scheme of energy levels of any crystal is often spoken of as its band structure. Thus for a complex line of atoms with m AOs in a unit cell the band structure is defined by the m branches of the dispersion law (2.19). Each of the functions (2.19), specified at points of the first Brillouin zone, can be represented in the form of a curve, with any of the curves corresponding to its own band on the ε axis. In the special case when $m = 1$ (a simple line of atoms), the secular equation (2.18) is reduced to the relation

(2.11), which defines a single branch of the dispersion law with the one band corresponding to it.

2.1.5 Genetic Relationship between Bands and Atomic Levels

Since the number of bands m for a complex line of atoms is determined by the number of AOs in a unit cell, these bands are related to the corresponding atomic levels. Thus in the case of a line of hydrogen atoms, for which the $1s$ and $2s$ AOs are taken into consideration, one speaks of the $1s$ and $2s$ bands (see Fig. 2.2), and in the case of a line of alternating hydrogen and lithium atoms one speaks of the $1s$ band of the hydrogen and the $2s$ band of the lithium.

To understand the meaning of this relationship, let us first assume that each jth AO χ_j in any unit cell interacts only with these same functions in the adjacent unit cells, but does not interact with any of the other types of AOs. Then the nondiagonal elements of the secular determinant (2.18) become zero and the determinant assumes the form

$$\mathrm{Det}\begin{Vmatrix} \langle \psi_1 | \hat{H} | \psi_1 \rangle - \varepsilon & 0 & \ldots & 0 \\ 0 & \langle \psi_2 | \hat{H} | \psi_2 \rangle - \varepsilon & \ldots & 0 \\ \cdots & \cdots & \cdots & \cdots \\ 0 & 0 & \ldots & \langle \psi_m | \hat{H} | \psi_m \rangle - \varepsilon \end{Vmatrix} = 0 \qquad (2.20)$$

and its roots will be the functions

$$\varepsilon = \varepsilon_1(\mathbf{k}) = \langle \psi_1 | \hat{H} | \psi_1 \rangle, \ldots, \varepsilon = \varepsilon_m(\mathbf{k}) = \langle \psi_m | \hat{H} | \psi_m \rangle \qquad (2.21)$$

Each of the functions (2.21) has the appearance of the dispersion law for the simple line of atoms containing only one type of AO, and in this case one can in fact speak of the individual χ_1, χ_2, etc., bands. Thus, if the interaction of the $1s$ AO of each hydrogen atom with the $2s$ AO of the adjacent hydrogen atom is ignored for a line of N hydrogen atoms, the determinant (2.20) assumes the form

$$\mathrm{Det}\begin{Vmatrix} \langle \psi_{1s} | \hat{H} | \psi_{1s} \rangle - \varepsilon & 0 \\ 0 & \langle \psi_{2s} | \hat{H} | \psi_{2s} \rangle - \varepsilon \end{Vmatrix} = 0 \qquad (2.22)$$

Then the dispersion law is given by the formulas

$$\varepsilon = \varepsilon_1(\mathbf{k}) = \langle \psi_{1s} | \hat{H} | \psi_{1s} \rangle; \quad \varepsilon = \varepsilon_2(\mathbf{k}) = \langle \psi_{2s} | \hat{H} | \psi_{2s} \rangle \qquad (2.23)$$

which have the form of (2.14). Correspondingly, in the band structure of such a line of atoms one can separate the individual $1s$ and $2s$ bands, the midlevels of which are equal to the energies of the atomic 1*s* and 2*s* levels α_{1s} and α_{2s} and the widths of which are $4|\beta_{1s}|$ and $4|\beta_{2s}|$ (Fig. 2.2).

In reality, however, each function χ_j interacts not only with these same AOs in the adjacent unit cells, but also with the other types of AOs in this same cell as well as in the other unit cells. Therefore the nondiagonal elements of the determinant (2.20) do not become zero in reality and the division of the band structure into $\chi_1, \chi_2, \ldots$ bands is rather arbitrary. Actually, the BF, corresponding to the *j*th band,.contains not only the AOs χ_j, but also other AOs $\chi_{j'}$ with $j' \neq j$. For example, the BFs corresponding to the 1*s* band in the H_N line of atoms contain the impurity 2*s* AOs (and vice versa), while the BFs belonging to the 2*s* band of lithium in the chain . . . H—Li—H—Li . . . contain the impurity 1*s* AOs of hydrogen. Such "mixing" of states is called band "hybridization," and in this case one speaks of "hybrid" bands.

In conjunction with band hybridization let us point out one more fact. As follows from the form of the BFs (2.16), all the matrix elements of the determinant (2.18) are functions of **k**. Therefore **k** values, for which the nondiagonal elements become zero, are possible. At these points of the first zone a division into purely atomic bands occurs even in the strict sense of the word.

Let us take, for example, a line consisting of atoms with one *s* and one *pσ* orbital (the σ system of the cumulene . . . C=C=C=C . . .), and let us consider the center of the **k** = 0 band. For **k** = 0 all the exponential terms are $\exp(i\mathbf{kR}) = 1$ and the corresponding basic BFs (2.16) have the form

$$\begin{aligned} \psi_1 &= \psi_s(0 \mid \mathbf{r}) = \frac{1}{\sqrt{N}} \sum_{\mathbf{R}} s(\mathbf{r}-\mathbf{R}) \\ \psi_2 &= \psi_p(0 \mid \mathbf{r}) = \frac{1}{\sqrt{N}} \sum_{\mathbf{R}} p(\mathbf{r}-\mathbf{R}) \end{aligned} \tag{2.24}$$

It is easy to see that the nondiagonal elements $\langle\psi_s|\hat{H}|\psi_p\rangle$ with the BFs (2.24) become zero since the interaction of each *pσ* AO with an *s* AO to the right is compensated by the interaction with that *s* AO that is at the left, which is equal in magnitude but opposite in sign. Therefore at the center of the band the levels are strictly divided into *s*- and *p*-type levels. Such a compensation does not occur for the other **k** values and, consequently, the other **k** values correspond to hybrid states.

2.1.6 Band Structure and Physical Properties of a Crystal

Finally, let us consider, for the line of atoms example, how the band structure is related to the physical properties of a crystal.

According to the one-electron model, the electrons in crystals fill, pairwise, the levels of each energy band, starting from the bottom level of the lowest band. A dual nature situation is possible then.

(a) The electrons are insufficient to fill the entire number of bands and one of them is only partially filled. For example, in a line of N hydrogen atoms the number of electrons is equal to N and they can fill only the $N/2$ lower levels of the 1s band. Then empty levels are located above the uppermost filled level to a distance of the order of $|\beta|/N$ (as $N \to \infty$, it approaches zero). An electron can move into these levels with an insignificant expenditure of energy. A one-dimensional metal will be such a line of atoms.

(b) The electrons in the crystal only suffice to fill several of the lower bands (the "valence band"), and the rest of the bands (the "conduction band") are separated from the filled by an energy gap (the "forbidden band"). This gap does not approach zero as $N \to \infty$, so that for the excitation of an electron into an empty level a finite energy must be expended. A one-dimensional insulator will be such a line of atoms. For example, in a line of N helium atoms there are $2N$ electrons that completely fill the 1s band. In order for the $(He)_N$ line of atoms to be able to conduct an electrical current, an electron must be excited into a level of the 2s band. This, as a minimum, requires the expenditure of an energy equal to the distance between the upper level of the 1s band (the "top of the valence band") and the lower level of the 2s band (the "bottom of the conduction band").

According to Eq. (2.14) this distance (the "forbidden band gap," E_g) is equal to

$$E_g = \alpha_{2s} - \alpha_{1s} + 2\beta_{2s} + 2\beta_{1s} \tag{2.25}$$

which amounts to ~20 eV for helium.

2.2 BAND THEORY FOR THREE-DIMENSIONAL CRYSTALS

2.2.1 Translational Symmetry and the Bloch Functions of a Three-Dimensional Lattice

For a real three-dimensional crystal the translational symmetry consists of the following: three linearly independent vectors (the basis vectors of the lattice; Fig. 2.3) can be selected in it

$$\mathbf{a}_1,\ \mathbf{a}_2,\ \mathbf{a}_3 \tag{2.26}$$

such that all the atoms of the crystal coincide with identical atoms upon a displacement along any of the vectors $\mathbf{a}_i$ or along their arbitrary linear combination with integer-number coefficients:

$$\mathbf{R} = m_1\mathbf{a}_1 + m_2\mathbf{a}_2 + m_3\mathbf{a}_3 \tag{2.27}$$

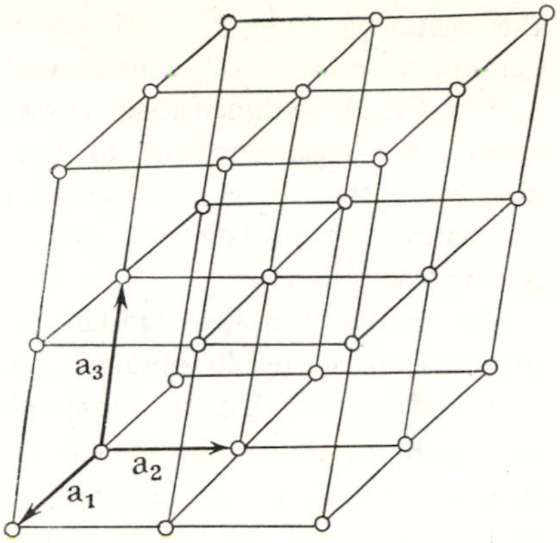

FIG. 2.3. Crystal, unit cell, and basis vectors of lattice.

The parallelepiped, formed by the vectors (2.26), is the unit cell of the crystal.*

The simplest examples of crystal lattices are shown in Fig. 2.4.

A real crystal does not possess complete translational symmetry since the translation of an atom at the surface of the specimen takes it outside the boundaries of the crystal. Therefore for simplicity in band theory a fictitious "cyclic" crystal in the form of a parallelepiped with formally identical opposite faces (the cyclic Born-Karman conditions) is considered instead of the actual crystal. Such a cyclic crystal is analogous to the closed line of atoms considered above. To use the cyclic conditions the crystal must be in the form of a parallelepiped with edges parallel to the vectors (2.26). Let us assume this parallelepiped has N_1, N_2, N_3 lattice periods in the directions $\mathbf{a}_1$, $\mathbf{a}_2$, $\mathbf{a}_3$, respectively. For a displacement, for example, by the vector $\mathbf{a}_1$ all the unit cells of the crystal coincide with other cells, except for the layer that consists of the cells located at the front face of the crystal. When the cyclic conditions are met, it is assumed that this layer, upon translation, moves to the location that was previously occupied (before the translation) by the layer at the rear face of the crystal. The situation is similar with translations along the vectors $\mathbf{a}_2$ and $\mathbf{a}_3$, so that for N_i successive translations in the direction of

*The geometry and theory of the symmetry of crystal lattices are described in all books on solid state physics, crystallography, and crystal chemistry; see, for example, [47, 48].

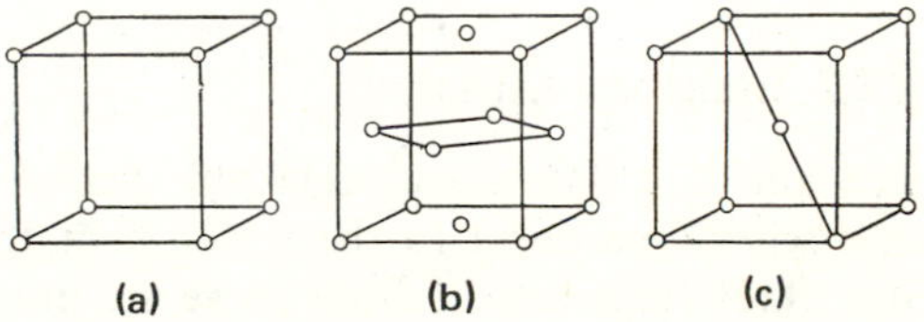

FIG. 2.4. Cubic lattices: (a) simple, (b) face-centered, (c) body-centered.

the vector $\mathbf{a}_i$ (i = 1, 2, 3) all the atoms of the crystal, having made a "complete revolution," coincide with themselves.

For the three-dimensional crystal, just as for the one-dimensional line of atoms, the eigenfunctions of the one-electron Hamiltonian (BFs) can be chosen so that they will belong to the irreducible representations of the translation group. This fact makes it possible to find how the BFs behave during translations.

Let us first consider translations in the direction of the vectors (2.26) and make use of the results obtained for the one-dimensional line.

It is obvious that the ensemble of translations along the vector $\mathbf{a}_1$ is equivalent to the group of translations of a line consisting of N_1 units and having a period $\mathbf{a}_1$. Accordingly, the irreducible representations of the group of translations of the crystal along $\mathbf{a}_1$ and of the group of translations of the line are identical and, consequently, for a translation along $\mathbf{a}_1$ the BFs of the crystal are multiplied by $\exp(2\pi i n_1/N_1)$, where n_1 is an integer. Similarly for a translation along the vector $\mathbf{a}_2$ the BFs of the crystal are multiplied by $\exp(2\pi i n_2/N_2)$, and for a translation along $\mathbf{a}_3$ by $\exp(2\pi i n_3/N_3)$.

Hence, for a translation along the arbitrary vector

$$\mathbf{R}_0 = p_1\mathbf{a}_1 + p_2\mathbf{a}_2 + p_3\mathbf{a}_3 \tag{2.28}$$

where p_1, p_2, p_3 are integers, the BF is multiplied by the factor

$$\exp\left[2\pi i\left(\frac{n_1 p_1}{N_1} + \frac{n_2 p_2}{N_2} + \frac{n_3 p_3}{N_3}\right)\right] \tag{2.29}$$

(the Bloch condition).

Let us rewrite the Bloch condition in vector form, after writing the exponent in the form

$$2\pi\left(\frac{n_1 p_1}{N_1} + \frac{n_2 p_2}{N_2} + \frac{n_3 p_3}{N_3}\right) = \mathbf{k}\mathbf{R}_0 \tag{2.30}$$

Then each BF for the three-dimensional lattice is characterized by the value of the vector $\mathbf{k}\psi = \psi(\mathbf{k}/\mathbf{r})$, and the transformation of the BF upon a translation along the vector $\mathbf{R}_0$ $[\psi(\mathbf{k}/\mathbf{r}) \rightarrow \psi(\mathbf{k}/\mathbf{r} - \mathbf{R}_0)]$ is written in the form of (2.7), where $\mathbf{R}_0$ and $\mathbf{k}$ are three-dimensional vectors.

2.2.2 Reciprocal Lattice

For the one-dimensional line the period b of the reciprocal lattice was inversely proportional to the period a of the direct lattice (see Sec. 2.1.2). In the three-dimensional case, however, it makes no sense to define the reciprocal lattice as the lattice with the basis vectors $\mathbf{b}_1, \mathbf{b}_2, \mathbf{b}_3$ such that the

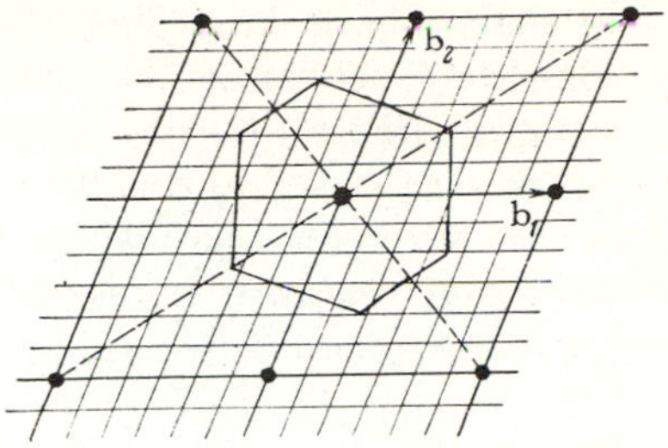

FIG. 2.5. Reciprocal lattice, k space, and first Brillouin zone (two-dimensional model).

direction of each vector $\mathbf{b}_i$ coincides with the direction of the corresponding vector $\mathbf{a}_i$ of the direct lattice, and the length b_i is inversely proportional to a_i.

Such a definition would not satisfy Eq. (2.30) since the vectors (2.26) in the general case are not orthogonal, while in an oblique-angled coordinate system the scalar product of two vectors is not written in the form of the sum of the products of the components of the vectors. Therefore in the three-dimensional case the reciprocal lattice is defined in a somewhat more complicated manner.

Let us consider the vectors

$$\mathbf{b}_1, \ \mathbf{b}_2, \ \mathbf{b}_3 \tag{2.31}$$

for which the equations

$$\begin{gathered} \mathbf{a}_1\mathbf{b}_1 = 2\pi; \quad \mathbf{a}_2\mathbf{b}_2 = 2\pi; \quad \mathbf{a}_3\mathbf{b}_3 = 2\pi \\ \mathbf{a}_1\mathbf{b}_2 = \mathbf{a}_1\mathbf{b}_3 = 0; \quad \mathbf{a}_2\mathbf{b}_1 = \mathbf{a}_2\mathbf{b}_3 = 0; \quad \mathbf{a}_3\mathbf{b}_1 = \mathbf{a}_3\mathbf{b}_2 = 0 \end{gathered} \tag{2.32}$$

are satisfied, or, more succinctly,

$$\mathbf{a}_i\mathbf{b}_j = 2\pi\delta_{ij} \tag{2.33}$$

(i.e., the vector $\mathbf{b}_1$ is orthogonal to the vectors $\mathbf{a}_2$ and $\mathbf{a}_3$, and the scalar product of $\mathbf{b}_1$ and $\mathbf{a}_1$ is equal to 2π, etc.).* The vectors (2.31) are also the basis vectors of the reciprocal lattice (Fig. 2.5). They can also be written in explicit form

$$\mathbf{b}_1 = \frac{2\pi\,[\mathbf{a}_2\mathbf{a}_3]}{\mathbf{a}_1\,[\mathbf{a}_2\mathbf{a}_3]}; \qquad \mathbf{b}_2 = \frac{2\pi\,[\mathbf{a}_3\mathbf{a}_1]}{\mathbf{a}_2\,[\mathbf{a}_3\mathbf{a}_1]}; \qquad \mathbf{b}_3 = \frac{2\pi\,[\mathbf{a}_1\mathbf{a}_2]}{\mathbf{a}_3\,[\mathbf{a}_1\mathbf{a}_2]} \tag{2.34}$$

where $[\mathbf{a}_i\mathbf{a}_j]$ is the vector product, and $\mathbf{a}_i[\mathbf{a}_j\mathbf{a}_k]$ is the scalar triple product of the vectors $\mathbf{a}_i$, $\mathbf{a}_j$, and $\mathbf{a}_k$.

*Let us point out that for the special case of orthogonal vectors (2.26) the vectors (2.31) are also orthogonal, and each vector (2.31) coincides in direction with the corresponding vector (2.26).

As follows from the relations (2.32), the geometry of the reciprocal lattice is completely determined by the geometry of the real lattice. Thus for a simple cubic lattice the reciprocal lattice is also simple cubic. In this case the length of each vector $\mathbf{b}_i$ is equal to $2\pi/a$, where a is the period of the direct lattice the reciprocal lattice will be body-centered cubic.

For future reference let us also point out that for a face-centered cubic lattice the reciprocal lattice will be body-centered cubic.

2.2.3 k Space and First Brillouin Zone for Three-Dimensional Crystals

Now it is not hard to determine what is meant by **k** space in the three-dimensional case. Let us divide the period b_1 of the reciprocal lattice into N_1 parts, the period b_2 into N_2 parts, and the period b_3 into N_3 parts so that each unit cell of the reciprocal lattice is divided into $N_1 = N_2 = N_3 = N$ small parallelepipeds (see Fig. 2.5). The ensemble of the sites of the fine lattice formed in this manner will be the **k** space for the three-dimensional case. It is obvious that each vector of **k** space is written in the form

$$\mathbf{k}=\frac{n_1}{N_1}\mathbf{b}_1+\frac{n_2}{N_2}\mathbf{b}_2+\frac{n_3}{N_3}\mathbf{b}_3 \tag{2.35}$$

where n_1, n_2, n_3 are integers.

Then the scalar product of the vector $\mathbf{R}_0$ (2.28) by the vector (2.35) will be exactly equal to the exponent (2.30) of the exponential term by which the BF $\psi(\mathbf{k}/\mathbf{r})$ is multiplied upon a translation along $\mathbf{R}_0$.

Let us recall that the vector **k** is needed here only to characterize the behavior of the BF during translations.* Therefore each vector **k** is determined within the accuracy of an arbitrary integral linear combination of the basis vectors of the reciprocal lattice:

$$\mathbf{K}=q_1\mathbf{b}_1+q_2\mathbf{b}_2+q_3\mathbf{b}_3 \tag{2.36}$$

where q_1, q_2, q_3 are integers.

Actually, in view of the equations (2.32) we have

$$\mathbf{K}\mathbf{R}_0=2\pi\,(q_1p_1+q_2p_2+q_3p_3)=2\pi\times\text{ an integer} \tag{2.37}$$

so that the vectors **k** and $(\mathbf{k} + \mathbf{K})$ correspond to identical values of the exponent (2.29).

*In addition to the scheme of so-called reduced bands used here, the "scheme of expanded bands" ("large bands," "Jones's bands") is also used. In the scheme of expanded bands the vector **k** also uniquely characterizes the energy of the BF and is defined in all **k** space and not just in the first zone [41].

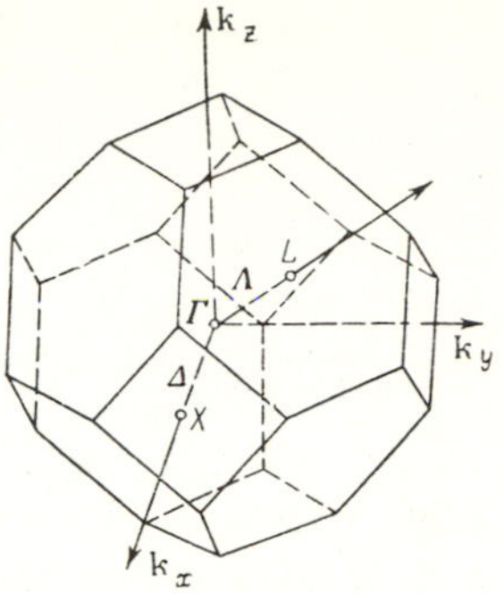

FIG. 2.6. First Brillouin zone, symmetry directions, and points for face-centered cubic lattice, as well as the diamond, zinc sulfide, and sodium chloride lattices.

Thus in the three-dimensional case, just as in the one-dimensional, it is sufficient to consider only the **k** vectors whose ends fill any volume in **k** space containing $N = N_1N_2N_3$ points of the fine lattice. As an example one can take a unit cell of the reciprocal lattice as this volume. It is more convenient, however, to use a polyhedron of this same volume that possesses all of the symmetry elements of the reciprocal lattice. Such a polyhedron is obtained if the coordinate origin of the reciprocal lattice is connected to the nearest sites of this lattice and planes drawn, to bisect these line segments.

The polyhedron constructed in this manner is called the first Brillouin zone (or simply the first zone; see Fig. 2.5). It is obvious that for a simple cubic (direct) lattice the first Brillouin zone is a cube with the center at the coordinate center of the reciprocal lattice and with an edge equal in length to the period of the reciprocal lattice.

Figure 2.6 shows the first Brillouin zone for a face-centered cubic lattice. For this lattice the reciprocal lattice is body-centered cubic; therefore the first zone can be obtained from a cube by cutting its eight corners with planes.

2.3 CALCULATION OF BAND STRUCTURE IN LCAO METHOD

2.3.1 Definition of Dispersion Law

The calculation of the band structure of three dimensional crystals in the LCAO method (this method is often called the "tight binding approximation") is accomplished in the same way as for the one-dimensional line of atoms.

If the unit cell of the direct lattice of the crystal contains one atom possessing a single AO, then the form of the BF is wholly determined by the translational symmetry. In this case the BFs are written in the form (2.5). Here **R** and **k** are three-dimensional vectors; the vector **R** denotes translationally equivalent sites of the direct lattice at which the atoms are, while **k** assumes values within the boundaries of the first Brillouin zone.

The dispersion law $\varepsilon(\mathbf{k})$ in this case is found from Eq. (2.11), where $\hat{H}$ is the effective one-electron Hamiltonian of the three-dimensional crystal.

Let us now consider a complex lattice containing several atoms in its unit cell, with each of them being able to have several AOs. Let us first make use of the translational symmetry, after constructing the basic BFs of the form (2.16) for each AO of each atom. In this case the eigenfunctions of the Hamiltonian will be linear combinations (2.17) of the basic BFs (2.16) and the finding of the dispersion laws reduces to solving the secular equation (2.18), the order m of which is equal to the total number of AOs to be considered in a unit cell of the crystal; the m roots of the secular equation (2.18) define the m branches of the dispersion laws that we seek. The only difference from the one-dimensional line is in the form of the functions $\varepsilon_i(\mathbf{k})$ themselves, which are scalar functions of the three-dimensional vector argument $\mathbf{k}$. Accordingly, the geometrical representation of the band structure of a three-dimensional crystal is given by a series of three-dimensional "hypersurfaces" in a four-dimensional space with axes corresponding to the three components $\mathbf{k}_x$, $\mathbf{k}_y$, $\mathbf{k}_z$ of the vector $\mathbf{k}$ and the energy ε.

For the graphical representation of the band structure the function $\varepsilon(\mathbf{k})$ is not considered within the entire first zone, but at isolated "symmetrical" points or along isolated "symmetrical" directions. For the crystals to be considered below with the structure of diamond and zinc sulfide (sphalerite) the directions $\Delta = [100]$ and $\Lambda = [111]$ are usually taken as these directions, i.e., vectors of the form $\mathbf{k} = \{k_x, 0, 0\}$ or $\mathbf{k} = \{k_x, k_x, k_x\}$ (see Fig. 2.6). In view of the so-called point symmetry of these crystals the dispersion laws are the same for each of all the six directions (symmetrically equivalent to the [100] direction) and for the eight directions equivalent to the [111] direction (see Sec. 2.5.2). Therefore the dispersion laws for the [100] and [111] directions give a good representation of the band structure within the entire first zone. There is also another reason that often forces one to consider only the symmetrical directions: the use of the point symmetry of the crystal in these directions makes it possible to lower the order of the secular equation (2.18).

It is obvious that along each such line in the first zone the function $\varepsilon(\mathbf{k})$ is transformed from a hypersurface to a conventional curve. Examples of such a graphical representation of the band structure along the symmetrical directions are given in Fig. 3.2.

2.3.2 Energy Bands and Band-to-Band Transitions in a Three-Dimensional Crystal

A characteristic property of the energy level structure in a three-dimensional crystal, just as for the one-dimensional model, is the presence of isolated bands (energy bands).

Let us actually consider a crystal with m AOs in the unit cell and let us choose a certain branch of the dispersion law $\varepsilon = \varepsilon_i(\mathbf{k})$. By taking all of the allowable **k** values from the first zone for it and marking the corresponding ε values on the energy axis, we obtain a band of N levels. The number of such bands is equal to the number of branches of the dispersion law. In the general case all the bands are hybrid and the BFs, corresponding to each of them, contain all the AOs of the crystal. However, for selected **k** values, falling at the symmetry elements of the reciprocal lattice, some of the bands or even all the bands can be "pure" in the sense that the BFs belonging to them are composed of only a certain type of AO.

The structure of the bands determines the physical properties of the crystal, and everything said above for the one-dimensional line is valid also for real three-dimensional crystals: the crystal has the properties of a metal when the uppermost band of those filled with electrons is only partially filled. Conversely, it will be an insulator when the valence band is separated from the conduction band by an energy gap (the forbidden band).

Crystals with a band structure typical of an insulator for which the value of E_g is less than 2-3 eV are segregated (arbitrarily) into the separate class called semiconductors. Thus diamond with E_g = 5.4 eV is considered to be an insulator, while silicon (E_g = 1.2 eV) or germanium (E_g = 0.8 eV) are semiconductors.

When an insulator or semiconductor is excited, an electron from some level ε_i^v of the valence band can be transferred to a level ε_j^c of the conduction band. The energy difference

$$\Delta\varepsilon_{ij} = \varepsilon_j^c - \varepsilon_i^v \tag{2.38}$$

is called the energy of the band-to-band transition $\varepsilon^v \to \varepsilon^c$.

Since certain **k** values belong to the levels in the crystal, both levels ε_i^v and ε_j^c can be associated with one and the same **k** value:

$$\Delta\varepsilon_{ij} = \varepsilon_j^c(\mathbf{k}) - \varepsilon_i^v(\mathbf{k}) \tag{2.39}$$

Such a transition $\varepsilon_i^v(\mathbf{k}) \to \varepsilon_j^c(\mathbf{k}')$ is called a direct transition (Fig. 2.7a). If the levels ε_i^v and ε_j^c correspond to different **k** values, one speaks of the indirect transition $\varepsilon_i^v(\mathbf{k}) \to \varepsilon_j^c(\mathbf{k}')$ with an energy

$$\Delta\varepsilon_{ij} = \varepsilon_j^c(\mathbf{k}') - \varepsilon_i^v(\mathbf{k}) \tag{2.40}$$

In different crystals the forbidden band can correspond both to direct and to indirect transitions (Fig. 2.7b, c). Thus in diamond, silicon, germanium, or certain $A^{III}B^V$ compounds (such as GaP) E_g corresponds to indirect transitions, while in GaAs and InSb, let us say, it corresponds to direct transitions.

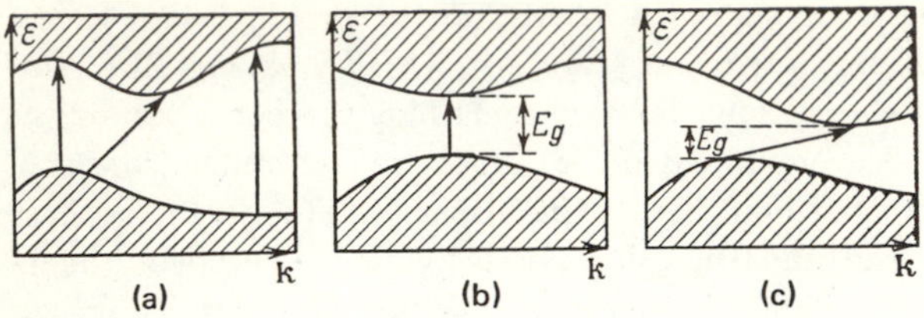

FIG. 2.7. Band-to-band transitions: (a) direct and indirect transitions, (b) forbidden band gap equal to direct transition energy, (c) forbidden band gap equal to indirect transition energy.

In band theory (in contrast to molecular quantum chemistry) there are bases for identifying the difference $\varepsilon_j-\varepsilon_i$ of the one-electron energies with the actual expenditure of energy during excitation of the system. This follows from the Hartree–Fock scheme and follows from Eq. (1.46)* since for a three-dimensional system of $\sim 10^{23}$ atoms $J_{ij} = K_{ij} = 0$. In fact, let us point out that the inequalities [7]

$$0 \leqslant K_{ij} \leqslant J_{ij} \leqslant \frac{1}{2}(J_{ii}+J_{jj}) \tag{2.41}$$

are valid for these quantities. Therefore it is sufficient to prove that for any i

$$J_{ii}=\int \frac{\rho_i(1)\,\rho_i(2)}{r_{12}}\,dv_1\,dv_2=0 \tag{2.42}$$

Here ρ_i is the electron density corresponding to the ith eigenfunction. For a crystal the distribution ρ_i has the periodicity of the lattice so that (2.42) is equal to the electrostatic "eigen" energy of a single charge (the eigenfunctions are normalized), distributed quasiuniformly throughout the volume of the crystal. In the general case this energy is of the order of q^2/a, where q is the total charge and a is the characteristic linear dimension of the volume within which it is distributed. For a real crystal specimen $a \sim 10^8$ (in atomic units). Since $q = 1$ for normalized functions, then $J_{ii} \sim 10^{-8}$, which was to be proved. Strictly speaking it must be borne in mind here, of course, that the Hartree–Fock method ignores correlation and also that the difference in total energies is equal to the difference in one-electron energies only for the exact Hartree–Fock Hamiltonian. In conjunction with the latter let us mention reference [50], where the question of the "Koopmans' corrections" when different approximations for the exchange potential are used is discussed.

2.3.3 Matrix Elements for Three-Dimensional Crystal in the LCAO Method

Since the LCAO approximation in band theory is a special case of the general LCAO scheme, everything said in Sec. 1.3.3 concerning the matrix elements is

*See [49] for another proof (for the case of a filled valence band and an extra electron in the conduction band).

also valid for crystals. It must be remembered, however, that the secular equation (2.18) does not have exactly the same meaning as Eq. (1.38). It is not written for the AOs as is Eq. (1.38), but for linear combinations of the AOs, symmetrized with respect to the irreducible representations of the group of translations. Therefore the elements of the determinant (2.18) will not be Coulomb and resonance integrals of the type (1.49) and (1.53). Taking into consideration, however, that the basic BFs (2.17) are linear combinations of the AOs, they can be expressed in terms of the matrix elements in the basis of the AOs, i.e., in terms of the integrals (1.49) and (1.53), to which everything that has been said in Sec. 1.3.3 applies completely.

2.3.4 Coefficients for Bloch Functions and the Effective Charges in Crystals

Solving the secular equation (2.18) also makes it possible to find the eigenfunctions of the one-electron Hamiltonian of the crystal, i.e., the coefficients $c_1, c_2, \ldots, c_m$ in the expansion (2.17) of the eigenfunctions in terms of the basic BFs. To do this one must formulate the system of equations

$$\begin{gathered}
[\langle \psi_1 | \hat{H} | \psi_1 \rangle - \varepsilon(\mathbf{k})] c_1 + \langle \psi_1 | \hat{H} | \psi_2 \rangle c_2 + \ldots + \langle \psi_1 | \hat{H} | \psi_m \rangle c_m = 0 \\
\langle \psi_2 | \hat{H} | \psi_1 \rangle c_1 + [\langle \psi_2 | \hat{H} | \psi_2 \rangle - \varepsilon(\mathbf{k})] c_2 + \ldots \\
+ \langle \psi_2 | \hat{H} | \psi_m \rangle c_m = 0 \\
\cdots\cdots\cdots\cdots\cdots\cdots\cdots\cdots \\
\langle \psi_m | \hat{H} | \psi_1 \rangle c_1 + \langle \psi_m | \hat{H} | \psi_2 \rangle c_2 + \ldots + [\langle \psi_m | \hat{H} | \psi_m \rangle - \varepsilon(\mathbf{k})] c_m = 0
\end{gathered} \tag{2.43}$$

The solutions of the system (2.43) define the unknown coefficients c_i.

Since the elements of the determinant (2.18) are written in the basis of BFs, they, along with the roots of Eq. (2.18), depend on **k**. Therefore the c_i will also be functions of **k**. To find these coefficients one can first pick a specific branch of the dispersion law $\varepsilon = \varepsilon_i(\mathbf{k})$ and then substitute different values of **k** in turn into the function $\varepsilon_i(\mathbf{k})$ and into the matrix elements $\langle\psi_i|\hat{H}|\psi_j\rangle$. Then for each branch $\varepsilon_i(\mathbf{k})$ and each value of **k** there will be a system of equations (2.43) with numerical coefficients, the solution of which will be the set of numbers $c_1(\mathbf{k}), c_2(\mathbf{k}), \ldots, c_m(\mathbf{k})$. By considering all N values of **k** from the first zone, we obtain N sets of numerical coefficients c_i, i.e., one set of the functions $c_1(\mathbf{k}), \ldots, c_m(\mathbf{k})$ for the selected branch $\varepsilon_i(\mathbf{k})$. Continuing this operation m times, we arrive at m sets of functions $c_1(\mathbf{k}), \ldots, c_m(\mathbf{k})$ for all the branches $\varepsilon_i(\mathbf{k})$.

Let us assume there is a set $c_1, \ldots, c_m$ of coefficients of the expansion of some eigenfunction in terms of the basic BFs (2.16). Then the quantity

$$\frac{1}{N} | c_i(\mathbf{k}) |^2 \tag{2.44}$$

is equal to the partial population of each of the AOs in the ith basic BF for a given **k** value.

To obtain a formula analogous to (1.56) for the total population of a given AO, it is necessary to sum the partial populations, i.e., to sum (2.44) over all **k** values (over the entire first Brillouin zone), as well as over all the filled eigenfunctions (over all the bands of the valence band). Thus for each AO the total population is expressed by the formula:

$$\nu = \frac{2}{N} \sum_{\substack{\text{over filled} \\ \text{bands}}} \sum_{\substack{\text{over first} \\ \text{zone}}} |c_i(\mathbf{k})|^2 \tag{2.45}$$

Then the effective number of electrons in each atom A is calculated from the formula

$$n_{\mathrm{A}} = \sum_{\substack{\text{over all AOs} \\ \text{of atom A}}} \nu = \frac{2}{N} \sum_{\substack{\text{over all AOs} \\ \text{of atom A}}} \sum_{\substack{\text{over filled} \\ \text{bands}}} \sum_{\substack{\text{over first} \\ \text{zone}}} |c_i(\mathbf{k})|^2 \tag{2.46}$$

and the effective charge is calculated from Eq. (1.58).

In practice, of course, it is impossible to use Eqs (2.45) and (2.46) verbatim since an actual summation over $N \approx 10^{23}$ **k** values is impossible. In practice the eigenfunctions are found for only some comparatively small number of points in the first zone for computing the effective charges. There is, however, another method to avoid the difficulty with the summation in Eqs. (2.45) and (2.46). This method involves the so-called "equivalent orbitals" (see Sec. 2.6.1).

2.4 OTHER METHODS OF CALCULATING THE BAND STRUCTURE*

2.4.1 Preliminary Comments

As mentioned, the LCAO approximation (or EO LCAO; see below) makes it possible to relate the properties of the crystal to the properties of the atoms and to the interaction of the atoms.

If this is not the desired goal, then along with the basis of AOs or instead of it one can use functions in the form of plane or diverging waves. This provides an advantage in accuracy compared with the nonempirical LCAO method. Such an idea serves as the basis of the four methods of calculating the band structure that are the most widely used along with the LCAO method†: the associated plane wave (APW) method, Green

*Ziman [51] recently published a complete and critically written review of all the main calculation methods; see also [38–40, 42].

†The LCAO method suggested by Bloch in 1928 for a long time occupied an inferior role to the other methods. Recently, however, there has been a rekindling of interest in the LCAO method and a considerable increase in the number of papers devoted to its application to a variety of materials; see, for example, [52–54] (diamond), [55] (diamond, Si, and Na), [56] (graphite), [57] (lithium halides, NiS, and VO).

function method [the Korringa-Kohn-Rostoker (KKR) method], the orthogonalized plane wave (OPW) method, and the empirical pseudopotential (EP) method. Here we shall briefly describe these methods, the results of which are continually referred to below, and analyze them from the viewpoint of using them in the theory of the chemical bond. We shall first consider the methods that basically allow a nonempirical treatment, although none of them, in reality, represents an approach from the viewpoint of "first principles," and empirical correction parameters very often are introduced even in the "nonempirical" methods (see Sec. 2.4.5).

2.4.2 Associated Plane Wave Method

The assumption [58] that the electrons in a crystal are subject to the action of the potential of the atomic core near the atoms (ions) of the crystal and move like free particles in the space between atoms serves as the basis of the APW method ([38, 40, 42, 51]; see also [58–60] and the reviews [61, 62]). In accordance with this the volume of the crystal is divided into parts by means of nonintersecting "APW spheres," drawn around the atoms. Inside each sphere the potential V of the crystal is assumed to be spherically symmetrical, while in the rest of the space it is assumed to be equal to a constant. The basis wave functions (APWs), which have the form of plane waves between the spheres and within each sphere and are obtained by solving the Schroedinger equation in the corresponding spherical potential well, are responsible for this "MT potential" (muffin tin potential). The individual APWs obtained in this manner are considered to be the basis functions, in terms of which the solutions of the Schroedinger equation for the entire crystal are expanded; a variational method is used for finding the coefficients in this expansion together with the corresponding energy.

In modern APW calculations the number of basis functions is ~50 for homoatomic and ~100 for heteroatomic diatomic crystals [62]. In this situation the solutions of the Schroedinger equation inside the APW spheres are expanded in terms of the spherical harmonics Y_e^m up to l = 10–15 (see for example, [63]; a calculation of a Ag crystal with l = 13); moreover, from 50 to 100 iterations are required for each **k** value [62] so that the use of the APW method is a cumbersome procedure that cannot be done without computers.

An important problem in the APW method is the selection of the radius of the APW spheres and the finding of the potential V in these spheres. In most cases the atomic potential is used as this potential, to which (usually for ionic crystals) is added the potential of the surrounding atoms, for example, in the form of the Madelung potential of the lattice $\pm MZ/R$ [64], where M is the Madelung constant, Z is the effective charge, and R is the smallest interatomic distance. The atomic potential is most often calculated from the Hartree–Fock atomic functions as computed by Herman and Skillman [9]; the exchange portion of the potential is taken in the form of the Slater $\rho^{1/3}$ potential for an electron gas [8].

Let us point out that even within the limits of all these approximations the determination of the potential in the crystal is based to a certain extent on intuition since it is based on certain concepts of the mechanism of the chemical bond.* Thus in the case of covalent crystals (diamond) in the calculation of the atomic potentials it is assumed *a priori* that the atoms are in the $nsnp^3$ valence state [67], whereas for crystals

*With the recent advent of more powerful computers a procedure was developed for the self-consistent calculation of the potential in a crystal by computing the electron density over several selected states in the zone, followed by an integration of the Poisson equation (see, for example, [62], [65], and also [66], where the APW method is extended to the case of a potential, different from its MT form, and is applied to the case of Li).

with a marked degree of ionicity (AgCl, AgBr) one usually starts from a purely ionic model ($Ag^+ Hal^-$, [64]). This comment also applies to the selection of the radius of the APW spheres, which is based on the assumption that the atomic radius corresponds to the position of the maximum on the curve representing the superposition of the atomic potentials [64, 68, 69].

2.4.3 Green Function Method

The Green function method is quite similar to the APW method ([70, 71]; see also the review [72]), extended to the case of several atoms in a unit cell [73]. The basis of this method is the integral form of the Schroedinger equation (1.7) for the Bloch functions

$$\psi(\mathbf{k}\,|\,\mathbf{r}) = \int_{\tau_0} G(\mathbf{r}, \mathbf{r}')\, V(\mathbf{r}')\, \psi(\mathbf{k}\,|\,\mathbf{r}')\, dv \tag{2.47}$$

where the Green function $G(\mathbf{r},\mathbf{r}')$, with the periodicity of the potential in the crystal taken into account, has the form

$$G(\mathbf{r}, \mathbf{r}') = -\frac{1}{\tau_0} \sum_{\mathbf{K}} \frac{\exp[i(\mathbf{k}+\mathbf{K})(\mathbf{r}-\mathbf{r}')]}{(\mathbf{k}+\mathbf{K})^2 - \varepsilon} \tag{2.48}$$

(where **K** are vectors of the reciprocal lattice, τ_0 is the volume of a unit cell of the direct lattice) and depends on the lattice symmetry, on the position in the first Brillouin zone of the point for which we want to know the energy, and on the unknown energy ε itself.

In the practical use of the method it is assumed that the potential V of the crystal has the MT form, just as in the APW method, and the potential between the spheres can be considered to be zero. In this case integration over the unit cell reduces to an integration over the volume of all the spheres within the confines of the cell. Accordingly, to find the dispersion laws it is sufficient to specify the BF in each sphere by the set of coefficients of its expansion in terms of spherical harmonics. Expansion of the unknown function ψ reduces the integral equation (2.47) to a system of homogeneous linear algebraic equations for the expansion coefficients. Thus the entire procedure leads to a secular equation, the solution of which gives (in implicit form) the desired dispersion law $\varepsilon = \varepsilon(\mathbf{k})$.

The Green function method is simpler than the APW method in the sense that for good convergence it is usually sufficient to consider terms with $l \leqslant 2$ in the expansion for ψ [72]. However, the KKR method, like the APW method, is limited by the assumption of the MT form of the crystal potential, the finding of which is not free of arbitrariness here either. Thus for $A^N B^{8-N}$ crystals with a diamond or ZnS lattice [74-78] the sum of the Herman-Skillman atomic potentials plus some "exchange" potential is usually taken as the initial potential. In this situation the resulting potential of the atoms A, B is quite far from spherical symmetry. To impart the MT form to it, the superposition of the atomic potentials V_{sup} along the line A-B is considered. Then the portion of the curve from atom A or B to the maximum point of the V_{sup} curve is considered as the radial part of the potential in the A and B spheres, respectively [74, 75]. A similar method within the framework of the relativistic modification of the KKR method was also used to find the MT potential in lead chalcogenides [79].

For ionic crystals with an NaCl lattice the spherically symmetrical approximation of the potential at the lattice sites is suitable to a larger extent than for diamond-like crystals. Nevertheless, here too the finding of the MT potential is not a rigorous and unambiguous procedure. In [80], for example, for KI the Liberman [81] potentials (relativistic) of the K^+ and I^- ions were used with an added Madelung term of $\pm MZ/R$, and the radii of the K^+ and I^- spheres are found from the maximum of the superposition of the corresponding atomic potentials [64].

2.4.4 Orthogonalized Plane Wave Method

The most reliable of the methods listed in Sec. 2.4.1 is the OPW method, recently developed by Herman and others employing two approaches–nonempirical (the self-consistent OPW (SC OPW) method [82–97]) and semiempirical [the empirically corrected OPW (EC OPW) method, see below]. In the OPW method* all the electron states in the crystal are divided into those belonging to the atomic cores and into valence states. The core states in the OPW method are described in the LCAO approximation by Bloch functions of the form

$$\psi_{i,\mathrm{core}}(\mathbf{k}\,|\,\mathbf{r})=\frac{1}{\sqrt{N}}\sum_{\mathbf{R}} e^{i\mathbf{k}\mathbf{R}}\chi_{i,\mathrm{core}}(\mathbf{r}-\mathbf{R}) \tag{2.49}$$

where $\chi_{i,\ \mathrm{core}}$ are the AOs of the atomic cores.

The functions (2.49) are used to find the OPWs, by which we mean functions of the form

$$\pi\,(\mathbf{k}\,|\,\mathbf{r})=e^{i\mathbf{k}\mathbf{r}}-\sum_{i} b_i\psi_{i,\mathrm{core}}(\mathbf{k}\,|\,\mathbf{r}) \tag{2.50}$$

where the coefficients b_i are determined by the condition of orthogonality of the functions (2.50) to the functions (2.49).

Then the wave functions of the valence electrons in the OPW method are sought in the form of combinations of the individual OPWs (2.50), the expansion coefficients in which are determined by a variational method so that the finding of the dispersion laws is reduced to that of solving the secular equation.

Self-consistency in the OPW method is achieved by a successive recalculation of the core and valence BFs [86, 97]. In each cycle of the iterations a spherical symmetry is given to the Coulomb potential of the valence electrons and the electron density, and the core functions are recalculated for fixed values of this potential and the density. The new potential is found from the previous values of the potential and the density of the valence electrons and from the new values of the basis functions. The coefficients b_i in the relations (2.50) and the coefficients of the expansion of the eigenfunctions in terms of the OPWs are recalculated. Each cycle of the iterations is concluded with a calculation of the valence electron levels ε_i, and the iterations are continued until the change in the energies ε_i from cycle to cycle is less than 0.01 eV [93]. In the first cycle the potential is taken in the form of a sum of the nonrelativistic self-consistent Herman-Skillman atomic potentials (as in the earlier non-self-consistent OPW calculations [100]). To find the electron density and the potential all the Bloch functions are taken each time at a certain number of **k** points of the first zone. Usually four (Γ, X, L, and W [88]) or six

*See [98] for a review of the OPW method and a method of implementing it on a computer. The localized orbital method, using a mixed basis (of AOs and plane waves) and vigorously developed in recent years by Kunz [99], is similar to the OPW method.

(Γ, X, L, W, Δ, and Σ [92]) points are taken and the "partial" densities belonging to them are averaged with specially selected weighting factors W_i. The electron density is calculated at 650 points of space, covering 1/24 of a unit cell [92, 93], and in recent work up to 259 OPWs are used as the basis for homopolar (Si [93]) and up to 537 OPWs for heteropolar (BAs [93]) crystals. This leads to the necessity for solving secular equations up to the 270th order, inclusively.

Compared with the APW method and the Green function method, the OPW method has the definite advantage that it not related with a specific MT form of the one-electron potential in the crystal. The MT model obviously is a rather poor representation of the true potential distribution in the quite brittle structures of semiconductors with the diamond or zinc sulfide lattice since for them only 20-30% of the crystal volume can be described more or less satisfactorily by means of a constant potential (between the spheres). At the same time the "directional character" of the tetrahedral bonds agrees poorly with the assumption of a spherically symmetrical potential inside the individual atomic spheres.

Therefore the OPW method (together with the empirical pseudopotential method, see below) has been used especially often and effectively for calculating the energy bands in the tetrahedral semiconductors that we are going to consider.

Nevertheless, from a critical point of view it is necessary to point out that the OPW method, even in its nonempirical SC OPW form, is by no means an *ab initio* method in the full sense of the word and retains a certain freedom of parameter choice, the criterion of which in the final analysis is trial and error. This is primarily true in the choice of the exchange potential in a crystal, for which three forms are now in use [94]: Slater [8]; the Kohn–Sham potential $V_{\text{exch K-S}} = 2/3\ V_{\text{exch Slat}}$; and the Liberman potential [101] (see Sec. 2.4.5 concerning the effect of the form of the potential on the band structure).

2.4.5 Accuracy of Nonempirical Calculations and the Parameter Fitting Method

Although the APW, OPW, and Green function methods discussed above include certain assumptions borrowed from atomic calculations and rely on experiment, they should be considered to be primarily nonempirical. Experience in their application to the calculation of the band structure of various crystals indicates the possibility of a successful nonempirical treatment of the electron structure of solids. Unfortunately, however, these methods do not yet always give the correct band structure, and in many cases lead to even a distorted qualitative picture of the band structure. This is well apparent from Table 1, in which the results of a calculation of ZnS crystals by the APW method [68], a calculation of Si, Ge, GaP, GaAs, and InP by the KKR method [77], as well as similar data for silicon, taken from a calculation by the SC OPW method [92], are listed.

As seen from Table 1, the energies of the $\Gamma^v_{25'} \rightarrow \Gamma^c_{2'}$ transitions in elementary semiconductors, obtained by the Green function method [77], deviate from experiment by more than 2 eV. For silicon, according to the calculation of [77], the $\Gamma^c_{2'}$ level is erroneously found to be narrower than the Γ^c_{15} level, while germanium turns out to be a conductor for which the principal minimum of the conduction band is at the Γ point 1 eV below the top of the valence band $\Gamma^v_{25'}$. This same distortion of the band structure is also observed in the SC OPW calculation [92] for silicon, in which the use of the Slater exchange potential leads to an incorrect level arrangement: $\varepsilon(\Gamma^c_{2'}) < \varepsilon(\Gamma^c_{15})$, while the use of the Kohn-Sham exchange potential leads to a forbidden band gap of $E_g = 0.3$ eV [$E_g(\text{Si})_{\text{exp}} = 1.2$ eV]. Let us also point out in connection with the latter that the results of calculations by the SC OPW method are generally quite sensitive with respect to the

TABLE 1. Band-to-Band Transition Energies for Some Crystals with the Diamond Lattice of Zinc Sulfide from the Data of Different Methods of Band Structure Calculation[a]

Transition	APW [68]	KKR [77]					SC OPW [92]
	ZnS	Si	Ge	GaP	GaAs	InP	Si
$\Gamma_{15}^{v} \longrightarrow \Gamma_{1}^{c}$	4.7	1.6	−1.1 (!)	1.7	0.8	—	2.75
	3.7	3.8	1.0	2.8	1.55		3.8
$\Gamma_{15}^{v} \longrightarrow \Gamma_{15}^{c}$	10.1	—	—	—	4.8	5.8	2.79
	8.4	—	—	—	4.2	3.8	3.5
$X_{5}^{v} \longrightarrow X_{1}^{c}$	8.1	—	—	4.1	4.0	3.9	—
	7.0	—	—	5.2	4.6	4.8	—
$X_{5}^{v} \longrightarrow X_{3}^{c}$	9.6	—	—	4.6	4.5	4.2	—
	7.4	—	—	5.5	5.1	5.1	—
$L_{3}^{v} \longrightarrow L_{1}^{c}$	6.6	2.4	1.4	2.8	—	—	2.78
	5.7	3.3	2.2	3.4	—	—	3.3
$L_{3}^{v} \longrightarrow L_{3}^{c}$	11.1	—	—	—	—	—	5.01
	9.7	—	—	—	—	—	5.4

[a]The transitions listed are those for which the maximum deviation from experiment is observed. The theoretical data for each transition are given in the upper line. The experimental data (lower line) are taken from the review of Phillips [146]. The transition energies are stated in eV.

exchange potential. Thus the use of the Liberman potential instead of the Slater and Kohn–Sham potentials in ZnS and ZnSe leads to a change of ~1–2 eV and ~1.5–2.5 eV, respectively, in the energy of the $\Gamma_{15}^{v} \to \Gamma_{1}^{c}$ and $\Gamma_{15}^{v} \to X_{1}^{c}$ transitions and a change of as much as 3.5–4.5 eV in the energy of the $\Gamma_{15}^{v} \to \Gamma_{15}^{c}$ transition [94].

The unique features of the electron structure of the most important semiconductors, for which the width of the energy gap and the energy of several band-to-band transitions are measured in a few electron volts or even in fractions of an electron volt, impose stringent requirements on the calculation accuracy. On the other hand any calculation within the framework of the one-electron model unavoidably contains errors since it ignores the many-electron effects. Just as in the theory of molecules, the often used method of compensating this disadvantage (and at the same time the inaccuracies of the calculation itself too) consists of introducing adjustment parameters in explicit form into the nonempirical scheme. These parameters are chosen such that the energy of one or several band-to-band transitions is identical to the corresponding experimental values. In the APW method the role of the adjustment parameter is usually played by the constant potential V_0 between the spheres* or by the coefficient in the expression for the exchange potential.† A similar approach is used in the Green function method, where a

*See, for example, the calculation in [67] for diamond, where v_0 is chosen from the $\Gamma_{25'}^{v} \to \Gamma_{15}^{c}$ transition, the calculations in [62] and [69] for AgCl, AgBr, and CdO, where v_0 is chosen from the energy of the $\Gamma_{15}^{v} \to \Gamma_{1}^{c}$ transition, and also the calculation in [79] for PbS, PbSe, and PbTe, where the direct transition at point L plays the role of the adjustment parameter.

†Thus, in [102] in the calculation of LiF, NaCl, and KBr the exchange potential is written in the form $V_{\text{exch}} = 6\alpha\{(3\pi/8)^{1/2}\}$, where the coefficient α is chosen from the E_g values.

nonzero potential v_0 between the spheres is also introduced. The latter is chosen from the forbidden band gap E_g (the cubic modification of ZnS, ZnSe [74, 75]; PbS, PbSe, PbTe [79]; KI [80]) or from the $\Gamma^v_{25'} \rightarrow \Gamma^c_2$ transition (Si, Ge [77]) although in the last case the empirical adjustment of the parameters leads to further degradation [77] even without unsatisfactory agreement with experiment (see Table 1) and also to an unreal value of v_0 [77].

The empirically corrected OPW (EC OPW) method suggested by Herman [84, 85, 88, 103], which includes two steps [103], is apparently the most successful example of supplementing the nonempirical calculation with an adjustment of parameters. In the first step the preliminary structure of the bands is calculated with a potential in the form of a superposition of the atomic potentials. Then the band structure calculated in the first step is corrected in accordance with experimental data on the band-to-band transitions. To do this in the covalent crystal case the corrections ΔV_{111}, ΔV_{220}, and ΔV_{331} (the adjustment parameters) are added to the Fourier coefficients V_{111}, V_{220}, and V_{331} of the crystal potential V. The procedure is altered for partially ionic $A^N B^{8-N}$ crystals. In this case the role of the adjustment parameters is played by the correction ΔV_{111}, as well as the two corrections $\Delta\varepsilon_{core}(A)$ and $\Delta\varepsilon_{core}(B)$ to the energies of the core states of the cations (A) and anions (B), determined from the $\Gamma^v_{15} \rightarrow \Gamma^c_1$, $\Gamma^v_{15} \rightarrow X^c_1$ and $\Gamma^v_{15} \rightarrow X^c_3$ (or $L^v_{15} \rightarrow L^c_1$) transitions [49]. As shown by calculations of the band structure of covalent crystals [84, 85], GaP, GaAs, GaSb [49], as well as ZnS, ZnSe, CdS, and CdSe [88], the EC OPW method determines the structure with adequate accuracy (to 0.5 eV). This is reasonable since the three most characteristic transitions in the crystal are used for the adjustment (together with a large number of basis functions–200 OPWs [103]).

2.4.6 Empirical Pseudopotential Method and the Fourier Expansion of the Dispersion Laws Method

Methods in which all the parameters are found from experiment are a logical extension of the methods using parameter adjustment. Among these the EP method should receive first mention [104-109].*

Formally, the basis of the EP method [104] follows from the OPW method. It can be shown that the Schroedinger equation (1.7) for a function expanded in terms of OPWs is written in the form of the Schroedinger equation for a "pseudowave" function, composed of plane waves only. The role of the potential term in such an equation is then played by the "pseudopotential," i.e., the operator $\hat{V}_{eff} = V + \hat{V}_R$, consisting of the true crystal potential V and the (nonlocal) operator $\hat{V}_R$, depending on the core functions (2.49) and on the energy ε that is being sought, but compensating to a considerable extent the action of the true crystal potential. In practice the nonlocal nature of the potential $\hat{V}_{eff}$ is ignored by assuming it is approximated by the usual potential $\hat{V}_{eff}$; it is no longer equal to the true one-electron potential in the crystal and is not calculated, as in the OPW method, but is chosen in the following manner from experimental data.

It is assumed that V_{eff} is the sum of the spherically symmetrical atomic pseudopotentials V_{eff}^{atom}. Then the Fourier coefficients $V_{eff}(\mathbf{K})$ of the potential V_{eff} will have the form:

*For a review of the pseudopotential method and the procedure for determining the parameters from empirical data see Heine, Cohen, and Ware [104] and Brust [105]. See [106–109] for the application of this method to group IV elements and other $A^N B^{8-N}$ crystals.

$$V_{\text{eff}}(\mathbf{K})=\sum_j V_{\text{eff}}^{\text{atom}}(\mathbf{K})_j e^{i\mathbf{K}\mathbf{r}_j} \tag{2.51}$$

Here the subscript j enumerates the atoms in the unit cell, so that, for example, for the tetrahedral $A^N B^{8-N}$ crystals [107]

$$V_{\text{eff}}(\mathbf{K})=V_{\text{sym}}(K)\, 2\cos \mathbf{Kb}+iV_{\text{anti}}(K)\, 2\sin \mathbf{Kb} \tag{2.52}$$

where $b = \{a/8, a/8, a/8\}$, a is the edge of the cubic unit cell,

$$\begin{aligned} V_{\text{sym}}(K)&=\frac{1}{2}\{V_{\text{A}}(K)+V_{\text{B}}(K)\} \\ V_{\text{anti}}(K)&=\frac{1}{2}\{V_{\text{A}}(K)-V_{\text{B}}(K)\} \end{aligned} \tag{2.53}$$

$V_{\text{A}}(K)$, $V_{\text{B}}(K)$ are the Fourier transformations of the atomic pseudopotentials:

$$V_{\text{A or B}}(K)=\frac{1}{\text{cell volume}}\int V_{\text{A or B}}(\mathbf{r})\, e^{-i\mathbf{K}\mathbf{r}}\, dv \tag{2.54}$$

(the last, in view of the spherical symmetry of $V_{\text{eff}}^{\text{atom}}(\mathbf{r})$, depends only on the modulus $|\mathbf{K}| = \text{K}$).

In this case $V_{\text{eff}}(\mathbf{r})$ is determined by the values of $V_{\text{A}}(K)$, $V_{\text{B}}(K)$ or, what is the same thing, by the values of $V_{\text{sym}}(K)$, $V_{\text{anti}}(K)$. These values are also considered as parameters in the empirical pseudopotential method,* which should be chosen so as to ensure the best possible agreement with experiment. To construct the $V_{\text{eff}}(\mathbf{r})$, of course, they must be specified only for the **K**, which are the vectors of the reciprocal lattice.

The independent values of the corresponding vectors are determined on the basis of the symmetry properties of the (reciprocal) lattice. Since the Fourier coefficients of the pseudopotential decrease quite rapidly with distance from the coordinate origin of the reciprocal space, only those that belong to the shortest vectors make a noticeable contribution to the pseudopotential.

In practice it is sufficient to take a small number of **K** values (three values for the diamond lattice [107] and six [107] or twelve [109] values for the ZnS lattice), belonging to the first several vectors of the reciprocal lattice; the squares of the lengths of these vectors in units of $1/a$ are equal to 3, 4, 8, 11, 12, and 16. In this situation for homopolar crystals the adjustment parameters are $V_{\text{sym}}(3)$, $V_{\text{sym}}(8)$, and $V_{\text{sym}}(11)$, which, for example, were determined in [107] from seven band-to-band transitions so that on the average agreement was assured between the calculated and experimental data. For $A^{\text{III}}B^{\text{V}}$ crystals the parameters $V_{\text{sym}}(K)$ were taken in [107] to be the same as for isoelectronic A^{IV} crystals and the parameters $V_{\text{anti}}(3)$, $V_{\text{anti}}(4)$, and $V_{\text{anti}}(11)$ were chosen from seven transitions.

Finally, for $A^{\text{II}}B^{\text{VI}}$ crystals the parameters $V_{\text{sym}}(K)$ were retained and the Fourier coefficients $V_{\text{anti}}(K)$ were taken from data for $A^{\text{III}}B^{\text{V}}$ crystals and multiplied by an empirical factor of 2.3.

*In particular, for the diamond lattice $V_{\text{A}}(K) = V_{\text{B}}(K)$ and $V_{\text{eff}}(K) = 2V_{\text{A}}(K)\cos \mathbf{Kb}$.

A somewhat different approach was used for heteropolar crystals in [109]. In it the values of the Fourier coefficients $V_C(K)$ and $V_{Si}(K)$ were used for the cubic modification of SiC. These coefficients were found from the band structure of diamond and silicon [106, 107] and were supplemented by two parameters found directly from data on the band structure of β-SiC.

Specification of the potential $V_{eff}(r)$ by means of the Fourier coefficients is completely reasonable in the EP method, where the pseudowave functions are expanded in terms of plane waves. The corresponding expansion coefficients are determined by variational means. This leads to the secular equation for finding the dispersion laws, and the set of basis functions usually numbers 20 plane waves.

It follows from the data of EP calculations [106, 109] that this method, just like the EC OPW method, generally yields a good* representation of the band structure although the EP method must be considered as an extrapolation method, rather than a calculation method, making it possible to find from certain band structure data other band structure data for the same crystal.

The "Fourier expansion of the dispersion laws" method, suggested by Dresselhaus [110] and used to describe the band structure of Si and Ge, is apparently an extreme example of this approach. Using only symmetry considerations, the Dresselhauses obtained the Fourier expansions of the dispersion laws, including 13 undefined parameters for each case. These were chosen from a variety of experimental data (optical transitions, effective masses, and the imaginary part of the dielectric constant). The dispersion laws obtained actually provide a good description of the uppermost portion of the valence band and the lower portion of the conduction band (to which the cited experimental data apply). However, the other portions of the band structure are incorrectly described. Thus, for both crystals the minimum of the valence band is obtained at the X_1^v point (and not Γ_1^v) and the $L_{2'}^v$–Γ_1^v–X_1^v branch of the dispersion law has a curvature opposite to that normally found. The total width E_v of the valence band for Si and Ge was found to be equal to $\approx$8 eV, whereas the experimental value of E_v is equal to 13–16 eV (see Sec. 3.5.3).

2.4.7 Relationship between Band Structure and Atomic Structure in the APW, Green Function, OPW, and EP Methods

As seen from Secs. 2.4.2–2.4.6, in a certain sense a relationship between the chemical composition and the band structure of the crystal is established in each of the methods considered. Thus in the APW and Green function methods the nature of the atoms is characterized by the MT potential and the radii of the APW or KKR spheres, while in the OPW method it is characterized by the atomic potential and the functions of the atomic cores. Of special interest in this sense is the EP method, where the atom is characterized by the form factor of the atomic pseudopotential. Calculations of binary and ternary semiconductors ($A^N B^{8-N}$ [107, 108], $A^{II}B^{IV}C_2^V$ [111]) show that the form factors, found for one crystal, can be used in the calculation of other crystals; consequently, each plays the role of a fairly stable characteristic of the atom.

Nevertheless, from the viewpoint of the problems of chemical bond theory (as they were formulated in the Introduction) these methods have a number of drawbacks, the most important of which are the following: (a) The use of the potential or its Fourier transform as "atomic" parameters is inconvenient since they are not directly related to

*Except for isolated transitions, such as the $\Gamma_{1s}^v \rightarrow \Gamma_{1s}^c$ transition for GaSb, ZnS, and ZnTe, where the deviation from experiment amounts to $\approx$1–2 eV.

the atomic energy spectrum or to other observable properties of the atoms; (b) a mixed basis of atomic-type functions and plane or divergent waves does not permit an interpretation of the dependence of the band structure on the atomic characteristics by a uniform method.

Methods of this type have received little practical use in the study of molecules.* On this basis points of contact with the corresponding molecular theory are still not possible.

It is desirable to investigate problems 1–3 (see "Introduction") in analytic form, having, if possible, explicit formulas for the dispersion laws and band-to-band transitions, whereas the methods named are essentially "numerical" and, in particular, involve the solution of high-order secular equations.

For this reason the APW, KKR, OPW, EP, or other similar methods can now hardly be considered as a basis for constructing a theory of the chemical bond in crystals, including parallels with molecular quantum chemistry, whereas the LCAO method (or EO LCAO method, see below) is free of these four drawbacks.

2.5 POINT SYMMETRY IN BAND THEORY

2.5.1 Point and Space Groups of a Crystal

In considering the electron structure of crystals, up to now we have taken account of only the translational symmetry of the lattice. For most crystals of interest, however, the symmetry transformations are not limited to translations, but also include rotations and reflections.

For example, after taking any atom in a diamond or sphalerite type crystal as the coordinate origin (see Fig. 3.1), it is easy to prove that all the atoms of the crystal coincide with others of these same atoms upon rotations and reflections of the T_d symmetry group of a tetrahedron. Moreover, a homoatomic diamond-type crystal has centers of symmetry in the middle of any of the A–B bonds (a heteroatomic crystal like ZnS does not have such centers of symmetry). Such symmetry, supplementing the translational symmetry, is called the "point" symmetry of a crystal.

The point symmetry operations together with the translations produce a "spatial" ("Fedorov") symmetry group of the crystal,† consisting of all the translations, all the transformations of the point group, and all the combination transformations, each of which includes a translation plus an operation of the point group.

Here let us mention one more subtle fact that is particularly true for a diamond-type crystal (see Fig. 3.1). Let us take a line, passing through the center of any atom parallel to a coordinate axis, for example, the x axis. This line is a second-order symmetry axis C_2 since rotation by π about the C_2 axis the structure coincides with itself. This line, however, is also a fourth-order screw axis since the diamond structure coincides with itself upon a $\pi/2$ rotation about this line and a simultaneous translation of $a/2$ along it,

*See footnote on page 17.

†Point and space symmetry of crystals are discussed, for example, in references [42, 47, 48].

where a is the lattice period. Thus these rotations (and reflections), which themselves are not symmetry operations of the crystal, can be associated with the space group of the crystal. Such crystals and the space groups belonging to them are called nonadditive in contrast to additive, where all rotations and reflections, entering into the Fedorov group of a crystal, are themselves symmetry operations too. The example given shows that the diamond-type crystal is nonadditive, whereas for example the simple cubic lattice depicted in Fig. 2.4, *a* is additive.

By definition, all rotations and reflections, associated with the space group—both "pure" and those encountered only at the same time as translations—are included in the point group. Therefore for additive crystals the point group is a subgroup of the space group; for nonadditive crystals it is not. In particular, the point group of diamond will be the octahedron group O_h, although the rotations and reflections, transforming the structure depicted in Fig. 3.1 into itself form only the tetrahedron group T_d.

The symmetry, defined by the space group, is the complete symmetry of the crystal; therefore it serves as the basis of the complete classification of the levels and eigenfunctions of the one-electron Hamiltonian of the crystal.*

2.5.2 Symmetry of Dispersion Law and the k-Vector "Star"

A rigorous explanation of the classification of states in terms of irreducible representations of the space group can be found in the cited literature; here we will limit our discussion to certain key points, needed below.

Let us point out that the first Brillouin zone of a crystal is a symmetrical polygon; it possesses all the symmetry elements of a point group. This fact is clearly seen for the example of the Brillouin zone for the diamond or zinc sulfide lattice (see Fig. 2.6). It follows from this that to each vector $\mathbf{k} = \mathbf{k}_1$ in the first zone there belong several other vectors $\mathbf{k}_2, \mathbf{k}_3, \ldots$, equivalent to them in terms of symmetry. All these vectors together with the original one comprise the so-called **k**-vector "star," and the dispersion law $\varepsilon(\mathbf{k})$ has the same form for all the directions belonging to the vectors of one and the same star. This is due to the fact that the BFs for the vectors of one and the same star are transformed into one another during the operations of the point group. Thus an examination of the dispersion laws in the entire first zone is split into two problems, the first of which is to find the various types of **k**-vector stars. This problem is solved trivially: there are "general type" stars for which the **k** vectors do not lie on any symmetry element of the first zone, and special types of stars for which the **k** vectors lie on symmetry axes or planes of the first zone. It is obvious that for the Brillouin zone (see Fig. 2.6) the general type star numbers 48 vectors. Of the special type stars let us call attention to the star of the vector $\mathbf{k} = \{k_x, 00\}$, consisting of six vectors, and the star of the vector $\mathbf{k} = \{k_x, k_y, k_z\}$, consisting of eight vectors. The

*See [24, 44–46] as well as [40–42] for the application of the theory of space group representations to band theory.

isolated "star," consisting of the one point $\mathbf{k} = 0$, forms the center of the Brillouin zone–point Γ.

2.5.3 k-Vector Group

Let us now consider the dispersion law $\varepsilon(\mathbf{k})$ along any ray of a star. It was noted above that the BF $\psi(\mathbf{k}/\mathbf{r})$, in the presence of certain point group transformations, becomes a BF with other $\mathbf{k}$ vectors of the same star. For a given $\mathbf{k}$ in the point group there will be, however, those operations that transform a BF with this $\mathbf{k}$ into a BF with this same $\mathbf{k}$ (although not necessarily into the same BF itself; a BF constructed, for example, from p_x functions can be transformed into a BF constructed from p_y functions, etc.). The transformations that preserve the $\mathbf{k}$ vector unchanged are spoken of as the $G(\mathbf{k})$ group of a given wave vector $\mathbf{k}$.

Let us take the complete set of basis BFs for this $\mathbf{k}$ and find the irreducible representations of the group $G(\mathbf{k})$ realized in this basis. Then the number of different branches of the dispersion law $\varepsilon(\mathbf{k})$ in the $\mathbf{k}$ direction will be equal to the number of irreducible representations of $G(\mathbf{k})$ (counting, as usual, different examples of one and the same representation). The degeneracy multiplicity of each branch is the dimensionality of the corresponding irreducible representation.

Let us consider, for example, a crystal with a diamond or ZnS lattice, where two atoms are present in the unit cell and, consequently, eight valence AOs–one s and three p AOs from each atom. For the general type direction the group $G(\mathbf{k})$ reduces to an identity transformation; in this case each basis BF belongs to a one-dimensional irreducible representation and $\varepsilon(\mathbf{k})$ contains eight single branches $\varepsilon_i(\mathbf{k})$.

On the other hand at the center of the zone Γ any operation of the point group transforms the vector $\mathbf{k} = 0$ into this same vector such that $G(\mathbf{k})$ coincides with the complete point group of the crystal. It is obvious that the operations of the point group transform the s AOs of the atoms of the crystal only into s AOs, and the p AOs only into p AOs (although, generally speaking, of other atoms). Hence, for the center Γ of the zone there are two one-dimensional and two three-dimensional irreducible representations of the group $G(\mathbf{k})$ and this means two single and two triply degenerate levels (let us remember that in the crystals being considered there are two basis BFs constructed from s AOs and six basis BFs constructed from p AOs).

In a similar manner it is easy to classify the representations and levels for the symmetry directions $\Delta = [100]$ and $\Lambda = [111]$. It turns out that for each of these directions in the basis of eight basis BFs four one-dimensional and two two-dimensional irreducible representations of the group $G(\mathbf{k})$ are realized and, consequently, four single and two double levels (see Sec. 3.1.1 for more details).

The possibility of factorization of the secular equation (2.18) for the symmetry directions and points of the Brillouin zone also follows from what has been said. Thus at point Γ for ZnS type crystals the eighth-order Eq. (2.18) can be replaced by two second-order equations–for the one-dimensional and for the three-dimensional representations and for the Δ and Λ directions–by one fourth-order equation and one second-order equation. [For diamond-type crystals the situation is even simpler since because of the homoatomicity all the representations of $G(\Gamma)$ are different and, consequently, the energies of the levels at Γ can be found without solving the secular equations.]

Finally, let us point out that the classification in terms of the irreducible representations of the space group follows from the classification of the states in terms of the irreducible representations of the group $G(\mathbf{k})$. In fact, let us assume there are BFs belonging to some irreducible representation of dimensionality n of the group $G(\mathbf{k})$. Let us act on these BFs with those transformations of the point group that convert $\mathbf{k} = \mathbf{k}_1$ into the rest of the vectors $\mathbf{k}_2, \mathbf{k}_3, \ldots$ of the k-vector star. As a result that set of BFs is obtained in which all the transformations of the point group [and not just from the group $G(\mathbf{k})$] no longer yield functions outside the set. Since translations in this situation simply multiply the BFs by a numerical coefficient, this set defines the irreducible representation of the complete space group, and the dimensionality of the representation will be equal to the product {dimensionality of the irreducible representation of $G(\mathbf{k})$} $\times$ {the number of vectors in the $\mathbf{k}$ = vector star}. Thus the irreducible representation of the space group is defined by the vector $\mathbf{k}$ (it specifies the irreducible representation of the group of translations and the type of star), as well as by the corresponding irreducible representation of the group $G(\mathbf{k})$.

Besides the point symmetry operations there is an operation that leads to an additional degeneracy of the levels. Let us take the complex conjugates of both sides of the Schroedinger equation, taking into consideration that the Hamiltonian (1.9) and the energy ε are real. Then along with (1.8) the equality

$$\hat{H}\psi^* = \varepsilon\psi^* \tag{2.55}$$

is obtained, which means that the complex conjugate function also belongs to this same level along with each eigenfunction. For a BF characterized by some value of the $\mathbf{k}$ vector the complex conjugate converts it into the BF corresponding to the opposite vector $(-\mathbf{k})$. Therefore for each branch $\varepsilon_i(\mathbf{k})$ of the dispersion law in any crystal we have

$$\varepsilon_i(-\mathbf{k}) = \varepsilon_i(\mathbf{k}) \tag{2.56}$$

Thus the energy, as a function of $\mathbf{k}$, is always even regardless of whether the crystal has a center of symmetry. As shown by Wigner, the complex conjugate operation is closely related to the time-reversal operation: $t \to -t$ [24, 40, 41, 44–46]. Therefore this symmetry is called symmetry with respect to time reversal.

2.6 EQUIVALENT ORBITALS IN BAND THEORY

2.6.1 Bloch Functions and Localized Orbitals in Crystals

The apparent contradiction between the one-electron model and the classical structure theory, mentioned in Sec. 1.5.1, is especially prominent in the case of crystals for which the eigenfunctions (BFs) are delocalized over the entire macrocrystal. At the same time, for example, the lengths and energies of the $A^{IV}-A^{IV}$ bonds in crystals of group IV elements are practically no different from the lengths or energies in saturated $A_n^{IV}H_{2n+2}$-type molecules (see Table 2), so that, let us say, diamond is traditionally included in the classical structure theory of organic chemistry* with its additive scheme.

The relationship between the concepts of delocalized BFs and localized two-center functions is given in this case by the LO theory† (see chapter 1). In this situation all the bonds in a diamond- or sphalerite-type crystal are identical, as in the CH_4 molecule, and the LOs can be expressed such that they will be EOs. In addition (see Fig. 3.1), in the diamond lattice there is an inversion center at the middle of each bond. Therefore the bonding EO $\varphi_i^{(+)}$ is symmetrical and the antibonding EO $\varphi_i^{(-)}$ is antisymmetrical with respect to the middle of the corresponding bond (for the same reason as for a diatomic homoatomic molecule, see Sec. 1.3.5). Accordingly, in the LCAO approximation they are written in the form‡

$$\begin{aligned} \varphi^{(+)} &= \frac{1}{\sqrt{2}}\{\chi_A + \chi_B\} \\ \varphi^{(-)} &= \frac{1}{\sqrt{2}}\{\chi_B - \chi_A\} \end{aligned} \tag{2.57}$$

where χ_A, χ_B are the sp^3 orbitals of the valence bound atoms A, B, directed along the A–B bond.

For heteroatomic sphalerite-type $A^N B^{8-N}$ crystals all the A–B bonds are also identical; however, there are no inversion centers at the middles of the

*The length of the C–C bond (1.542 Å) in diamond is usually taken as the standard of the ordinary $C(sp^3)$–$C(sp^3)$ σ bond.

†The unitary equivalence of the BFs and the atomic type LOs was first mentioned by Wannier [112]. For crystals with the diamond structure the LOs (EOs) were introduced by Hall immediately after this as the EO method made it possible for him to study the ionization potentials in a number of paraffins (see [113], paraffins; [114, 115], diamond). The general LO theory for any crystals with an insulator character of the band structure is given by Koster [115].

‡In the form (2.57) the EOs are no longer strictly unitarily equivalent to the BFs since they are not completely orthogonal. However, the corresponding integrals overlap little so that this equivalency is nearly satisfied.

bonds. The bonding and antibonding EOs in such crystals are similar to the bonding and antibonding MOs in a heteroatomic diatomic molecule (see Sec. 1.3.5). They are written in the form similar to (2.57) but more general:

$$\varphi^{(+)} = \frac{1}{\sqrt{1+\lambda^2}}\{\chi_B + \lambda\chi_A\}$$
$$\varphi^{(-)} = \frac{1}{\sqrt{1+\lambda^2}}\{\lambda\chi_B - \chi_A\} \qquad (2.58)$$

where $0 < \lambda \leqslant 1$ is a parameter defining the degree of covalency and is related to the effective charge by a simple formula.

In fact, the conversion from the eigenfunctions (the BFs here) to the EOs leaves the invariant electron density:

$$\rho(\mathbf{r}) = 2\sum_i \sum_k |\psi_i(\mathbf{k}\,|\,\mathbf{r})|^2 \qquad (2.59)$$

so that

$$\rho(\mathbf{r}) = 2 \sum_{\text{over the bonds}} [\varphi_i^{(+)}]^2 \qquad (2.60)$$

In (2.59) the summation is carried out over all **k** in the first Brillouin zone (see Fig. 2.6) and over all four filled bands in the valence band (see Fig. 3.2).

It follows from (2.58) that

$$[\varphi^{(+)}]^2 = \frac{1}{1+\lambda^2}\{\chi_B^2 + \lambda^2\chi_A^2\} \qquad (2.61)$$

(as before, we ignore the overlap distribution). Then the effective numbers of electrons at the atoms A, B are given by the formulas

$$n_A = \frac{8\lambda^2}{1+\lambda^2} \qquad n_B = \frac{8}{1+\lambda^2} \qquad (2.62)$$

and the effective charge on the atoms $Z = Z_A = -Z_B$ is determined by the formula [117] (see Sec. 1.3.5):

$$Z = \frac{N - (8-N)\lambda^2}{1+\lambda^2} \qquad (2.63)$$

All that has been said in chapter 1 with regard to the relationship between the EOs and the one-electron properties are also valid for solids, except for the more specialized terminology. Thus instead of the quantities $\langle\varepsilon^{(+)}\rangle$ and

$\langle \varepsilon^{(-)} \rangle$, here we are dealing with the average level of the valence band $\langle \varepsilon^{v} \rangle$ and the average level of the conduction band $\langle \varepsilon^{c} \rangle$, which are defined by the relations

$$\langle \varepsilon^{v} \rangle = \frac{1}{\text{total number of states}} \sum_{i} \sum_{\mathbf{k}} n_i^{v}(\mathbf{k})\, \varepsilon_i^{v}(\mathbf{k})$$
$$\langle \varepsilon^{c} \rangle = \frac{1}{\text{total number of states}} \sum \sum n_i^{c}(\mathbf{k})\, \varepsilon_i^{c}(\mathbf{k}) \tag{2.64}$$

Then Eqs. (1.96) and (1.101) for a crystal can be rewritten in the form [118]:

$$\langle \varepsilon^{v} \rangle = \langle \varphi^{(+)} | \hat{H} | \varphi^{(+)} \rangle$$
$$\langle \varepsilon^{c} \rangle = \langle \varphi^{(-)} | \hat{H} | \varphi^{(-)} \rangle \tag{2.65}$$

$$\langle \Delta\varepsilon \rangle = \langle \varepsilon^{c} \rangle - \langle \varepsilon^{v} \rangle = \langle \varphi^{(-)} | \hat{H} | \varphi^{(-)} \rangle - \langle \varphi^{(+)} | \hat{H} | \varphi^{(+)} \rangle \tag{2.66}$$

2.6.2 Equivalent Orbital Method

The EOs in crystals can also be used for studying individual energy levels, i.e., the band structure. To do this it is necessary to form the BFs [114, 115] from the EOs $\varphi_i^{(+)}$ or $\varphi^{(-)}$ in the same manner as was done in Sec. 2.3.1 for the AOs (see Sec. 3.1.2 for more details). In this situation the BFs of the valence band must include only the bonding EOs $\varphi_i^{(+)}$. In fact, in the sense of the LO theory the set $\{\varphi_i^{(+)}\}$ is determined as the result of applying a unitary transformation to the BFs of this band, so that the latter in turn must be expressed only in terms of $\{\varphi_i^{(+)}\}$ (by means of an inverse transformation).

Similarly, the BFs of the conduction band include only the EOs $\varphi_i^{(-)}$.

This method of writing the BFs is usually called the "equivalent orbital method" (although the name "LCBO method"–linear combination of bound orbitals–that is sometimes used may be better). The EO method has a whole series of advantages. It makes it possible to reduce the order of the secular equations encountered to such an extent that for diamond- and sphalerite-type crystals and for the fundamental directions Δ and Λ in the first zone it becomes possible to express the dispersion laws in analytic form.

Another advantage of this method appears when one is studying partially ionic $A^N B^{8-N}$ crystals. In this case the EO method makes it possible to relate the wave functions to the amount of effective charge on the atoms and, consequently, to ascribe a "self-consistent" character* to the theory in the

*Let us point out that the tight binding method does not possess this self-consistency; it can be achieved only by iteration means.

sense that Z determines both the potential in the Hamiltonian of the crystal as well as the form of the BFs.

In addition, for tetrahedral crystals the EOs correspond to the "bonds" of classical structure theory, so that use of the EO method makes it possible to include the voluminous empirical data that has been accumulated within the framework of structure chemistry.

3

Chemical Bond and Structure of Energy Bands in Covalent Crystals with the Diamond Lattice. Equivalent Orbital Method and Valence Band

In this chapter we shall proceed to a systematic investigation of the relationship between the band structure and the nature of the chemical bond for a number of the most widely known crystals in accordance with the problems that were formulated in the Introduction. It is natural to begin this investigation with the simplest case of purely covalent crystals with the diamond lattice, which include diamond, Si, Ge, and α-Sn (gray tin). We shall consider the valence band separately.

The structure of the valence band is quite insensitive to the individual features of the chemical bond in crystals of various group IV elements. This makes it possible to describe the valence band by the simplest means. Such an approach makes it possible to develop a preliminary viewpoint, on the basis of which a number of more subtle problems, associated with the structure of the conduction band and the total band structure, will be investigated in the next chapter. A simple method of describing the valence band is also of interest in itself since knowledge of the structure of only this band is sufficient for interpreting a considerable amount of experimental data from X-ray and X-ray–electron spectroscopy.

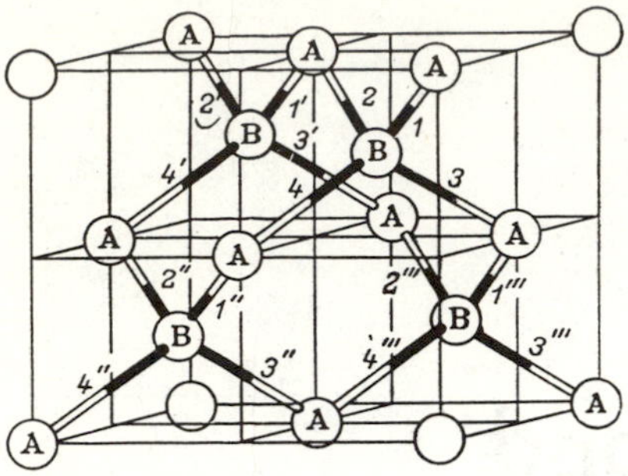

FIG. 3.1. Atomic structure of diamond- and zinc sulfide-type crystal. In the diamond-type crystal the atoms A and B are identical, while in the ZnS-type crystal they are different. The "rods" joining the atoms form four families of translationally equivalent A–B bonds and can also serve as representations of the eight EO families. The rods with black and white halves correspond to the four families of antibonding EOs $\varphi_j^{(-)}$, where the "negative" halves are white. To obtain the four bonding EOs $\varphi_j^{(+)}$, both halves of each rod must be colored black.

3.1 GENERAL DESCRIPTION OF CRYSTALS WITH THE DIAMOND STRUCTURE

The atomic structure of the diamond-type crystal (shown in Fig. 3.1) can be considered as two interpenetrating face-centered cubic sublattices of A^{IV} atoms, so that there are two A^{IV} atoms in a unit cell of the crystal. Each such atom in the ground state has an ns^2np^2 valence electron configuration, where for C (diamond), Si, Ge, and Sn n is equal to 2, 3, 4, and 5 and, consequently, its valence AOs are one ns and three np functions. Accordingly, when a basis of valence AOs is used, the number of branches of the dispersion law $\varepsilon(\mathbf{k})$ in the first Brillouin zone (Fig. 2.6) for the diamond-type crystal is equal to $2 \times 4 = 8$, and these branches in the LCAO method are the solutions of an eighth-order secular equation (see Sec. 2.3.1).

The structure of the energy bands* for C (diamond), Si, Ge, and α-Sn is shown in Fig. 3.2 (for the symmetry directions $\Delta = [100]$ and $\Lambda = [111]$). As seen from these figures, these eight branches of the dispersion law decompose into two groups, with four branches in each (two branches in each direction are doubly degenerate and are counted twice, see below). The four lower branches correspond to four filled bands and constitute the valence band, while the four upper branches are empty bands and constitute the conduction band. For C, Si,

*A detailed description of the band structure of group IV elements can be found also in [40] and especially in [39]. The band structure of diamond specifically is discussed in the review [119].

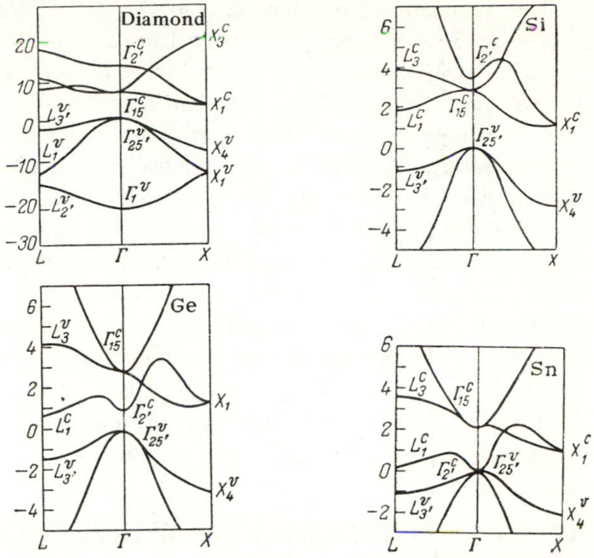

FIG. 3.2. Band structure of diamond, silicon, germanium, and gray tin according to Herman and Kortum [84]. The complete band structure is shown for diamond. For the rest of the crystals only the conduction band and the upper part of the valence band are shown. On the whole the valence band of Si, Ge, and α-Sn is completely analogous to the valence band of diamond.

and Ge the valence band is separated from the conduction band by an energy gap; however, for the α-Sn there is no gap and the $\Gamma^c_{2'}$ level already lies somewhat below the top of the valence band $\Gamma^v_{25'}$ (by ~0.2 eV).

The band structure of the diamond-type crystal looks the simplest at the center of the first Γ band. In a basis of valence AOs the energy level diagram for this point consists of two single levels Γ^v_1 and $\Gamma^c_{2'}$ and two triply degenerate levels $\Gamma^v_{25'}$ and Γ^c_{15}; it resembles the level diagram for the AH_4 molecule (see Fig. 1.7). The single level Γ^v_1, forming the bottom of the valence band, belongs to a bonding combination of the *s* orbitals of the atoms of both sublattices, while the $\Gamma^c_{2'}$ level in the conduction band corresponds to an antibonding combination of these *s* orbitals. On the other hand the triple level $\Gamma^v_{25'}$ (the top of the valence band) and the level Γ^c_{15} in the conduction band correspond to bonding and antibonding combinations of the *p* orbitals of these atoms.

At other points of the Brillouin zone (when $\mathbf{k} \neq 0$) the threefold degeneracy of the *p* levels is completely or partially removed and the levels correspond simultaneously to both the *s* and the *p* orbitals of the atoms of both sublattices. Thus for a general-type point of the first Brillouin zone in the valence band of a diamond-type crystal there are four different (hybrid) *s*–*p* levels and there are the same levels in the conduction band.

The symmetry directions Δ and Λ in the first zone, where the level degeneracy is not completely removed, are an exception to this general rule. For the $\Delta = [100]$ direction the two branches Δ_1^v and $\Delta_{2'}^v$ of the dispersion law in the valence band and these same two branches Δ_1^c and $\Delta_{2'}^c$ in the conduction band are nondegenerate. They correspond to combinations of s and p_x AOs simultaneously–bonding for the valence band and antibonding for the conduction band. However, the Δ_5^v branch in the valence band and the Δ_5^c branch in the conduction band are doubly degenerate. They belong to combinations of only p functions of the atoms of both sublattices (p_y and p_z).

The situation is similar for the direction $\Lambda = [111]$. Here the two Λ_1^v branches in the valence band and these same two branches Λ_1^c in the conduction band are not degenerate and belong to combinations of the s and $\{p_x + p_y + p_z\}$ AOs of the atoms of the two sublattices. At the same time the Λ_3^v and Λ_3^c branches are doubly degenerate and correspond to the combinations of $\{p_x - p_y\}$ AOs and $\{2p_z - p_x - p_y\}$ AOs of these atoms.

3.2 EQUIVALENT ORBITAL (EO) METHOD FOR A DIAMOND-TYPE CRYSTAL

3.2.1 Bloch Functions and the Equivalent Orbital Method

As already mentioned in chapter 2, it is natural to use the LCAO approximation in the study of solid state quantum chemistry problems. However, for tetrahedral $A^N B^{8-N}$ crystals and, in particular, for diamond-type crystals it is convenient to use the equivalent orbital (EO) method beforehand, which in this case makes it possible to find the dispersion laws in analytic form. In addition, as also already mentioned, the EO method applied to tetrahedral $A^N B^{8-N}$ crystals makes it possible to trace the relationship between the band theory of solids and the classical structure theory in chemistry.

Let us point out that the use of the EO LCAO method instead of the tight binding approximation is not an important restriction of the theory since the results of using both methods differ only insignificantly (see Sec. 3.6.2).

For the application of the EO method to diamond-type crystals we will arbitrarily divide all the atoms of the diamond lattice into two face-centered cubic sublattices {A} and {B}; then each atom of one sublattice will be surrounded only by atoms of the other sublattice. Furthermore, we shall assume (also arbitrarily, of course) that each A–B bond belongs to that B atom that takes part in this bond. In that way we establish a one-to-one correspondence between the B atoms and the quadruplets of >B< bonds of the crystal, so that all the bonds in the crystal are naturally decomposed into four families of parallel and identically oriented bonds (Fig. 3.1). It is seen from Fig. 3.1 that the bonds belonging to one and the same family are translationally equivalent. They coincide with one another by means of translations of the atoms of the B sublattice into one another.

For each A–B bond let us now take the bonding EO $\varphi^{(+)}$ and antibonding EO $\varphi^{(-)}$ corresponding to it. Then for each of these four families one can formulate two Bloch functions of the form:

$$\psi_j^{(+)}(\mathbf{k}\,|\,\mathbf{r}) = \frac{1}{\sqrt{N}} \sum_{\substack{\text{over B} \\ \text{atoms}}} e^{i\mathbf{kr}} \varphi_j^{(+)}(\mathbf{r}-\mathbf{R}),\ j=1,2,3,4 \tag{3.1}$$

$$\psi_j^{(-)}(\mathbf{k}\,|\,\mathbf{r}) = \frac{1}{\sqrt{N}} \sum_{\substack{\text{over B} \\ \text{atoms}}} e^{i\mathbf{kr}} \varphi_j^{(-)}(\mathbf{r}-\mathbf{R}),\ j=1,2,3,4 \tag{3.2}$$

As a result, two independent quadruplets of basis Bloch functions are obtained. The Bloch functions of the first quadruplet are linear combinations of the bonding EOs and belong to the valence band, while the basis Bloch functions of the second quadruplet are linear combinations of the antibonding EOs and belong to the conduction band.

As stated in Sec. 2.6.2, the "true" Bloch functions $\psi^{(+)}(\mathbf{k}|\mathbf{r})$, $\psi^{(-)}(\mathbf{k}|\mathbf{r})$ for each of these bands,* i.e., the eigenfunctions of the effective one-electron Hamiltonian of the crystal $\hat{H}$, are unitarily equivalent to the corresponding EOs–bonding or antibonding. Therefore these Bloch functions of the valence band are linearly expressed only in terms of the basis Bloch functions $\psi_j^{(+)}(\mathbf{k}|\mathbf{r})$, whereas the "true" Bloch functions for the conduction band are expressed only in terms of the basis Bloch functions $\psi_j^{(-)}(\mathbf{k}|\mathbf{r})$:

$$\psi^{(+)}(\mathbf{k}\,|\,\mathbf{r}) = \sum_{j=1}^{4} c_j^{(+)}(\mathbf{k})\, \psi_j^{(+)}(\mathbf{k}\,|\,\mathbf{r}) \tag{3.3}$$

$$\psi^{(-)}(\mathbf{k}\,|\,\mathbf{r}) = \sum_{j=1}^{4} c_j^{(-)}(\mathbf{k})\, \psi_j^{(-)}(\mathbf{k}\,|\,\mathbf{r}) \tag{3.4}$$

Thus finding the corresponding dispersion law $\varepsilon = \varepsilon(\mathbf{k})$ at an arbitrary point $\mathbf{k}$ of the first Brillouin zone is reduced to finding the two dispersion laws $\varepsilon = \varepsilon^{(+)}(\mathbf{k})$ and $\varepsilon = \varepsilon^{(-)}(\mathbf{k})$ separately, i.e., to solving two fourth-order secular equations:

*Properly, four "true" BFs for the valence band and the same number of BFs for the conduction band belong to each k; the corresponding subscripts are omitted here in order not to confuse the "true" BFs with the basis ones.

$$\mathrm{Det}\begin{Vmatrix} H_{11}-\varepsilon(\mathbf{k}) & H_{12} & H_{13} & H_{14} \\ H_{21} & H_{22}-\varepsilon(\mathbf{k}) & H_{23} & H_{24} \\ H_{31} & H_{32} & H_{33}-\varepsilon(\mathbf{k}) & H_{34} \\ H_{41} & H_{42} & H_{43} & H_{44}-\varepsilon(\mathbf{k}) \end{Vmatrix}=0 \qquad (3.5)$$

where $H_{ij} = H_{ij}^{(+)}$ for the valence band and $H_{ij} = H_{ij}^{(-)}$ for the conduction band, and

$$H_{ij}^{(+)} = \langle \psi_i^{(+)} | \hat{H} | \psi_j^{(+)} \rangle, \quad H_{ij}^{(-)} = \langle \psi_i^{(-)} | \hat{H} | \psi_j^{(-)} \rangle \qquad (3.6)$$

3.2.2 Symmetry and Factorization of Secular Equations for a Crystal with a Diamond Lattice

To find the dispersion laws in analytic form we will restrict our consideration (as is usual in band structure calculations) to the principal directions $\Delta = [100]$ and $\Lambda = [111]$ in the Brillouin zone (see Fig. 2.6), making use of the symmetry considerations (see Sec. 2.5.1). Then it is not hard to see* that the wave vector group $G(\mathbf{k})$ for the Δ direction is the C_{4v} group, while for the Λ direction it is the C_{3v} group. Correspondingly, the functions belonging to the irreducible representations of the group $G(\mathbf{k})$ will be the following linear combinations of the basis Bloch functions [the (+) superscripts refer to the valence band and the (−) superscripts refer to the conduction band]:

$\Delta = [100]$ direction, $G(\mathbf{k}) = C_{4v}$:

$$\Psi_1^{(\pm)} = \frac{1}{\sqrt{2}} \{\psi_1^{(\pm)} + \psi_4^{(\pm)}\} \quad \text{representation } \Delta_1$$

$$\Psi_2^{(\pm)} = \frac{1}{\sqrt{2}} \{\psi_2^{(\pm)} + \psi_3^{(\pm)}\} \quad \text{representation } \Delta_1$$

$$\Psi_3^{(\pm)} = \frac{1}{\sqrt{2}} \{\psi_1^{(\pm)} - \psi_4^{(\pm)}\} \quad \text{representation } \Delta_5 \qquad (3.7)$$

$$\Psi_4^{(\pm)} = \frac{1}{\sqrt{2}} \{\psi_2^{(\pm)} - \psi_3^{(\pm)}\} \quad \text{representation } \Delta_5$$

$\Lambda = [111]$ direction, $G(\mathbf{k}) = C_{3v}$:

*Remember that the point group is the octahedron group O_h for the diamond-type (nonadditive) crystal.

$$\begin{aligned}
&\Psi_1^{(\pm)}=\psi_1^{(\pm)} && \text{representation } \Lambda_1\\
&\Psi_2^{(\pm)}=\frac{1}{\sqrt{3}}\{\psi_2^{(\pm)}+\psi_3^{(\pm)}+\psi_4^{(\pm)}\} && \text{representation } \Lambda_1\\
&\Psi_3^{(\pm)}=\frac{1}{\sqrt{2}}\{\psi_3^{(\pm)}-\psi_4^{(\pm)}\} && \text{representation } \Lambda_3\\
&\Psi_4^{(\pm)}=\frac{1}{\sqrt{6}}\{2\psi_2^{(\pm)}-\psi_3^{(\pm)}-\psi_4^{(\pm)}\} && \text{representation } \Lambda_3
\end{aligned} \tag{3.8}$$

Therefore the dispersion laws $\varepsilon = \varepsilon(\mathbf{k})$ in both directions and for both bands (valence and conduction) are given by the solutions of one quadratic secular equation:

$$\text{Det}\begin{Vmatrix} \mathcal{H}_{11}-\varepsilon(\mathbf{k}) & \mathcal{H}_{12} \\ \mathcal{H}_{21} & \mathcal{H}_{22}-\varepsilon(\mathbf{k}) \end{Vmatrix}=0 \tag{3.9}$$

and any of the first-degree secular "equations"

$$\mathcal{H}_{33}-\varepsilon(\mathbf{k})=0, \quad \mathcal{H}_{44}-\varepsilon(\mathbf{k})=0 \tag{3.10}$$

Here the matrix elements $\mathcal{H}_{ij}$

$$\mathcal{H}_{ij}^{(+)}=\langle \Psi_i^{(+)} | \hat{H} | \Psi_j^{(+)} \rangle ; \quad \mathcal{H}_{ij}^{(-)}=\langle \Psi_i^{(-)} | \hat{H} | \Psi_j^{(-)} \rangle \tag{3.11}$$

in the basis of symmetrized Bloch functions $\Psi_i^{(+)}$, $\Psi_i^{(-)}$ are expressed in the following manner in terms of the matrix elements H_{ij} in the basis of the functions $\psi_i^{(+)}$, $\psi_i^{(-)}$:

$\Delta = [100]$ direction:

$$\begin{aligned}
\mathcal{H}_{11}&=\frac{1}{2}\{H_{11}+H_{14}+H_{41}+H_{44}\}\\
\mathcal{H}_{12}&=\frac{1}{2}\{H_{12}+H_{42}+H_{13}+H_{43}\}\\
\mathcal{H}_{22}&=\frac{1}{2}\{H_{22}+H_{32}+H_{23}+H_{33}\}\\
\mathcal{H}_{33}&=\frac{1}{2}\{H_{11}+H_{44}+H_{41}+H_{14}\}
\end{aligned} \tag{3.12}$$

$\Lambda = [111]$ direction:

$$\begin{aligned}
\mathcal{H}_{11}&=H_{11}\\
\mathcal{H}_{12}&=\frac{1}{\sqrt{3}}\{H_{12}+H_{13}+H_{14}\}
\end{aligned} \tag{3.13}$$

$$\mathscr{H}_{22}=\frac{1}{3}\{H_{22}+H_{33}+H_{44}+H_{23}+H_{24}+H_{34}+H_{42}+H_{43}\} \quad (3.13)$$

(continued)

$$\mathscr{H}_{33}=\frac{1}{2}\{H_{33}+H_{44}-H_{34}-H_{43}\}$$

3.2.3 Dispersion Laws in Nearest Neighbor Approximation

Since the basis Bloch functions $\psi_i^{(+)}(\mathbf{k}|\mathbf{r})$, $\psi_i^{(-)}(\mathbf{k}|\mathbf{r})$ are in turn expanded in terms of the EOs $\varphi_j^{(+)}$, $\varphi_j^{(-)}$, the specific form of the matrix elements H_{ij} and $\mathscr{H}_{ij}$—and, therefore, the form of the dispersion laws—is determined by the number of matrix elements of the Hamiltonian $\hat{H}$ in the basis of EOs to be taken into consideration. Below we shall always use the approximation in which all possible types of EO interactions are considered, but only for those EOs that reflect the interaction of the valence-bound atoms (see Sec. 3.2.4). It is not hard to see that these atoms participate, first of all, in the formation of first-neighbor bonds (having one common A or B atom; for example, the 1′, 2, and 1, 2 bonds in Fig. 3.1). Second, the valence bound atoms can also take part in the formation of second-neighbor bonds (1′,1; 3,1′; and 4,1′ bonds in Fig. 3.1). Therefore the successive application of this approximation requires consideration of the interaction of both cited types of EOs.

Accordingly, the system of matrix elements, defining the dispersion laws, includes eight integrals of the following four types (Fig. 3.3).

(1) Coulomb integrals $\alpha^{(+)}$, $\alpha^{(-)}$ that correspond to the average value of the Hamiltonian $\hat{H}$ in the EOs $\varphi^{(+)}$ and $\varphi^{(-)}$:

$$\alpha^{(-)}=\langle\varphi^{(+)}|\hat{H}|\varphi^{(+)}\rangle \quad (3.14)$$

$$\alpha^{(-)}=\langle\varphi^{(+)}|\hat{H}|\varphi^{(-)}\rangle \quad (3.15)$$

(2) Resonance integrals $\beta_A^{(+)}$, $\beta_A^{(-)}$ that belong to the interaction of two EOs having one common atom (the EOs corresponding to first-neighbor bonds):

$$\beta_A^{(+)}=\langle\varphi^{(+)\prime}|\hat{H}|\varphi^{(+)\prime\prime}\rangle \quad (3.16)$$

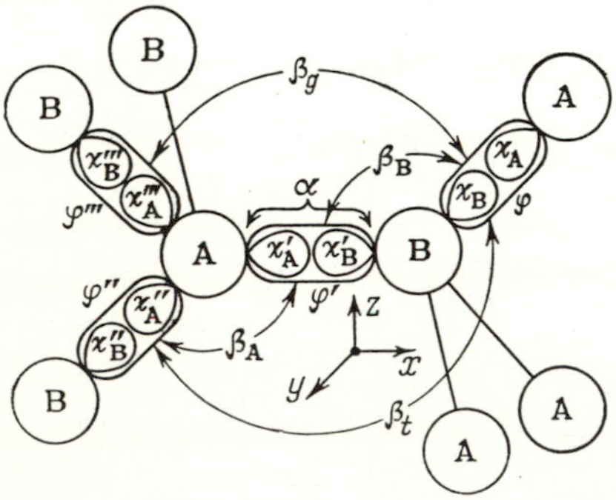

FIG. 3.3. Matrix elements in basis of equivalent orbitals for diamond and zinc sulfide type crystals.

$$\beta_A^{(-)} = \langle \varphi^{(-)\prime} | \hat{H} | \varphi^{(-)\prime\prime} \rangle \tag{3.17}$$

(3) Resonance integrals $\beta_t^{(+)}$, $\beta_t^{(-)}$ that belong to the interaction of the EOs for second-neighbor bonds transposed with respect to each other (EOs φ, φ'' in Fig. 3.3):

$$\beta_t^{(+)} = \langle \varphi^{(+)} | \hat{H} | \varphi^{(+)\prime\prime} \rangle \tag{3.18}$$

$$\beta_t^{(-)} = \langle \varphi^{(-)} | \hat{H} | \varphi^{(-)\prime\prime} \rangle \tag{3.19}$$

(4) Resonance integrals $\beta_g^{(+)}$, $\beta_g^{(-)}$ that belong to the interaction of the EOs for second-neighbor bonds gauche-positioned with respect to one another (EOs φ, φ''' in Fig. 3.3):

$$\beta_g^{(+)} = \langle \varphi^{(+)} | \hat{H} | \varphi^{(+)\prime\prime\prime} \rangle \tag{3.20}$$

$$\beta_g^{(-)} = \langle \varphi^{(-)} | \hat{H} | \varphi^{(-)\prime\prime\prime} \rangle \tag{3.21}$$

Correspondingly, the matrix elements $H_{ij}^{(\pm)}$ in this case are written in the form

$$\begin{aligned}
H_{11}^{(\pm)} &= \alpha^{(\pm)} + 2\beta_t^{(\pm)} \{\cos(q_x+q_y) + \cos(q_x+q_z) + \cos(q_y+q_z)\} \\
H_{22}^{(\pm)} &= \alpha^{(\pm)} + 2\beta_t^{(\pm)} \{\cos(q_x+q_y) + \cos(q_x-q_z) + \cos(q_y-q_z)\} \\
H_{33}^{(\pm)} &= \alpha^{(\pm)} + 2\beta_t^{(\pm)} \{\cos(q_x+q_z) + \cos(q_x-q_y) + \cos(q_z-q_y)\} \\
H_{44}^{(\pm)} &= \alpha^{(\pm)} + 2\beta_t^{(\pm)} \{\cos(q_y+q_z) + \cos(q_y-q_x) + \cos(q_x-q_z)\} \\
H_{12}^{(\pm)} &= \beta_A^{(\pm)} \{1 + e^{i(q_x+q_y)}\} + 2\beta_g^{(\pm)} \{e^{iq_x} + e^{iq_y}\} \cos q_z \\
H_{13}^{(\pm)} &= \beta_A^{(\pm)} \{1 + e^{i(q_x+q_z)}\} + 2\beta_g^{(\pm)} \{e^{iq_x} + e^{iq_z}\} \cos q_x \\
H_{14}^{(\pm)} &= \beta_A^{(\pm)} \{1 + e^{i(q_y+q_z)}\} + 2\beta_g^{(\pm)} \{e^{iq_y} + e^{iq_z}\} \cos q_x \\
H_{23}^{(\pm)} &= \beta_A^{(\pm)} \{1 + e^{i(q_z-q_y)}\} + 2\beta_g^{(\pm)} \{e^{iq_z} + e^{-iq_y}\} \cos q_x \\
H_{24}^{(\pm)} &= \beta_A^{(\pm)} \{1 + e^{i(q_z-q_x)}\} + 2\beta_g^{(\pm)} \{e^{iq_z} + e^{-iq_x}\} \cos q_y \\
H_{34}^{(\pm)} &= \beta_A^{(\pm)} \{1 + e^{i(q_y-q_x)}\} + 2\beta_g^{(\pm)} \{e^{iq_y} + e^{-iq_x}\} \cos q_z
\end{aligned} \tag{3.22}$$

By solving the secular equations (3.9) and (3.10) with the matrix elements (3.12) and (3.13), where the elements $H_{ij}^{(\pm)}$ are expressed in terms of the parameters (3.14)-(3.21) by means of Eqs. (3.22), it is easy to find the dispersion laws for the Δ and Λ directions. They have the following form [114, 115]:

$\Delta = [100]$ direction:

$$\begin{aligned}
\varepsilon_{1,2}^{(\pm)}(q) = \alpha^{(\pm)} &+ 2\{\beta_A^{(\pm)} + \beta_t^{(\pm)}\} + 4\{\beta_g^{(\pm)} + \beta_t^{(\pm)}\}\cos q \\
&\pm 4\{\beta_A^{(\pm)} + 2\beta_g^{(\pm)}\}\cos\frac{q}{2}
\end{aligned} \tag{3.23}$$

$$\varepsilon_{3,4}^{(\pm)}(q) = \alpha^{(\pm)} - 2\{\beta_A^{(\pm)} - \beta_t^{(\pm)}\} - 4\{\beta_g^{(\pm)} - \beta_t^{(\pm)}\} \cos q \tag{3.24}$$

$\Lambda = [111]$ direction:

$$\varepsilon_{1,2}^{(\pm)}(q) = \alpha^{(\pm)} + 2\{\beta_A^{(\pm)} + \beta_g^{(\pm)} + \beta_t^{(\pm)}\} + 2\{\beta_g^{(\pm)} + 2\beta_t^{(\pm)}\} \cos 2q$$

$$\pm 2\sqrt{\{(\beta_A^{(\pm)} + \beta_g^{(\pm)} + \beta_t^{(\pm)}) + (\beta_g^{(\pm)} - \beta_t^{(\pm)})\cos 2q\}^2 + 3\{\beta_A^{(\pm)} + 2\beta_g^{(\pm)}\}^2 \cos^2 q} \tag{3.25}$$

$$\varepsilon_{3,4}^{(\pm)}(q) = \alpha^{(\pm)} - 2\{\beta_A^{(\pm)} + \beta_g^{(\pm)} - 2\beta_t^{(\pm)}\} - \{\beta_g^{(\pm)} - \beta_t^{(\pm)}\} \cos 2q \tag{3.26}$$

Here the superscripts (+), as usual, refer to the valence band, and the superscripts (—) to the conduction band, while q denotes the quantity $q_x = (a/2)k_x$, i.e., the xth component of the vector $(a/2)\mathbf{k}$.

3.2.4 Common Characteristic of First and Second Nearest Neighbor Bonds

To conclude this section let us discuss the approximation of first and second nearest neighbor bonds that we have adopted. As already indicated, this is equivalent to considering only the valence-bound atoms.

It is known that in organic molecules both the lengths as well as the energies of ordinary sp^3-sp^3 C–C bonds are nearly constant and independent of the molecule structure as a whole (additive scheme; see Sec. 1.5.1). Thus for *n*-butane (I) and isobutane (II) the heats of formation are, respectively, 1,221 and 1,222.6 kcal/mol although in the first case there are two and in the second case three nonvalence interacting C atoms. An even clearer example of this is *n*-pentane (III) (heat of formation of 1,498.1 kcal/mol) and neopentane (IV) (heat of formation of 1,502.1 kcal/mol), where the numbers of nonvalence interacting C atoms are, respectively, three and six. (For thermochemical data see, for example, [26].)

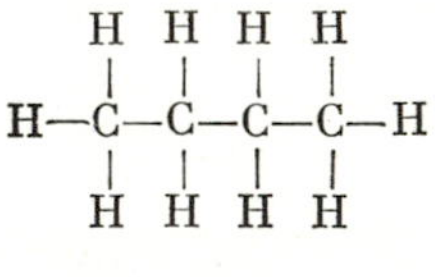

(I)

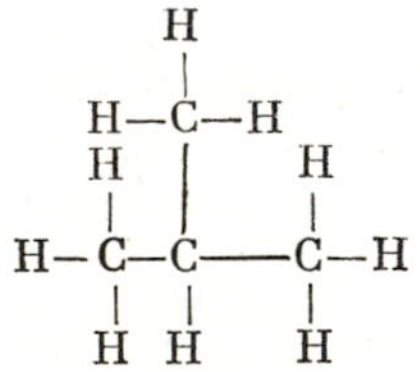

(II)

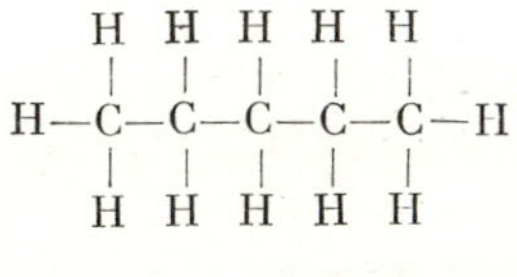

(III)

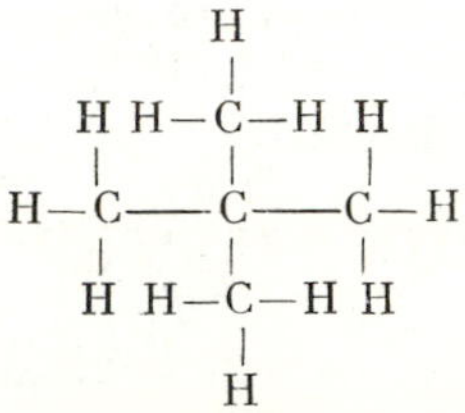

(IV)

Finally, the same is seen from a comparison of paraffin chains and the diamond crystal, for which the binding energies differ by ~2%, although in the C_nH_{2n+2} molecule there are two nonvalence interacting C atoms for each C atom and in the diamond crystal there are 12 (second-neighbor atoms). These examples show that for saturated carbon compounds, as well as the rest of the group IV elements (Table 2) only the interaction of the valence bound atoms is important, which is just what is taken into consideration in the approximation that has been adopted here.

Let us point out that such an approximation for our purposes is simultaneously both necessary and sufficient. Thus, the absence of the EO interaction integrals $\beta_t^{(\pm)}$, $\beta_g^{(\pm)}$ for the second-neighbor bonds (although these integrals are not large) leads, according to Eqs. (3.24) and (3.26), to a qualitatively incorrect picture of the band structure. In this band structure, for example, the dispersion law $\varepsilon_{3,4}(\mathbf{k})$ for the doubly degenerate p level would have the form $\varepsilon_{3,4}(\mathbf{k}) =$ const, which, of course, clearly contradicts the test data.

The other case—when the absence of the integrals $\beta_t^{(\pm)}$, $\beta_g^{(\pm)}$ could lead to considerable distortion of the band structure—applies to the structure of the conduction band (see chapter 4 for more detail).

As the experimental data show, in diamond and silicon crystals, unlike crystals of germanium and gray tin, the antibonding s level $\Gamma_{2'}^c$ lies above the antibonding p level Γ_{15}^c, and the absolute minimum of the conduction band is located near the point X. At the same time it is not hard to obtain the following relations from Eqs. (3.23)–(3.26) [127]:

$$\varepsilon(\Gamma_{2'}^c) - \varepsilon(\Gamma_{15}^c) = 8\beta_A^{(-)} + 16\beta_g^{(-)} \tag{3.27}$$

$$\varepsilon(\Gamma_{15}^c) - \varepsilon(X_1^c) = -4\beta_A^{(-)} + 8\beta_t^{(-)} \tag{3.28}$$

Therefore the inequalities

$$8\beta_A^{(-)} + 16\beta_g^{(-)} > 0 \tag{3.29}$$

$$-4\beta_A^{(-)} + 8\beta_t^{(-)} > 0 \tag{3.30}$$

TABLE 2. Lengths and Energies of Ordinary Bonds in Crystals and Molecules

Element	Distance in Crystal (Å [120])	Binding Energy in Crystal (kcal/mol [121])	Distance in Molecule (Å [122])	Binding Energy in Molecule (kcal/mol)
C	1.542	85.5	1.536 ($H_3C{-}CH_3$)	83—85 [123] ($H_3C{-}CH_3$)
Si	2.35	53	2.32 ($H_3Si{-}SiH_3$)	53 [124] (Si_nH_{2n+2})
Ge	2.45	39	2.41 ($H_3Ge{-}GeH_3$)	37.9 [125] ($H_3Ge{-}GeH_3$)
Sn	2.80	37	—	37 [126] (Sn_nH_{2n+2})

must be satisfied simultaneously for the conduction band in diamond and silicon. A necessary—although not sufficient—condition of the consistency of this system is the fulfillment of the inequality [127]

$$\beta_t^{(-)} + \beta_g^{(-)} > 0 \tag{3.31}$$

so that the condition $\beta_t^{(\pm)} = \beta_g^{(\pm)} = 0$ clearly leads to an incorrect description of the conduction band of diamond and silicon.

Moreover, as will be shown below, the successive consideration of all the interaction integrals for the valence-bound atoms makes it possible to explain all the important features of the band structure of the group IV elements. Consideration of, let us say, the third-neighbor bonds no longer leads to any significant changes and therefore it is superfluous. For this same reason we ignore the overlap integrals between different EOs [for the "true" equivalent orbitals obtained by a unitary transformation of the Bloch functions these overlap integrals are equal to zero; below, however, we will not use the true EOs, but the approximate expressions (2.57)]. In addition, these integrals are generally not large. For example, for diamond, when the Slater functions are used for the s and p AOs of the C atoms, they amount to $|S_A| = \langle\varphi|\varphi'\rangle \approx 0.1$; $|S_t| = \langle\varphi|\varphi''\rangle \approx 0.05$; and $|S_g| = \langle\varphi|\varphi'''\rangle = 0.03$. Therefore the entire procedure of finding the energy levels by the EO method fits into the framework of the "zero differential overlap" approximation employed in quantum chemistry (see the footnote on page 19 also).

Finally, let us point out another argument in favor of the restriction to the interaction of valence-bound atoms only. As already repeatedly indicated, the character of the problems we have formulated is due to a considerable extent to the semiempirical nature of the theory, in which the resonance integrals, describing the interaction of the atoms, are found from test data. These parameters can actually be found for the valence-bound atoms. For nonvalence-bound atoms, however, any estimate of such parameters would hardly have sufficient reliability.

3.3 EO LCAO METHOD AND THE COMPUTATION OF THE PARAMETERS IN THE DISPERSION LAWS

3.3.1 Matrix Elements in "Complete" Version of Theory

Thus far we have not used for the EOs their expression in LCAO form (2.57), so that Eqs. (3.22)-(3.26) are not related to this specific (and approximate) form of the EOs. Thus, in principle, one could generally ignore the LCAO form of the EOs and consider the "true" EOs as the primary basis, in terms of which the eigenfunctions of the Hamiltonian of the system are expanded. This is often done in the quantum chemistry of the C_nH_{2n+2} molecular chains. In this case the matrix elements of $\hat{H}$ in the basis of EOs—α, β_A, etc., which are determined from test data in the semiempirical version of the theory—play the role of the fundamental parameters of the theory. In view of the rigorous unitary equivalence of the "true" EOs and the Bloch functions such an approach would perhaps be even more rigorous and would not be needed in a comparison with the tight binding method (see Sec. 3.6.1). Nevertheless, such an approach to the formulation of a theory is of absolutely no use from the viewpoint of the problems we have formulated since we wish to investigate the dependence of band structure on the properties of the atoms and not on the properties of the bonds. In addition, crystals with two-center A–B bonds are only a special case of coordination crystals, so that such an approach would suffer the disadvantage of a lack of generality.

Finally, the realization of this approach in its semiempirical form is generally difficult in practice. To find the eight parameters $\alpha^{(\pm)}$, $\beta_A^{(\pm)}$, $\beta_t^{(\pm)}$, and $\beta_g^{(\pm)}$ from independent data would require similar information on the properties of the A_nH_{2n+2}-type molecular σ systems, where A = C, Si, Ge, and Sn, in particular. Data on the electron spectra of these molecules are available, whereas there is no adequately detailed information now available on the uv spectra even for the paraffins.*

Therefore we will consider the EO method only as a convenient method for obtaining the dispersion laws in analytic form, and we will use the matrix elements of the Hamiltonian $\hat{H}$ in the basis of the atomic functions as the fundamental parameters of the theory. It is obvious that to do this it is only necessary to use the expression of the EOs in the LCAO form (the EO LCAO method). It is seen below that this viewpoint also has a number of practical advantages besides the theoretical. It makes it possible to use the data of atomic spectroscopy and to evaluate the integrals of interaction between the atomic functions from the data for a larger variety of molecules and not just for the A_nH_{2n+2} σ systems.

Using the atomic orbitals as the basis and considering only the interaction of the AOs for the valence bound atoms, we obtain the six following molecular integrals as the principal parameters.

(1) Two Coulomb integrals $\alpha_A^{(s)}$ and $\alpha_A^{(p)}$ that correspond to the energy of one electron in the s and p levels of atom A:

$$\alpha_A^{(s)} = \langle s_A | \hat{H} | s_A \rangle \tag{3.32}$$

$$\alpha_A^{(p)} = \langle p_A | \hat{H} | p_A \rangle \tag{3.33}$$

(2) Four resonance integrals that belong to all possible versions of the interaction of the s and p orbitals of the adjacent atoms A and B (see Sec. 1.3.3):

*Only information about the parameters $\alpha^{(+)}$, $\beta_A^{(+)}$, and $\beta_t^{(+)}$ can be gathered directly from data on the ionization potentials for unbranched and branched paraffins [128, 129] now available; the value of $\beta_t^{(+)}$ is still insufficiently reliable [128].

$$\beta_{ss} = \langle s_A | H | s_A \rangle \tag{3.34}$$

$$\beta_{sp} = \langle s_A | \hat{H} | p_B \rangle = \langle p_A | \hat{H} | s_B \rangle \tag{3.35}$$

$$\beta_{pp} = \langle p_A | \hat{H} | p_B \rangle \quad p\sigma\text{–}p\sigma \text{ interaction} \tag{3.36}$$

$$\beta_{\pi} = \langle p_A | \hat{H} | p_B \rangle \quad p\pi\text{–}p\pi \text{ interaction} \tag{3.37}$$

To express the matrix elements (3.14)-(3.21) in the basis of EOs in terms of the matrix elements (3.32)-(3.37) in the basis of AOs, instead of the EOs we substitute their expressions (2.57) in the right sides of Eqs. (3.14)-(3.21). Using the notations of Fig. 3.3, it is not hard to see that [130]

$$\alpha^{(\pm)} = \langle \varphi' | \hat{H} | \varphi' \rangle = \langle \chi'_A | \hat{H} | \chi'_A \rangle \pm \langle \chi'_A | \hat{H} | \chi'_B \rangle \tag{3.38}$$

$$\beta_A^{(\pm)} = \langle \varphi^{(\pm)} | \hat{H} | \varphi^{(\pm)\prime} \rangle = \frac{1}{2} \{ \langle \chi_B | \hat{H} | \chi'_B \rangle \pm \langle \chi_B | \hat{H} | \chi'_A \rangle \pm \langle \chi'_B | \hat{H} | \chi_A \rangle \} \tag{3.39}$$

$$\beta_t^{(\pm)} = \langle \varphi^{(\pm)} | \hat{H} | \varphi^{(\pm)\prime\prime} \rangle = \pm \frac{1}{2} \langle \chi_B | \hat{H} | \chi''_A \rangle \tag{3.40}$$

$$\beta_g^{(\pm)} = \langle \varphi^{(\pm)} | \hat{H} | \varphi^{(\pm)\prime\prime\prime} \rangle = \pm \frac{1}{2} \langle \chi_B | \hat{H} | \chi'''_A \rangle \tag{3.41}$$

The term $\langle \chi_A | \hat{H} | \chi'_A \rangle$ in Eq. (3.39) and the terms $\langle \chi_A | \hat{H} | \chi''_A \rangle$, etc., in Eqs. (3.40) and (3.41) are dropped since they belong to the interaction of atoms that are not directly bound.

Now, taking into consideration that each of the functions χ_A, χ_B (as the hybrid sp^3 orbital) is in turn a linear combination of the atomic s and p functions

$$\chi'_A = \frac{1}{2} s^A + \frac{\sqrt{3}}{2} p_x^A \tag{3.42}$$

$$\chi'_B = \frac{1}{2} s^B - \frac{\sqrt{3}}{2} p_x^B \tag{3.43}$$

$$\chi_A = \frac{1}{2} s^A - \frac{1}{2\sqrt{3}} p_x^A - \sqrt{\frac{2}{3}} p_z^A \tag{3.44}$$

$$\chi_B = \frac{1}{2} s^B + \frac{1}{2\sqrt{3}} p_x^B + \sqrt{\frac{2}{3}} p_z^B \tag{3.45}$$

$$\chi''_A = \frac{1}{2} s^A - \frac{1}{2\sqrt{3}} p_x^A - \sqrt{\frac{2}{3}} p_z^A \tag{3.46}$$

$$\chi'''_A = \frac{1}{2} s^A - \frac{1}{2\sqrt{3}} p_x^A + \frac{1}{\sqrt{2}} p_y^A + \frac{1}{\sqrt{6}} p_z^A \tag{3.47}$$

we finally have [131]

$$\alpha^{(\pm)}=\frac{1}{4}\{\alpha_{\mathrm{A}}^{(s)}+3\alpha_{\mathrm{A}}^{(p)}\}\pm\frac{1}{4}\{\beta_{ss}+2\sqrt{3}\,\beta_{sp}+3\beta_{pp}\} \tag{3.48}$$

$$\beta_{\mathrm{A}}^{(\pm)}=\frac{1}{8}\{\alpha_{\mathrm{A}}^{(s)}-\alpha_{\mathrm{A}}^{(p)}\}\pm\frac{1}{4}\left\{\beta_{ss}+\frac{2}{\sqrt{3}}\beta_{sp}-\beta_{pp}\right\} \tag{3.49}$$

$$\beta_{g}^{(\pm)}=\pm\frac{1}{8}\left\{\beta_{ss}-\frac{2}{\sqrt{3}}\beta_{sp}+\frac{1}{3}\beta_{pp}\right\}\pm\frac{1}{6}\beta_{\pi} \tag{3.50}$$

$$\beta_{t}^{(\pm)}=\pm\frac{1}{8}\left\{\beta_{ss}-\frac{2}{\sqrt{3}}\beta_{sp}+\frac{1}{3}\beta_{pp}\right\}\mp\frac{1}{3}\beta_{\pi} \tag{3.51}$$

Let us note that in the derivation of relations (3.48) and (3.49) from (3.38) and (3.39) the relation $\langle s_{\mathrm{A}}|\hat{H}|p_{\mathrm{A}}\rangle = 0$ is used, which is satisfied, in view of the tetrahedral symmetry of the potential at any site of the diamond lattice, not only for an isolated atom but also for an atom in the crystal.

3.3.2 Simplified Expression of Matrix Elements for the Valence Band

As we shall see below, the valence band, unlike the conduction band, is slightly related to the individual characteristics of the group IV elements, and its structure is quite stable with respect to small variations in the matrix elements. Therefore the valence band is amenable to a simple description [130]. We will consider it separately as a preliminary version of the more general theory, presented in chapter 4, especially as the structure of the valence band is of interest in itself.

This is a simpler version in that instead of the three independent resonance integrals β_{ss}, β_{sp}, and β_{pp} we shall use only one, which, for convenience, we will define as a new parameter*:

$$\beta=\langle\chi_{\mathrm{A}}|\hat{H}|\chi_{\mathrm{B}}\rangle=\frac{1}{4}\{\beta_{ss}+2\sqrt{3}\,\beta_{sp}+3\beta_{pp}\} \tag{3.52}$$

i.e., the resonance integral belonging to the σ interaction of the two hybrid sp^3 orbitals of the adjacent atoms A, B, directed along their common bonds A–B (see Fig. 3.3).

To find the expression of the parameters (3.38)–(3.41) in terms of the parameters β, we use in the intermediate step [130] the Wolfsberg-Helmholz method [132] that is employed in quantum chemistry (see also [18]).

*This make it possible to use thermochemical data to find the scales of the resonance integrals since the thermochemical strength of the A–B bonds in saturated compounds depends on the values of the parameter β.

According to this method, one can write:

$$\beta_{ss} = KS_{ss}\alpha_A^{(s)} \tag{3.53}$$

$$\beta_{sp} = KS_{sp}\frac{1}{2}\{\alpha_A^{(s)} + \alpha_A^{(p)}\} \tag{3.54}$$

$$\beta_{pp} = KS_{pp}\alpha_A^{(p)} \tag{3.55}$$

$$\beta = KS\alpha_A \tag{3.56}$$

where K is some proportionality constant*; S_{ss}, S_{sp}, S_{pp}, and S are the overlap integrals; and α_A is the average value of the Hamiltonian in the hybrid sp^3-orbital:

$$\alpha_A = \frac{1}{4}\{\alpha_A^{(s)} + 3\alpha_A^{(p)}\} \tag{3.57}$$

Hence we obtain

$$\beta_{ss} = \beta\left\{\frac{S_{ss}\alpha_A^{(s)}}{S\alpha_A}\right\} \tag{3.58}$$

$$\beta_{pp} = \beta\left\{\frac{S_{pp}\alpha_A^{(p)}}{S\alpha_A}\right\} \tag{3.59}$$

$$\beta_{sp} = \beta\left\{\frac{S_{sp}(\alpha_A^{(s)} + \alpha_A^{(p)})}{2S\alpha_A}\right\} \tag{3.60}$$

In these formulas one can use any of the estimates employed in the literature for evaluating the ratio of the overlap integrals $S_{ij}{:}S$ since this is only slightly reflected in the structure of the valence band (see Sec. 4.3.4). Here we will use the estimate that follows from Pauling's well-known "rule of angular parts" [31]:

$$S_{ss} : S_{sp} : S_{pp} : S = 1 : \sqrt{3} : 3 : 4 \tag{3.61}$$

*In the theory of molecules (for example, in the theory of complex compounds) the constant K is usually assigned a value of 1.67 for σ bonds and a value of 2 for π bonds [132], although other values are also discussed [133, 134]. This ambiguity is a serious drawback of the Wolfsberg–Helmholz method. Here, however, we are not concerned with the choice of K since K generally does not enter into the final formulas. For this reason the absolute magnitude of the overlap integrals is unimportant to us; only their ratio is important. In view of the latter, however, let us note that for cyclic systems or for sufficiently long organic chains $K = 1$. Thus, for aromatic compounds and polyenes it is usually assumed that $\beta_\pi \approx -2.6$ eV $\approx S_\pi\alpha_A^{(p)}$ [135]. Similarly, for the paraffins $\beta \approx -6.5$ eV $\approx S\cdot\alpha$ [136–138].

Its advantage is that it makes it possible to simplify Eqs. (3.48)-(3.51) for the parameters (3.14)-(3.21) considerably. They are now rewritten in the form [130]

$$\alpha = \alpha_A + \beta = \frac{1}{4}\{\alpha_A^{(s)} + 3\alpha_A^{(p)}\} + \beta \tag{3.62}$$

$$\beta_A = \frac{1}{8}\{\alpha_A^{(s)} - \alpha_A^{(p)}\} + \frac{1}{8}\,\frac{\alpha_A^{(s)} - \alpha_A^{(p)}}{\alpha_A}\,\beta \tag{3.63}$$

$$\beta_t = -\frac{1}{3}\beta_\pi \tag{3.64}$$

$$\beta_g = \frac{1}{6}\beta_\pi \tag{3.65}$$

The superscripts (+) and (−) are temporarily omitted here since Eqs. (3.62)-(3.65) will only be used for the valence band.

Thus in the "simplified" version of the theory the valence band structure of the elements of the fourth group is expressed only in terms of four parameters—the two Coulomb integrals $\alpha_A^{(s)}$, $\alpha_A^{(p)}$ and two resonance integrals β and β_π.

3.4 EMPIRICAL DETERMINATION OF PARAMETERS

3.4.1 Coulomb Integrals and the Promotion Energy

Equations (3.48)-(3.51) or (3.62)-(3.65) obtained above together with the dispersion laws (3.23)-(3.26) express the structure of the energy bands of covalent crystals in terms of the energy levels of the atoms and in terms of the parameters describing the atom interactions. Now in order to find the structure of the bands it is necessary to substitute the values of the Coulomb and resonance integrals into the dispersion laws, and for our purposes—for the reasons stated above—it is desirable to use the semiempirical version of formulating the theory.*

Thus we will strive to evaluate the parameters of the theory (in this chapter—$\alpha_A^{(s)}$, $\alpha_A^{(p)}$, β, β_π) in the dispersion laws from independent test data, with the idea of establishing the dependence of the band structure on the energy levels of the atoms and on the character of the chemical bond. This is necessary,

*Let us emphasize that this approach is significantly different from the EP method or from the method of Fourier expansion of the dispersion laws [110]. Within the framework of this approach none of the empirical parameters is borrowed from the band structure of the crystal to be investigated.

of course, in order to ensure, in the class of crystals being considered and in those systems from which we use the data, the necessary parameters have the same physical meaning and numerical values that are as close as possible. In particular, let us recall that we are investigating the band structure within the framework of the one-electron model without an explicit consideration of the electron interaction. Therefore in the three sections that follow—as well as later—we will use similar versions of the atomic level theory or the theory of the electron structure of molecules for finding the parameters.

As follows from Eqs. (3.32) and (3.33), the Coulomb integrals $\alpha_A^{(s)}, \alpha_A^{(p)}$ are defined as the average values of the Hamiltonian of the crystal in the s and p AOs of an isolated atom. Since for purely covalent crystals the long-range Madelung potential of the lattice is absent, it can be assumed that the lattice field in the vicinity of each given atom is quite close to the potential of a free atom. In this case it can be assumed, as is often done in the semiempirical theory of molecular structure, that $\alpha_A^{(s)}$, $\alpha_A^{(p)}$ are equal to the energies of the corresponding atomic levels, i.e., the orbital ionization potentials for these levels, taken with the opposite sign.

Let us point out, moreover, that in reality such an assumption can be replaced by a less stringent one. It will be seen below that the structure of the bands, i.e., the relative location of the levels at different points **k** of the first zone, is significantly dependent on only the difference $\alpha_A^{(p)}-\alpha_A^{(s)}$, whereas the absolute values of $\alpha_A^{(p)}$, $\alpha_A^{(s)}$ primarily determine only the position of all the levels with respect to "vacuum." Therefore it is actually sufficient to assume that the field of the lattice shifts the atomic s and p levels in approximately the same manner.

Let us consider in more detail the question of the numerical estimate of the Coulomb integrals $\alpha_A^{(s)}, \alpha_A^{(p)}$. As just stated, we shall equate the Coulomb integral $\alpha_A^{(p)}$ (taken with the opposite sign), to the orbital ionization potential for the p level of the A^{IV} atom (ns^2np^2 electron configuration), and the Coulomb integral $\alpha_A^{(s)}$ (taken with this same sign) to the orbital ionization potential for the s level of this atom. Thus, the Coulomb integrals can be determined from the processes

$$\begin{aligned} &A\,(s^2p^2) \rightarrow A^+\,(s^2p) + \alpha_A^{(p)} + e \\ &A\,(s^2p^2) \rightarrow A^+\,(sp^2) + \alpha_A^{(s)} + e \end{aligned} \tag{3.66}$$

Equations (3.66), however, do not define the integrals $\alpha_A^{(p)}, \alpha_A^{(s)}$ completely unambiguously since there are different terms within the limits of one-electron configuration of the atom (see Sec. 1.3.4). Therefore to find the energy of the atomic s and p levels (the orbital ionization potentials) we make use of the method of averaging over the terms (suggested by Slater [14]). Then, as follows from (3.66), for group IV elements $\alpha_A^{(p)}$, $\alpha_A^{(s)}$ will be determined by means of the equations

$$\alpha_A^{(p)} = \langle E\,(As^2p^2)\rangle - \langle E\,(As^2p)\rangle = \frac{1}{15}\{E\,(^1S) + 9E\,(^3P) + 5E\,(^1D)\}_{s^2p^2} - E\,(^2P)_{s^2p} \tag{3.67}$$

$$\alpha_A^{(s)} = \langle E\,(As^2p^2)\rangle - \langle E\,(Asp^2)\rangle = \frac{1}{15}\{E\,(^1S)+9E\,(^3P)+5E\,(^1D)\}_{s^2p^2}$$

$$-\frac{1}{30}\{2E\,(^2S)+6E\,(^2P)+12E\,(^4P)+10E\,(^2D)\}_{sp^2} \quad (3.68)$$

in which the first set of braces refers to the neutral atom A (s^2p^2) and the second to the singly charged ion $A^+(s^2p)$ or $A^+(sp^2)$.

To reduce Eqs. (3.67) and (3.68) to a form suitable for specific calculations, for each state of the atom or ion let us replace its (average) total energy $E(^{2S+1}L)$ by the sum

$$E\,(^{2S+1}L) = E_0 + \mathscr{E}\,(^{2S+1}L) \quad (3.69)$$

where E_0 is the (average) total energy of the atom or ion in its ground state, corresponding to the lower term of the unexcited configuration, and $\mathscr{E}$ is the spectroscopic (average) energy of the corresponding term, measured from the average energy for the lower term of the unexcited configuration.* Let us also note that the experimentally measured first ionization potential I_1 is equal to the difference of the total energies of the neutral atom A (s^2p^2) and the positive ion A^+ in their ground states. In this case we finally obtain from Eqs. (3.67), (3.68), and (3.69)

$$\alpha_A^{(p)} = -I_1 + \frac{1}{15}\{\mathscr{E}\,(^1S)+5\mathscr{E}\,(^1D)\}_{s^2p^2} \quad (3.70)$$

$$\alpha_A^{(s)} = -I_1 + \frac{1}{15}\{\mathscr{E}\,(^1S)+5\mathscr{E}\,(^1D)\}_{s^2p^2}$$

$$-\frac{1}{30}\{2\mathscr{E}\,(^2S)+6\mathscr{E}\,(^2P)+12\mathscr{E}\,(^4P)+10\mathscr{E}\,(^2D)\}_{sp^2} \quad (3.71)$$

Let us write here another expression for the difference of the Coulomb integrals $\alpha_A^{(p)} - \alpha_A^{(s)}$, which determines the band structure to a considerable extent. As follows from (3.70) and (3.71):

$$\alpha_A^{(p)} - \alpha_A^{(s)} = \langle E\,(Asp^2)\rangle - \langle E\,(As^2p)\rangle$$

$$= \frac{1}{30}\{2\mathscr{E}\,(^2S)+6\mathscr{E}\,(^2P)+12\mathscr{E}\,(^4P)+10\mathscr{E}\,(^2D)\}_{sp^2} \quad (3.72)$$

or in abbreviated form (we use this notation)

$$\alpha_A^{(p)} - \alpha_A^{(s)} = \mathscr{E}\,(A^+s^2p \rightarrow A^+sp^2) \quad (3.73)$$

where the term-averaged energy of the transition between the two configurations s^mp^n and

*For example, for the unexcited configuration A (s^2p^2) or A^+ (s^2p) the energy E_0 is equal to the average energy of the atom in the ^3P and ^2P states, respectively, and $\mathscr{E}$ is measured from these terms. At the same time for the excited configuration sp^2 of the ion A^+ E_0 corresponds to the total energy of the ion A^+ not for the lower term ^4P of this configuration, but for the lower term ^2P of the unexcited s^2p configuration, so that in this case too $\mathscr{E}$ is measured from the term ^2P.

$s^{m-1}p^{n+1}$ is denoted by $\mathcal{E}(s^m p^n \to s^{m-1}p^{n+1})$. The Coulomb integrals $\alpha_A^{(p)}$, $\alpha_A^{(s)}$ for group IV elements, estimated in this manner, together with similar integrals for atoms (ions) of several other groups (see Sec. 5.3.1 for more details) are listed in Table 3, in which some of the data of Bash and Viste are used (see, for example, the appendix to [139]).

In conclusion we shall discuss the question of the effect of the valence state of the atom on the magnitude of the Coulomb integrals $\alpha_A^{(p)}$, $\alpha_A^{(s)}$. The estimation of the Coulomb integrals from Eqs. (3.70) and (3.71) assumes that the inherent potential of the atoms in a diamond-type crystal is the same as for the free atom in its ground state $ns^2\,np^2$. In fact, in the diamond lattice the atoms are in the sp^3 hybridization states. Therefore the potential for such an atom should resemble the potential for an excited atom with the configuration A^{IV} ($nsnp^3$), so that the Coulomb integrals $\alpha_A^{(p)}$, $\alpha_A^{(s)}$ should be determined from the processes

$$\begin{aligned} &A\,(sp^3) \to A^+\,(sp^2) + \alpha_A^{(p)} + e \\ &A\,(sp^3) \to A^+\,(p^3) + \alpha_A^{(s)} + e \end{aligned} \tag{3.74}$$

and not from the processes as given in Eqs. (3.66). To estimate these values of the Coulomb integrals, let us consider the following cycle (here the new value of the integral $\alpha_A^{(p)}$ is temporarily denoted by $\alpha_A^{*(p)}$, and the old value by $\alpha_A^{(p)}$ as before):

$$\begin{aligned} &A\,(s^2p^2) \to A\,(sp^3) - \mathcal{E}\,(As^2p^2 \to Asp^3) \\ &A\,(sp^3) \to A^+\,(sp^2) + \alpha_A^{*(p)} + e \end{aligned} \tag{3.75}$$

TABLE 3. Orbital Ionization Potentials for Elements That Form $A^N B^{8-N}$ Tetrahedral Crystals[a]

Element	$-\alpha_A^{(s)}(0)$	$-\alpha_A^{(p)}(0)$	$-\alpha_A^{(s)}(1)$	$-\alpha_A^{(p)}(1)$
C	19.5	10.7	—	—
Si	15.0	7.8	—	—
Ge	15.6	7.5	—	—
Sn	15—14.5	7	—	—
B	14.0	8.3	25.1	19.5
Al	11.3	6.0	18.8	13.5
Ga	12.7	6.0	20.5	13.8
In	11.9	5.9	18.9	12.3
N	25.5	13.1	12.5	—0.1
P	18.7	10.3	7.5	0.7
As	17.6	9.0	7.1	0.6
Sb	16.3	7.4	—	—
Be	9.3	6	18.2	14.2
Zn	9.4	4.9	18.0	11.9
Cd	9.0	4.8	16.9	11.3
O	32.3	13.2	17.1	1.5
S	20.7	9.8	11.0	2.1
Se	20.8	9.2	9.9	1.7

[a]The data for the neutral atoms (except Sn, In, Cd, and Sb) are taken from [139]; the rest of the potentials are estimated from spectroscopic data [13].

$$A^+(sp^2) \to A^+(s^2p) + \mathscr{E}(A^+s^2p \to A^+sp^2)$$
$$A^+(s^2p) + e \to A(s^2p^2) - \alpha_A^{(p)} \tag{3.75 (continued)}$$

Hence

$$\alpha_A^{*(p)} = \alpha_A^{(p)} + \{\mathscr{E}(As^2p^2 \to Asp^3) - \mathscr{E}(A^+s^2p \to A^+sp^2)\} \tag{3.76}$$

Similarly, from the cycle

$$\begin{gathered} A(s^2p^2) \to A(sp^3) - \mathscr{E}(As^2p^2 \to Asp^3) \\ A(sp^3) \to A^+(p^3) + \alpha_A^{*(s)} + e \\ A^+(p^3) \to A^+(sp^2) + \mathscr{E}(A^+s^2p \to A^+p^3) \\ A^+(sp^2) + e \to A(s^2p^2) - \alpha_A^{(s)} \end{gathered} \tag{3.77}$$

we have

$$\alpha_A^{*(s)} = \alpha_A^{(s)} + \{\mathscr{E}(A\,s^2p^2 \to A\,sp^3) - \mathscr{E}(A^+\,sp^2 \to A^+\,p^3)\} \tag{3.78}$$

Numerical estimates on the basis of spectroscopic data [13] show that the quantities inside the braces in Eqs. (3.76) and (3.78) are not large. For example, for carbon they amount to ~0.4 and ~0.8 eV, respectively. In this case the "new" ionization potentials coincide with the "old" to within about 95%.

The "promotion" (excitation) energy $\mathscr{E}_{promot}$ of an s electron into the p level also plays an important role in the discussion below. It is easy to show that

$$\mathscr{E}_{promot} \approx \alpha_A^{(p)} - \alpha_A^{(s)} \tag{3.79}$$

i.e., $\mathscr{E}_{promot}$ can be evaluated simply as the difference of the orbital ionization potentials for the s and p levels of free atoms in their ground states. In reality $\mathscr{E}_{promot}$ is usually taken to mean the energy of this transition in a neutral atom, which can be estimated within the framework of the Slater method [14] as the energy of the transition between the average (obtained by averaging over the terms) levels for the s^2p^2 and sp^3 configurations:

$$\mathscr{E}_{promot} = \mathscr{E}[A(s^2p^2) \to A(sp^3)] \tag{3.80}$$

As just stated, however, the energy of the transition between the s and p levels in the neutral atom is close to the corresponding energy of the transition in the positive ion. Thus

$$\mathscr{E}_{promot} \approx \mathscr{E}[A^+(s^2p) \to A^+(sp^2)] \tag{3.81}$$

Actually, for example, for carbon $\mathscr{E}_{promot} - \mathscr{E}(C^+s^2p \to C^+sp^2) \simeq 0.5$ eV. Hence using Eqs. (3.66), we have Eq. (3.79).

3.4.2 Resonance Integrals β

As already stated, an empirical evaluation of the resonance parameters makes it possible to relate the structure of the energy levels to the character of the interaction of the atoms in different systems, thereby making it possible to establish the relationship between the band structure and the electron structure of the individual molecules.

We also pointed out above that the interatomic distances and the thermochemical A^{IV}–A^{IV} bond energies in tetrahedral crystals of group IV elements are close to the corresponding quantities in saturated compounds A_nX_{2n+2} (see Table 2).* For example, for diamond the energy to break the C–C bond (85.5 kcal/mol) is the same as the energy to break this bond in the aliphatic hydrocarbons C_nH_{2n+2} (83-85 kcal/mol) within ~98%, and the interatomic distance (1.542 Å) in the diamond crystal is the same as the C–C distance in the aliphatic chains within 99.5%.

Thus it can be assumed that the nature of the interaction of the A^{IV} atoms in A^{IV} crystals is generally the same as in the molecules A_nX_{2n+2}. Therefore for our purposes the most satisfactory method of empirical determination of the resonance parameters and, in particular, the parameter β would be to estimate them from the electron structure of the A_nX_{2n+2} molecules–preferably from the uv spectra, such as is done in the semiempirical quantum chemistry of conjugated systems [135]. For carbon compounds one can actually obtain such an estimate of the parameter β and by two different methods. (Let us only state that neither is based on direct uv spectroscopy data of the alkanes because of the lack of sufficiently complete experimental information at present.)

In the version of the theory of paraffins C_nH_{2n+2}, which is similar in the nature of the approximation to the theory of covalent crystals to be considered, a value of -6.5 eV is usually assigned to the parameter β [136-138]. This leads to a satisfactory description of quite a wide circle of the properties of the paraffins (and, in particular, their ionization potentials), so that for diamond we can adopt the estimate of $\beta(\text{diamond}) \approx \beta(\text{paraffin}) \approx -6.5$ eV although a somewhat higher value ($\beta \approx -8$ to -8.3 eV) may be a somewhat better estimate.†

*Let us note, incidentally, that the analogy between, let us say, the diamond crystal and the paraffins is not limited by only the "static" properties of these systems. Thus shown in [140], the spectrum $\omega(\mathrm{k})$ of the oscillations of the diamond lattice can be calculated by borrowing the force constants for the C–C bonds from the oscillation spectra of saturated hydrocarbons.

†The choice of the estimate $\beta = -6.5$ eV in [136] is motivated by the fact that the overlap integral for the two hybrid sp^3 orbitals of the C atoms, located at a distance of 1.54 Å, is equal to 0.65, whereas the average value of the Hamiltonian in the sp^3 orbitals is of the order of -10 eV. A more precise value of this average is -12.8 eV, from which $\beta \approx S\alpha_A \approx -8.3$ eV (see footnote on page 100).

Another method of estimating the parameter β is based on uv spectroscopy data of conjugated systems in combination with the Wolfsberg-Helmholz method. In the Hückel method for π systems—this method is similar to our approximation—the resonance integral β_π describing the interaction of the $p\pi$ orbitals of the C atoms can be determined from the uv spectra. This integral has a value of $\beta_\pi = -2.6$ eV [135, 137]. Let us now state that

$$\beta_\pi \approx K_\pi S_\pi \alpha_A^{(p)}; \quad \beta \approx K_\sigma S \alpha_A \tag{3.82}$$

where K_π and K_σ are the coefficients for the π and σ interactions of the AOs in the Wolfsberg-Helmholz method (see comment on page 100). Then

$$\beta \approx (K_\sigma / K_\pi)(S/S_\pi)(\alpha_A / \alpha_A^{(p)}) \beta_\pi \tag{3.83}$$

Since the corresponding overlap potentials for the Slater atomic functions are $S = 0.65$ and $S_\pi = 0.25$, then we obtain from Eq. (3.83): $\beta \approx -8.1$ eV for $K_\sigma \approx K_\pi$ and $\beta \approx -6.7$ eV for $K_\sigma / K_\pi \approx 1.67{:}2.00$. The latter agrees with the previous estimate, and the result obtained is only slightly dependent on the ratio K_σ / K_π.

Let us now turn to an estimation of the parameter β for Si, Ge, and α-Sn. In the case of these crystals the second method of estimating the parameter β is not applicable since for Si, Ge, and Sn there are no conjugated systems (organic molecule analogs with conjugated bonds). Moreover, uv spectra data for the $A_n^{IV} X_{2n+2}$ compounds (A^{IV} = Si, Ge, Sn) are even scarcer* than for the paraffins, and the theory for these molecules, from which one could take the parameter β, has not been developed either. Therefore we turn to thermochemical data for finding the parameter β for silicon, germanium, and gray tin.

Since the stability of tetrahedral covalent crystals is determined by the exchange interaction of the hybrid sp^3 functions of the valence-bound atoms, it is natural to assume that the parameters β in the series C (diamond)-Si-Ge-α-Sn change approximately the same as the energy of the A^{IV}–A^{IV} bonds, or, in other words, proportionally to the parameters $\beta_{\text{thermochem}}$, which are defined by the relation

$$\beta_{\text{thermochem}} = \frac{1}{4}(-\Delta H_{\text{AT}}) = \frac{1}{2} Q_{\text{AT}} \tag{3.84}$$

Here $-\Delta H_{\text{at}}$ is the atomization energy of the crystal, and Q_{at} is the energy of the A^{IV}-A^{IV} bond in molecule or crystal (so that the value of $\beta_{\text{thermochem}}$ is equal to the values of $-\Delta H_{\text{at}}$ or Q_{at} in the calculation at one bonding electron). Thus one can assumed that

$$\beta(\text{element}) : \beta(\text{diamond}) \approx \beta_{\text{thermochem}}(\text{element}) : \beta_{\text{thermochem}}(\text{diamond}) \tag{3.85}$$

*The only investigation in this direction is apparently that in [141], in which, however, only the overall profiles of the absorption band are obtained without reference to the transitions.

or

$$\beta(\text{element}) \approx \beta(\text{diamond}) \cdot \frac{\Delta H_{\text{at}}(\text{element})}{\Delta H_{\text{at}}(\text{diamond})} \tag{3.86}$$

Then, using the already established values of $\beta(\text{diamond}) = -5.5$ to -8.3 eV for $\beta(\text{diamond})$, we obtain the scale of the empirical values of the parameter β for all the group IV elements (Table 4).

Finally, let us make one more comment concerning the choice of β. Since the strength of the A–A bonds is determined by the interaction of the corresponding sp^3 orbitals, a question naturally arises as to whether it is possible to simply equate β to the quantity $\beta_{\text{thermochem}} \approx -(1/4)\Delta H_{\text{at}}$. These "thermochemical" values of the resonance integrals $\beta(\text{diamond}) = 1.86$ eV, $\beta(\text{Si}) = -1.15$ eV, $\beta(\text{Ge}) = -0.85$ eV, $\beta(\alpha\text{-Sn}) = -0.75$ eV were used in the works of Coulson, Redei, and Stocker [117], Stocker [142], and Doggett [143]. They lead, however, to a markedly poorer agreement between theory and experiment since they are drastically reduced in magnitude [130, 144]. The latter is explained by a number of reasons and, in particular, by the fact that the thermochemical value of $\beta_{\text{thermochem}} = -(1/4)\Delta H_{\text{at}}$ does not include the expenditures of energy for excitation of the atoms into the valence sp^3 state. [The gain in energy due to the exchange interaction should not only ensure the observed strength of the bonds, but also compensate the loss in energy during the transition of an atom from the ground state A^{IV} (ns^2np^2) into the valence state A^{IV} (*tetetete*).]

From a more general point of view it must be noted that in the semiempirical methods the "spectroscopic" values of the resonance parameters, serving to describe the structure of the energy levels, should generally not necessarily agree with their "thermochemical" values necessary to describe the bond energy. This fact is well known, for example, in the quantum chemistry of conjugated systems, where (within the framework of the Hückel method) the value of $\beta_\pi \approx -1$ eV is used to describe the heat of formation and the value of $\beta_\pi \approx -2.6$ eV is used to explain the spectra [135, 137]. Thus the assumption that β_{spectr} and $\beta_{\text{thermochem}}$ are equal is unjustified although it is logical to assume, as was done, that the larger thermochemical values of the parameter β correspond to the larger spectroscopic values.

3.4.3 Resonance Integrals β_π

For the empirical evaluation of the parameters β_π we again make use of the scheme described above, by first determining β_π for diamond and only then for

TABLE 4. Resonance Parameters (in eV) for Covalent Crystals from Data for Hydrocarbons

Parameters		Diamond	Silicon	Germanium	Gray Tin
$-\beta$		6.5—8.3	4—5.2	3—3.8	2.7—3.5
$-\beta_\pi$		1.7	1.0	0.8	0.7
$-\beta_{ss}$	$-\beta_{pp} \approx -\beta_0$	5.1	3.2	—	—
$-\beta_{sp}$		2.55	1.6	—	—

the rest of the covalent crystals. At first glance the empirical evaluation of β_π for diamond from data for individual molecules encounters the obvious difficulty that the $p\pi$-$p\pi$ interaction of the C atoms is not directly evident in the properties of such typical σ systems as the hydrocarbons C_nH_{2n+2}. In this case, however, to find the spectroscopic values of β_π one can make use of reliable experimental data for organic molecules with conjugated bonds, in which the $p\pi$-$p\pi$ interaction of the C atoms, conversely, is evident in a very direct manner.

As is well known from the quantum chemistry of conjugated systems, the dependence of the absorption frequencies in the uv region on the energy difference of the upper filled and lower empty molecular orbitals makes it possible to find, within the framework of the Hückel method (similar to our approximation), the spectroscopic value of the parameter β_π although, of course, only for the distance of 1.4 Å. This dependence for the series ethylene-butadiene-hexatriene-octatetraene leads to a value of $\beta_\pi(1.4\ \text{Å}) \approx -2.6$ eV [135], which is easily extrapolated to a distance of 1.54 Å by making use of any standard extrapolation method. Thus, for example, an extrapolation in terms of the ratio of the overlap integrals [130] $\beta_\pi(1.54\ \text{Å}):\beta_\pi(1.4\ \text{Å}) = S_\pi(1.54\ \text{Å}):S_\pi(1.4\ \text{Å})$ gives a value of $\beta_\pi(1.54\ \text{Å}) = -2.0$ eV. A similar extrapolation using the Morse curve [130] leads to an adequately close value of $\beta_\pi(1.54\ \text{Å}) = -1.5$ eV.

Using data from the uv spectra for a number of aromatic hydrocarbons, on the basis of which a value of -2.3 eV is usually assigned to the parameter $\beta_\pi(1.4\ \text{Å})$, also gives approximately these same results [135]. An extrapolation in terms of the ratio of the overlap integrals gives a value of $\beta_\pi(1.54\ \text{Å}) = -1.8$ eV in this case, whereas an extrapolation from the Morse curve leads to a similar value of $\beta_\pi(1.54\ \text{Å}) = -1.4$ eV.

The good agreement of all such estimates (Table 5) indicates, to all appearances, the adequately reliable nature of the extrapolation procedure of finding β_π from uv spectra data for the conjugated systems. Therefore as the resulting value of β_π(diamond) one can adopt the average value of

TABLE 5. Various Methods for Evaluating β_π for Diamond

Estimation Method	$-\beta_\pi$ (eV)
From overlap integrals, starting from value of $\beta = -6.5$ eV for paraffins	2.0
Extrapolation from Morse curve, starting from value of β_π for polyenes	1.5
Extrapolation from overlap integrals, starting from β_π for polyenes	2.0
Extrapolation from Morse curve, starting from β_π for aromatic hydrocarbons	1.4
Extrapolation from overlap integrals, starting from β_π for aromatic hydrocarbons	1.8

β_π(diamond) = −1.7 eV from the range of estimates obtained (−1.4 to −2.0 eV). This value (as pointed out in Sec. 2.3.3) also agrees well with the previously adopted estimate of β(diamond) = −6.5 to −8.3 eV for the resonance integral.

Finally, let us point out that the value of $\beta_\pi(1.54$ Å$) = -1.7$ eV obtained can be adjusted by using uv spectra data for toluene $C_6H_6-CH_3$. The C atom of the $-CH_3$ methyl group is 1.51 Å from the C atom of the benzene ring bound to it. Then, assuming that $\beta_\pi(1.54$ Å$) = -1.7$ eV and $\beta_\pi(1.4$ Å$) = -2.3$ eV, a value of $\beta_\pi(1.54$ Å$) = -1.8$ eV $= 0.78\beta_\pi(1.4$ Å$)$ should be chosen for the interaction integral of the $p\pi$ orbitals of the benzene and methyl C atoms. In fact, in explaining the bathochromic shift of the absorption band in the uv spectrum of toluene for $\beta_\pi(1.51$ Å$)$ a value of $0.7\beta_\pi(1.4$ Å$)$ is usually assumed. This agrees well with the value of $0.78\beta_\pi(1.4$ Å$)$.

It is impossible to extend the method described for determining β_π directly to the rest of the group IV elements, for which there are no compounds that are analogs of the polyenes or aromatic hydrocarbons. Therefore to construct the scale of the parameters β_π let us again assume that the resonance integrals β_π vary approximately the same as the parameters β or $\beta_{\text{thermochem}}$ throughout the entire series C(diamond)-Si-Ge-α-Sn. Then

$$\beta_\pi(\text{element}) = \frac{\Delta H_{\text{at}}(\text{element})}{\Delta H_{\text{at}}(\text{diamond})}\,\beta_\pi(\text{diamond}) \tag{3.87}$$

which leads to the desired scale of the parameters β_π (Table 4). We shall see below how these values can be compared with experimental data and prove that they agree well with experiments.

3.5 STRUCTURE OF THE VALENCE BAND OF COVALENT CRYSTALS. COMPARISONS WITH EXPERIMENT AND QUALITATIVE BEHAVIOR

3.5.1 Fundamental Parameters of the Valence Band Structure

Having obtained an empirical system of parameters for describing the band structure, let us turn to a discussion of some of the results of the theory.

The typical nature of the structure of the valence band for group IV elements can be understood from Fig. 3.2. As seen from this figure, the structure of the valence band is determined in its basic aspects by the location of the levels X_4^v, X_1^v, L_3^v, L_1^v, L_2^v, and Γ_1^v with respect to the top of the valence band $\Gamma_{25'}^v$. Thus the fundamental parameters of the band structure for the valence band are

total width of valence band $E_v = \varepsilon(\Gamma_{25'}^v) - \varepsilon(\Gamma_1^v)$

$$E_v = -8\beta_A^{(+)} - 16\beta_g^{(+)} \tag{3.88}$$

width of p band $\Delta E_p = \varepsilon(\Gamma_{25'}^{v}) - \varepsilon(X_4^{v})$

$$\Delta E_p = 8\{\beta_t^{(+)} - \beta_g^{(+)}\} \tag{3.89}$$

width of upper s-p band $\Delta E_1 = \varepsilon(\Gamma_{25'}^{v}) - \varepsilon(X_1^{v})$

$$\Delta E_1 = -4\beta_A^{(+)} + 8\beta_t^{(+)} \tag{3.90}$$

width of lower s-p band $\Delta E_2 = \varepsilon(X_1^{v}) - \varepsilon(\Gamma_1^{v})$

$$\Delta E_2 = -4\beta_A^{(+)} - 8\beta_t^{(+)} - 16\beta_g^{(+)} \tag{3.91}$$

difference in the levels $\delta E_p = \varepsilon(\Gamma_{25'}^{v}) - \varepsilon(L_{3'}^{v})$

$$\delta E_p = 4\{\beta_t^{(+)} - \beta_g^{(+)}\} \tag{3.92}$$

difference in the levels $\varepsilon(\Gamma_{25'}^{v}) - \varepsilon(\overline{\overline{L}}^{v}) = \varepsilon(\Gamma_{25'}^{v}) - \varepsilon(L_1^{v})$

$$\varepsilon(\Gamma_{25'}^{v}) - \varepsilon(\overline{\overline{L}}^{v}) = -2\beta_A^{(+)} - 4\beta_g^{(+)} + 12\beta_t^{(+)} \tag{3.93}$$

difference in the levels $\varepsilon(\Gamma_{25'}^{v}) - \varepsilon(\overline{L}^{v}) = \varepsilon(\Gamma_{25'}^{v}) - \varepsilon(L_2^{v})$

$$\varepsilon(\Gamma_{25'}^{v}) - \varepsilon(\overline{L}^{v}) = -6\beta_A^{(+)} - 4\beta_g^{(+)} + 4\beta_t^{(+)} \tag{3.94}$$

Here, the right sides of Eqs. (3.88)–(3.94) give expressions for the parameters E_v, ΔE_p, etc., in terms of the matrix elements in a basis of EOs and are valid for any version of the theory, either "complete" or "simplified."

By making use of Eqs. (3.48)–(3.51) or Eqs. (3.62)–(3.65) we can express the parameters of the band structure in terms of "atomic" parameters—the Coulomb and resonance integrals. In particular, for the "simplified" version of the theory* the corresponding expressions have the following form:

$$E_v = \alpha_A^{(p)} - \alpha_A^{(s)} + \frac{\alpha_A^{(p)} - \alpha_A^{(s)}}{\alpha_A}\beta - \frac{8}{3}\beta_\pi \tag{3.95}$$

$$\Delta E_p = -4\beta_\pi \tag{3.96}$$

$$\Delta E_1 = \frac{1}{2}\{\alpha_A^{(p)} - \alpha_A^{(s)}\} + \frac{1}{2}\frac{\alpha_A^{(p)} - \alpha_A^{(s)}}{\alpha_A}\beta - \frac{8}{3}\beta_\pi \tag{3.97}$$

$$\Delta E_2 = \frac{1}{2}\{\alpha_A^{(p)} - \alpha_A^{(s)}\} + \frac{1}{2}\frac{\alpha_A^{(p)} - \alpha_A^{(s)}}{\alpha_A}\beta \tag{3.98}$$

*Let us remember that in this chapter we are considering only this preliminary version; the more complicated or "complete" version is covered in chapter 4.

$$\delta E_p = -2\beta_\pi \qquad (3.99)$$

$$\varepsilon(\Gamma^v_{25'}) - \varepsilon(L^v_1) = \frac{1}{4}\{\alpha^{(p)}_A - \alpha^{(s)}_A\} + \frac{1}{4}\,\frac{\alpha^{(p)}_A - \alpha^{(s)}_A}{\alpha_A}\,\beta - \frac{14}{3}\beta_\pi \qquad (3.100)$$

$$\varepsilon(\Gamma^v_{25'}) - \varepsilon(L^v_2) = \frac{3}{4}\{\alpha^{(p)}_A - \alpha^{(s)}_A\} + \frac{3}{4}\,\frac{\alpha^{(p)}_A - \alpha^{(s)}_A}{\alpha_A}\,\beta - 2\beta_\pi \qquad (3.101)$$

Let us now examine how Eqs. (3.95)–(3.101)–as yet without a numerical evaluation of the parameters–make possible a qualitative explanation of a number of the characteristic features of band structure. As follows, for example, from Eq. (3.95), the total width E_v of the valence band (i.e., the distance between the bonding s and p levels in a crystal) consists of the distance between the s and p levels for a free atom $\alpha^{(p)}_A$-$\alpha^{(s)}_A$ and an additional resonance term. Therefore E_v should always be greater than the promotion energy $\mathscr{E}_{promot}$ for the free atom. This is the actual case (compare Tables 3 and 6).

On the basis of Eq. (3.95) it is also easy to predict the behavior of E_v in the series C(diamond)-Si-Ge-α-Sn. Since the values of $\mathscr{E}_{promot}$ and the resonance parameters become smaller as one goes from diamond to α-Sn, the total width E_v of the valence band should also vary in this same manner. The largest decrease in E_v should be observed as one goes from C to Si since in this case the values of $\mathscr{E}_{promot}$ and $|\beta|$ also decrease simultaneously by a considerable amount. On the other hand, as one goes from Si to Ge the value of $|\beta|$ decreases whereas the value of $\mathscr{E}_{promot}$ increases, so that the increments in the Coulomb

TABLE 6. Values of the Fundamental Parameters (in eV) of the Valence Band of Diamond, Silicon, and Germanium, Obtained by Different Methods

Crystal	Parameter	Experiment	Semi-empirical EO LCAO	OPW [62]	EC OPW [83–85]	EP [106, 107]	APW [67]	LCAO [52]	LCAO [55]
C (diamond)	ΔE_p	6.7—7.3	6.5	6.8	5	7	5.2	5.3	5.5
	δE_p	—	3.3	5	1.5	3	2.4	2.4	2.4
	ΔE_1	—	12	14	13	18	11.5	11.6	11.4
	ΔE_2	—	6	8	8	10	8.1	8.0	7.8
	E_v	16—36	18—19	22	21	28	19.6	19.6	19.2
Si	ΔE_p	2.2—3.1	4	2.7	2.5	2.5	—	—	—
	δE_p	—	2	2.4	1.2	1.3	—	—	—
	ΔE_1	~8	8	6.8	7.3	—	—	—	—
	ΔE_2	~6.5	5	4.2	4.2	—	—	—	—
	E_v	12—16.5	13—14	11	11.5	—	—	—	—
Ge	ΔE_p	2.4	3	2.7	2.8	2.5	—	—	—
	δE_p	1.4	1.5	1.7	1.2	1.1	—	—	—
	ΔE_1	~8.5	6	7.5	8.5	—	—	—	—
	ΔE_2	~4.5	5	3.5	3.5	—	—	—	—
	E_v	12.3—13	12—12.5	11	12	—	—	—	—

and resonance terms in Eq. (3.95) compensate one another to a certain extent. Thus one should expect that for the pair Si and Ge the corresponding values of E_v are quite close. This also is confirmed by the experimental data.

As another example of the qualitative explanation of the band structure, let us take the character of the variation of the width of the p band ΔE_p as well as the difference in levels δE_p in the series C(diamond)-Si-Ge-α-Sn. According to Eqs. (3.96), (3.99), and (3.87), the values of ΔE_p and δE_p decrease in this series just like the corresponding values of the resonance integral β_π, i.e., like the heat of atomization ΔH_{at}:

$$\Delta E_p\,(\mathrm{C}) : \Delta E_p\,(\mathrm{Si}) : \Delta E_p\,(\mathrm{Ge}) : \Delta E_p\,(\mathrm{Sn}) = \beta_\pi\,(\mathrm{C}) : \beta_\pi\,(\mathrm{Si}) : \beta_\pi\,(\mathrm{Ge}) : \beta_\pi\,(\mathrm{Sn}) \tag{3.102}$$

$$\delta E_p\,(\mathrm{C}) : \delta E_p\,(\mathrm{Si}) : \delta E_p\,(\mathrm{Ge}) : \delta E_p\,(\mathrm{Sn}) = \beta_\pi\,(\mathrm{C}) : \beta_\pi\,(\mathrm{Si}) : \beta_\pi\,(\mathrm{Ge}) : \beta_\pi\,(\mathrm{Sn}) \tag{3.103}$$

$$\beta_\pi\,(\mathrm{C}) : \beta_\pi\,(\mathrm{Si}) : \beta_\pi\,(\mathrm{Ge}) : \beta_\pi\,(\mathrm{Sn}) = \Delta H_{\mathrm{aT}}\,(\mathrm{C}) : \Delta H_{\mathrm{aT}}\,(\mathrm{Si}) : \Delta H_{\mathrm{aT}}\,(\mathrm{Ge}) : \Delta H_{\mathrm{aT}}\,(\mathrm{Sn}) = 1 : 0.6 : 0.45 : 0.4 \tag{3.104}$$

It is not hard to see that experiment gives very close values for these ratios (see Sec. 3.5.2; there are no corresponding experimental data for Sn):

$$\beta_\pi\,(\mathrm{C}) : \beta_\pi\,(\mathrm{Si}) : \beta_\pi\,(\mathrm{Ge}) \approx 1 : 0.5 : 0.4 \tag{3.105}$$

A second result, which can be extracted from Eqs. (3.96)-(3.99), is that the ratio δE_p:ΔE_p is constant and equal to 0.5 throughout the entire series C-Si-Ge-α-Sn. It is impossible to prove this statement by direct experiment since only ΔE_p can be determined from optical data for C and Si and only δE_p for Ge (see Sec. 4.2). However, from the data obtained by EP [106-108] and EC OPW [83-85] methods it follows that for this series the ratio δE_p:ΔE_p actually does have approximately the same value and it is 0.4-0.5.

3.5.2 Width of p Band. Theoretical and Experimental Values of β_π

Let us now examine how the previously found system of Coulomb and resonance integrals makes it possible to describe qualitatively the structure of the valence band of crystals with a diamond lattice, and let us also compare the values obtained for the band structure parameters (3.97)-(3.103) with experimental data. We start from the width ΔE_p of the p band and the level difference δE_p since the comparison with experiment can be done most simply for these parameters.

The values of ΔE_p and δE_p, calculated using the parameters β_π from Table 4, are listed in Table 6. At the same time, as seen from Fig. 3.2, the width of the p band $\Delta E_p = \varepsilon(\Gamma_{25'}^v) - \varepsilon(X_4^v)$ for diamond and silicon can be found directly from spectroscopic data. It is known that for these crystals the minimum of the conduction band is located on the line Δ near the point X and the energy of the level X_1^c is close to the value of the energy $\varepsilon(\Delta)_{\min}$ at the minimum point. In this case the experimental value of ΔE_p can be estimated as the energy difference of the band-to-band transition $X_4^v \to X_1^c$ and the forbidden band gap E_g:

$$(\Delta E_p)_{\exp} \approx \varepsilon\,(X_4^v \to X_1^c) - E_g \tag{3.106}$$

The energy of the transition $X_4^v \to X_1^c$ is equal to 12.2 [145]-12.7 eV [146, 147] and 4.3-4.5 eV [146, 147] for diamond and silicon, respectively, while the forbidden band gap E_g amounts to 5.4-5.5 eV [148] for diamond and 1.2 eV [149] for silicon. Hence for the experimental value of the width of the p band we have $\Delta E_p(\mathrm{C})_{\exp} = 6.7\text{-}7.3$ eV and $\Delta E_p(\mathrm{Si})_{\exp} = 3.1\text{-}3.3$ eV. These do not agree well with the "theoretical" estimates of $\Delta E_p(\mathrm{C}) = 6.8$ eV and $\Delta E_p(\mathrm{Si}) = 4$ eV obtained by using the parameters from Table 5.

For germanium the method of comparing theory with experiment must be altered somewhat. Although the "theoretical" value of ΔE_p for germanium is also determined from Eq. (3.96), the experimental value of $\Delta E_p(\mathrm{Ge})_{\exp}$ can no longer be calculated from Eq. (3.106) since for the Ge crystal the absolute minimum of the conduction band does not lie on the Δ line but at point L. Thus, in this case the "theoretical" value of δE_p must be taken instead of ΔE_p and compared with the corresponding experimental value

$$(\delta E_p)_{\exp} = \varepsilon\,(L_{3'}^v \to L_1^c) - E_g \tag{3.107}$$

Since for germanium the energy of the transition $L_{3'}^v \to L_1^c$ is equal to 2.2 eV [146] and the forbidden band gap E_g amounts to 0.8 eV, then $\delta E_p(\mathrm{Ge})_{\exp} = 1.4$ eV. This is also close to the "theoretical" value ($\delta E_p = 1.6$ eV).

It is also useful to consider these results from a somewhat different point of view. As Eqs. (3.96) and (3.98) show, the value of ΔE_p and δE_p depend only on one parameter–the resonance integral β_π. Therefore a comparison of the "theoretical" values of $\Delta E_p = -4\beta_\pi$ and $\delta E_p = -2\beta_\pi$ with the experimental values $(\Delta E_p)_{\exp}$ and $(\delta E_p)_{\exp}$ is, in essence, a direct experimental verification of the β_π scale obtained above (see Table 5). It is convenient to perform such a verification after the "experimental" values of β_π for diamond and silicon have been determined from the formula

$$(\beta_\pi)_{\exp} = \frac{1}{4}(\Delta E_p)_{\exp} \approx \frac{1}{4}\{\varepsilon\,(X_4^v \to X_1^c) - E_g\} \tag{3.108}$$

and for germanium from the formula

$$(\beta_\pi)_{\exp} = \frac{1}{2}(\delta E_p)_{\exp} = \frac{1}{2}\{\varepsilon\,(L_{3'}^v \to L_1^c) - E_g\} \tag{3.109}$$

The experimental values of β_π obtained in this manner are equal to ~1.7 eV for diamond, ~0.8 eV for Si, and ~0.7 eV for Ge. They coincide nearly exactly with our "theoretical" values of β_π from Table 4, although it is likely that one should not attach an extremely large amount of significance to such a close numerical agreement.

Nevertheless, the agreement between β_π and $(\beta_\pi)_{exp}$ is real in the sense that it can be considered as a confirmation of both the validity of the extrapolation method of finding β_π(1.54 Å) for diamond and the applicability of the general procedure of scaling the resonance integrals from diamond to the other crystals by using the heat of atomization.

Since the system of parameters β is matched to the system of parameters β_π, the nearness of β_π and $(\beta_\pi)_{exp}$ also attests to the validity of the choice of the scale of the parameters β.

Special mention must be made, however, of the agreement between the "theoretical" and experimental values of the parameters β_π for diamond. We recall that the "theoretical" value of β_π in this case was obtained from the electron spectra of the molecules, whereas the experimental value was obtained from the electron spectra of crystals. Therefore the agreement of β_π(C) with $\beta_\pi(C)_{exp}$ indicates an analogy between the electron structure of molecules and the band structure of solids and says that there is a real possibility of using the parameters, obtained for individual molecules, for the calculation of crystals.

In winding up the discussion of the question of the width of the p band, let us point out that there are other possibilities besides optical data for evaluating ΔE_p experimentally. Let us mention here the recent studies on the photoelectron (Grobman and Eastman [150]) and X-ray–electron (Ley [151]) spectra of silicon and germanium. According to the data of [150, 151], $\Delta E_p(\mathrm{Si})_{exp} =$ 2.2–3 eV and $\Delta E_p(\mathrm{Ge})_{exp} = 2.4$ eV. The "photoelectron" and "X-ray–electron" values of ΔE_p obtained for germanium agree well with out "theoretical" estimate (≈3 eV) although the analogous data for silicon are lower than the corresponding "theoretical" value of ≈4 eV.

3.5.3 Width E_v of Valence Band and Others of Its Subbands

Of the rest of the parameters determining the valence band structure, experimental data are now available for the total width of the valence band and in some cases for others of its subbands. For silicon the experimental estimates of E_v were obtained by investigating the X-ray spectra of Wiech [152]. More detailed information about the valence band structure was obtained from the already cited papers [150, 151] (Table 6). The values obtained for $E_v(\mathrm{Si})_{exp}$ lie in the interval from 11.8 to 16.5 eV, whereas the corresponding "theoretical" values of E_v using the parameters from Table 4 are equal to ≈13 eV for $\beta(\mathrm{Si}) =$ -4 eV and ≃14 eV for $\beta(\mathrm{Si}) \simeq -5.2$ eV.

There is a similar correspondence between the "theoretical" and experimental data (see Table 6) for other parameters of the valence band of Si as well as Ge* (the valence band of Ge has been discussed in detail in [150, 151]).

Experimental data on the width of the valence band for diamond are given in [154-158] and are, perhaps, the least definite in nature. Thus, in Holliday's paper [154] a value of 8.1 eV is obtained for the half width of the X-ray emission *CK* band for diamond, so that according to this paper $E_\nu(C)_{exp} = 16$ eV. At the same time Kleinman and Phillips [155], citing unpublished X-ray spectral data, give values for $E_\nu(C)$ in the interval from 16 to 30 eV, while Thomas et al. [157] even give values in the interval from 30 to 36 eV on the basis of X-ray-electron spectra, although this last estimate is probably much too high. This is supported in later papers: $E_\nu(C)_{exp} = 21$ eV. (The last value has been obtained by Gora et al. [158], also on the basis of X-ray-electron spectroscopy data.) As seen from Table 6 the "simplified" version of the theory within the framework of the EO LCAO method gives an estimate of $E_\nu = 18$ eV for $\beta(C) = -6.5$ and $E_\nu = 19$ eV for $\beta(C) = -8.3$ eV. These estimates agree with experimental estimates within the accuracy limits of the test data although the "simplified" version of the theory leads to reduced values of E_ν compared with those from the "complete" theory.† Besides, let us note that for diamond only there is an additional method that can be used to check the theoretical value of E_ν by means of a comparison of the theoretical and experimental values of E_ν for graphite. This check is made in Sec. 3.5.4, and its results support the theoretical estimates of E_ν presented above for diamond.

3.5.4 σ Band of Graphite and Width of the Valence Band in Diamond

In conjunction with the question of the width of the valence band in diamond let us consider the structure of the σ band in graphite [159], which is also of interest by itself as an example of the application of the equivalent orbital method to systems that are different from crystals with a diamond lattice.

Since graphite consists of separate layers, located a distance of 3.35 Å apart and bound by Van der Waals forces, we shall restrict our discussion–as is so often done in calculations of graphite–to an examination of one layer of C atoms.

The arrangement of the atoms in the layer and the choice of the basis vectors $\mathbf{a}_1$, $\mathbf{a}_2$ of the unit cell are shown in Fig. 3.4 along with the numbering of the bonds.

The analog of the equivalent orbitals (2.57) in graphite will obviously be the orbitals φ_j which, like in the diamond crystal, are localized along the σ bonds of C–C. Unlike the diamond crystal, however, in graphite the C atom does not occur in the sp^3 state, but in the sp^2 hybridization state. Thus, in the graphite layer the EOs are also expressed by Eq. (2.57), where for χ_A and χ_B we mean the sp^2 orbitals.

*For completeness let us point out that the X-ray spectral estimate of $E_\nu(Ge)_{exp} = 7$ eV given in [153] is significantly less than all these values. It is unreliable, however, just like the old estimate of $E_\nu = 10$ eV for diamond [156], since such small values of E_ν agree neither with new experimental data nor with the results of recent detailed calculations [83-85, 106, 108].

†For example, for diamond the "complete" version gives $E_\nu(C) = 27$ eV; see Sec. 4.3.4.

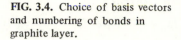

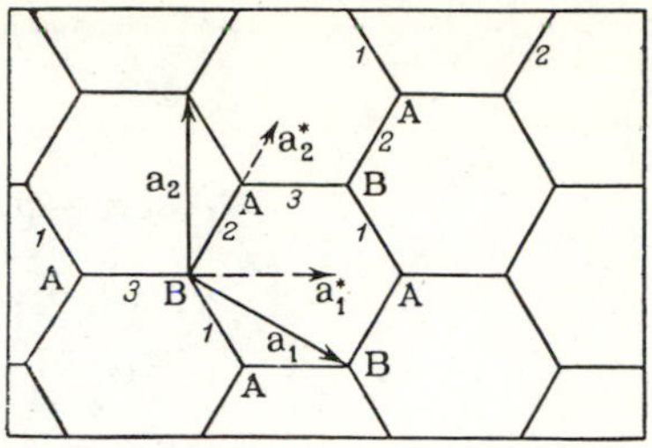

FIG. 3.4. Choice of basis vectors and numbering of bonds in graphite layer.

It is easy to see then that in the graphite layer, just as in the diamond crystal, one can identify two sublattices {A}, {B} and can arbitrarily refer to EOs to the corresponding atoms of one of them, let us say {B}. Then all the EOs in this layer are divided into three families of translationally equivalent EOs and the basis Bloch functions of the graphite layer are formally written in the form (3.1), (3.2), where the subscript j assumes values of $j = 1$, 2, 3. Again, taking into consideration the interaction of EOs of the first-neighbor bonds and the second-neighbor bonds, it is not hard to see that the matrix elements $H_{ij} = \langle\psi_i|\hat{H}|\psi_j\rangle$ in the problem at hand have the form:

$$
\begin{aligned}
H_{11} &= \alpha + 2\beta_t \{\cos q_1 + \cos q_2\} \\
H_{22} &= \alpha + 2\beta_t \{\cos (q_1 + q_2) + \cos q_2\} \\
H_{33} &= \alpha + 2\beta_t \{\cos q_1 + \cos (q_1 + q_2)\} \\
H_{12} &= \beta_A \{1 + e^{-iq_2}\} + \beta_g \{e^{iq_1} + e^{i(q_1+q_2)}\} \\
H_{13} &= \beta_A \{1 + e^{iq_1}\} + \beta_g \{e^{-iq_2} + e^{i(q_1+q_2)}\} \\
H_{23} &= \beta_A \{1 + e^{i(q_1+q_2)}\} + \beta_g \{e^{iq_1} + e^{iq_2}\}
\end{aligned}
\qquad (3.110)
$$

so that the dependence of the energy on the vector **k** for the σ band of graphite for an arbitrary value of **k** is given by the roots of a third-degree secular equation Det $\|H_{ij} - \varepsilon(\mathbf{k})\delta_{ij}\| = 0$ with the elements H_{ij} from the relations (3.110). Here $q_1 = \mathbf{ka}_1$, $q_2 = \mathbf{ka}_2$; see below concerning the rest of the parameters α, β_A, etc.

For factorization of this equation let us consider the dispersion laws in the [01] symmetry direction (see Fig. 3.4). It is easy to show that the symmetrized combinations of the basis Bloch functions in this direction will be the functions

$$\Psi_1 = \frac{1}{\sqrt{2}} (\psi_1 + \psi_3); \; \Psi_2 = \psi_2 \qquad (3.111)$$

$$\Psi_3 = \frac{1}{\sqrt{2}} (\psi_1 - \psi_2) \qquad (3.112)$$

In this case the three branches of the dispersion law are given by the roots of the second-order secular equation

$$\mathrm{Det} \begin{Vmatrix} \mathcal{H}_{11} - \varepsilon & \mathcal{H}_{12} \\ \mathcal{H}_{21} & \mathcal{H}_{22} - \varepsilon \end{Vmatrix} = 0; \; \mathcal{H}_{ij} = \langle \Psi_i | \hat{H} | \Psi_j \rangle \qquad (3.113)$$

and also by the relation

$$\varepsilon = \mathcal{H}_{33} = \langle \Psi_3 | \hat{H} | \Psi_3 \rangle \tag{3.114}$$

where

$$\begin{aligned} \mathcal{H}_{11} &= \alpha + 2\beta_A + 2\beta_t + 2(\beta_g + \beta_t)\cos q \\ \mathcal{H}_{22} &= \alpha + 4\beta_t \cos q \\ \mathcal{H}_{12} &= \sqrt{2}\,(\beta_A + \beta_g)(1 + e^{-iq}) \\ \mathcal{H}_{33} &= \alpha + 2\beta_A + 2\beta_t + 2(\beta_t - \beta_g)\cos q \end{aligned} \tag{3.115}$$

As follows from Eqs. (3.113) and (3.114), with relations (3.115) taken into consideration the dispersion law is defined by the formulas

$$\varepsilon_{1,2}(q) = \alpha + \beta_A + \beta_t + (\beta_g + 3\beta_t)\cos q$$
$$\pm \sqrt{\{(\beta_A + \beta_t) + (\beta_g - \beta_t)\cos q\}^2 + 8(\beta_A + \beta_g)^2 \cos^2 \frac{q}{2}} \tag{3.116}$$

$$\varepsilon_3(q) = \alpha - 2\beta_A + 2\beta_t + 2(\beta_t - \beta_g)\cos q \tag{3.117}$$

from which, in particular, we have

$$E_{v\sigma}\,(\text{graphite}) = \varepsilon_2(0) - \varepsilon_1(0) = -6\{\beta_A + \beta_g\} \tag{3.118}$$

for the width of the valence σ band.

Let us now find expressions for the matrix elements α, β_A, β_t, β_g, which enter into Eqs. (3.110)–(3.118) and for which–by analogy to diamond–the following notations are introduced (Fig. 3.5):

α is the average value of the Hamiltonian in the bonding EOs

$$\alpha = \langle \varphi | \hat{H} | \varphi \rangle \tag{3.119}$$

β_A is the interaction integral of the two EOs for the first-neighbor bonds

$$\beta_A = \langle \varphi | \hat{H} | \varphi' \rangle \tag{3.120}$$

β_t is the interaction integral of the EOs for the bonds of the second neighbors in the *trans* position

$$\beta_t = \langle \varphi | \hat{H} | \varphi'' \rangle \tag{3.121}$$

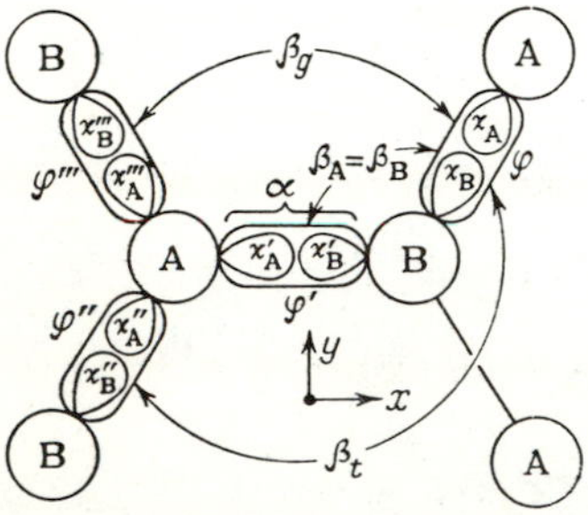

FIG. 3.5. Matrix elements in basis of equivalent orbitals for graphite layer.

β_g is this same integral for the bonds of the second neighbors in the *cis* position

$$\beta_g = \langle \varphi | \hat{H} | \varphi''' \rangle \tag{3.122}$$

To relate the structure of the valence σ band in graphite with the valence band structure in diamond, let us first express the matrix elements (3.119)–(3.122) in a form similar to (3.62)–(3.65). To do this we take account of the fact that the orbitals $\chi_B, \chi'_A, \chi'_B, \chi''_A, \chi''_B$ are hybrid sp^2 orbitals of the form (see Fig. 3.5):

$$\chi'_A = \frac{1}{\sqrt{3}} s^A + \sqrt{\frac{2}{3}} p_x^A \tag{3.123}$$

$$\chi'_B = \frac{1}{\sqrt{3}} s^B - \sqrt{\frac{2}{3}} p_x^B \tag{3.124}$$

$$\chi_B = \frac{1}{\sqrt{3}} s^B + \frac{1}{\sqrt{6}} p_x^B + \frac{1}{\sqrt{2}} p_y^B \tag{3.125}$$

$$\chi''_A = \frac{1}{\sqrt{3}} s^A - \frac{1}{\sqrt{6}} p_x^A - \frac{1}{\sqrt{2}} p_y^A \tag{3.126}$$

$$\chi'''_A = \frac{1}{\sqrt{3}} s^A - \frac{1}{\sqrt{6}} p_y^A + \frac{1}{\sqrt{2}} p_y^A \tag{3.127}$$

and we use the Wolfsberg–Helmholz method in combination with Pauling's rule of angular parts to evaluate the integrals $\beta_{ss}, \beta_{sp}, \beta_{pp}$ (see Sec. 3.2.2). Here by β we mean the interaction integral of the two trigonal sp^2 functions, directed along one and the same A–B bond; by S we mean the corresponding overlap integral. Then from Pauling's rule

$$S_{ss} : S_{sp} : S_{pp} = 1 : \sqrt{3} : 3, \quad S_{ss} : S = 1 : \left(\frac{1}{\sqrt{3}} + \sqrt{2}\right)^2 \tag{3.128}$$

and

$$\alpha = \alpha_A + \beta \tag{3.129}$$

$$\beta_A = \frac{1}{6}\{\alpha_A^{(s)} - \alpha_A^{(p)}\} + \frac{\alpha_A^{(s)}\left(2 - \sqrt{\frac{3}{2}}\right) - 3\alpha_A^{(p)}\left(1 - \sqrt{\frac{3}{2}}\right)}{2(1+\sqrt{6})\,\alpha_A} \tag{3.130}$$

$$\beta_t = \frac{\frac{\sqrt{6}}{3}\alpha_A^{(s)} + \frac{\sqrt{6}}{2}\alpha_A^{(p)}}{4(1+\sqrt{6})^2\,\alpha_A}\beta - \frac{1}{4}\beta_\pi \tag{3.131}$$

$$\beta_g = \frac{\frac{\sqrt{6}}{3}\alpha_A^{(s)} + \frac{\sqrt{6}}{2}\alpha_A^{(p)}}{4(1+\sqrt{6})^2\,\alpha_A}\beta + \frac{1}{4}\beta_\pi \tag{3.132}$$

where α_A denotes the average value of the Hamiltonian in the sp^2 orbitals:

$$\alpha_A = \langle \chi_A | \hat{H} | \chi_A \rangle = \frac{1}{3}\{\alpha_A^{(s)} + 2\alpha_A^{(p)}\} \tag{3.133}$$

To establish a correlation with the band structure of diamond let us now use the Coulomb integrals from Table 3 and the resonance integrals β and β_π from Table 4, after recalculating them for graphite. Since $\beta_\pi(1.39\ \text{Å}) \approx -2.3$ eV for the aromatic compounds and $\beta_\pi(1.54\ \text{Å}) \approx -1.7$ eV for diamond, then for any method of interpolation $\beta_\pi(1.42\ \text{Å}) = (-2.1)$-$(-2.2)$ eV and, to be specific, we can adopt $\beta_\pi(\text{graphite}) = -2.15$ eV. In a similar manner it is not hard to determine the value of the parameter β too, bearing in mind that the difference between the values of β in diamond and in graphite is caused, first, by the difference in the hybridization of the orbitals of the C atoms and, second, by the difference in the interatomic distances. To account for these two factors, we can write, for example,

$$\beta\ (\text{graphite}) \simeq \frac{S\,(sp^2 - sp^2)_{1.42\text{Å}}}{S\,(sp^3 - sp^3)_{1.54\text{Å}}} \cdot \beta\ (\text{diamond}) \tag{3.134}$$

or, what is practically the same thing:

$$\beta\ (\text{graphite}) = \frac{S\,(sp^2 - sp^2)_{1.42\text{Å}}}{S\,(sp^3 - sp^3)_{1.54\text{Å}}} \cdot \frac{\alpha_A\ (\text{graphite})}{\alpha_A(\text{diamond})} \cdot \beta\ (\text{diamond})$$
$$\left(\text{coefficient}\ \frac{\alpha_A\ (\text{graphite})}{\alpha_A(\text{diamond})} = 1.05\right) \tag{3.135}$$

Then $\beta(\text{graphite}) = -7.5$ eV for $\beta(\text{diamond}) = -6.5$ eV and $\beta(\text{graphite}) = -9.6$ eV for $\beta(\text{diamond}) = -8.3$ eV. Now, one can estimate the value of $E_{v\sigma}$ for graphite from the parameters previously used to calculate the band structure of diamond. It is not hard to obtain from Eqs. (3.118), (3.130), and (3.132) that for $\beta(\text{diamond}) = -6.5$ eV $E_{v\sigma}(\text{graphite}) \approx 12$ eV, and for $\beta(\text{diamond}) = -8.3$ eV $E_{v\sigma}(\text{graphite}) = 16.5$ eV, whereas X-ray spectral data give a value of 12.5–12.8 [154, 160] and 18–22 eV [161] for $E_{v\sigma}(\text{graphite})$.

Having the numerical estimates of the parameters $\alpha_A^{(s)}$, $\alpha_A^{(p)}$, β, and β_π one can calculate the complete structure of the valence σ band of graphite, as was done in [159].

3.6 STRUCTURE OF THE VALENCE BAND OF COVALENT CRYSTALS. COMPARISON WITH CALCULATED DATA AND SOME OTHER PROBLEMS

3.6.1 Comparison with Calculated Data

We shall now continue our discussion of the questions associated with the structure of the valence band of covalent crystals.

As shown above, the estimates of the parameters of the valence band structure of diamond-type crystals, obtained by the semiempirical EO LCAO method, agree quite well with experiment. However, direct experimental data, relating to the structure of the valence bands, are sometimes rather indeterminate in nature. Therefore for a more detailed evaluation of the calculational possibilities of the semiempirical EO LCAO method we should turn to a comparison of the data obtained by this method with other calculated data. To do this we use the results of the most thorough calculations of the structure of group IV elements by the OPW, APW, and EP methods (see Secs. 2.4.2, 2.4.4, 2.4.6).

During the past 10 years the band structure of diamond-type crystals has been investigated by the OPW method by Bassani and Yoshimine [162], the APW method by Keown [67], the "empirically corrected OPW" method in a number of papers by Herman et al. [83-85], as well as the EP method of Saslow, Bergstresser, and Cohen [106], Cohen and Bergstresser [107], and Van Haeringen and Junginger [108]. The results of these calculations as well as the results obtained by the semiempirical EO LCAO method are listed in Table 6. It is not hard to see that the data of the semiempirical EO LCAO method agree with the calculations from the OPW, APW, EC OPW, and EP methods to about the same extent as these latter agree among themselves, and this shows that the semiempirical EO LCAO method is also suitable for obtaining quite detailed quantitative information about the band structure. Below we shall see that this conclusion is extended to the complete band structure of covalent and partially covalent crystals and is also valid for the semiempirical LCAO method.

Let us emphasize that (as is usual in quantum chemistry) the successful application of semiempirical methods assumes, of course, a correct choice of all the parameters. Incorrect values for even some of them can drastically degrade the results. Thus, the use of "thermochemical" rather than "spectroscopic" values of the parameter β in Stocker's paper [142] led to the fact that the width of the valence band was much too small—by ~10 eV for diamond and 5-6 eV for Si and Ge (the conduction band was not considered in [142]).

3.6.2 Valence Band of Crystals with Diamond Lattice in the Tight Binding Method

As mentioned in Sec. 2.6.1, the EO LCAO method is not completely equivalent to the tight binding method in view of the incomplete orthogonality of the EOs $\varphi_j^{(+)}$ and $\varphi_j^{(-)}$. Here, and also in Sec. 4.3.5, we will examine how the picture of the band structure of covalent crystals is altered when one goes from the EO LCAO method to the LCAO method [163]. Beforehand we will write down several general formulas that are also necessary for the study of the conduction band. For a comparison with the EO LCAO method we will use the equivalent basis of bonding and antibonding EOs instead of the basis of atomic functions.

As already stated, in the EO LCAO method it is assumed that the valence band is formed only from the bonding orbitals $\varphi_j^{(+)}$ and the conduction band from only the antibonding orbitals $\varphi_j^{(-)}$. Thus the tight binding method is equivalent to considering an impurity of antibonding EOs $\varphi_j^{(-)}$ for the valence band and bonding $\varphi_j^{(+)}$ for the conduction band. Therefore now besides the matrix elements $H_{ij}^{(+)} = \langle\psi_i^{(+)}|\hat{H}|\psi_j^{(-)}\rangle$ and $H_{ij}^{(-)} = \langle\psi_i^{(-)}|\hat{H}|\psi_j^{(-)}\rangle$—here we will denote them by the symbols $H_{ij}^{(++)}$ and $H_{ij}^{(--)}$—it is necessary to also consider the matrix elements

$$H_{ij}^{(+-)} = \langle \psi_i^{(+)} | \hat{H} | \psi_j^{(-)} \rangle \tag{3.136}$$

$$H_{ij}^{(-+)} = \langle \psi_i^{(-)} | \hat{H} | \psi_j^{(+)} \rangle \tag{3.137}$$

that define the interaction of BFs composed of bonding EOs and BFs composed of antibonding EOs. Then for an arbitrary point **k** within the first Brillouin zone, when finding the function $\varepsilon(\mathbf{k})$, it is necessary to solve an eighth-order equation instead of two fourth-order secular equations Det $\|H_{ij}^{(++)} - \varepsilon\delta_{ij}\| = 0$ and Det $\|H_{ij}^{(--)} - \varepsilon\delta_{ij}\| = 0$

$$\text{Det} \left\| \begin{array}{c:c} H_{ij}^{(++)} - \varepsilon(\mathbf{k})\,\delta_{ij} & H_{ij}^{(+-)} \\ \hdashline H_{ij}^{(-+)} & H_{ij}^{(--)} - \varepsilon(\mathbf{k})\,\delta_{ji} \end{array} \right\| = 0 \tag{3.138}$$

For the symmetry directions $\Delta = [100]$ and $\Lambda = [111]$ Eq. (3.138) is factored by converting to symmetrized combinations of the basis Bloch functions (3.7) and (3.8). Accordingly, for the Δ and Λ directions the dispersion laws are obtained by solving a fourth-degree secular equation:

$$\text{Det} \left\| \begin{matrix} \mathcal{H}_{11}^{(++)} - \varepsilon & \mathcal{H}_{12}^{(++)} & \mathcal{H}_{11}^{(+-)} & \mathcal{H}_{12}^{(+-)} \\ \mathcal{H}_{21}^{(+-)} & \mathcal{H}_{22}^{(++)} - \varepsilon & \mathcal{H}_{21}^{(+-)} & \mathcal{H}_{22}^{(+-)} \\ \mathcal{H}_{11}^{(-+)} & \mathcal{H}_{12}^{(-+)} & \mathcal{H}_{11}^{(--)} - \varepsilon & \mathcal{H}_{12}^{(--)} \\ \mathcal{H}_{21}^{(-+)} & \mathcal{H}_{22}^{(-+)} & \mathcal{H}_{21}^{(--)} & \mathcal{H}_{22}^{(--)} - \varepsilon \end{matrix} \right\| = 0 \tag{3.139}$$

and one of two quadratic secular equations, for example, the equation

$$\text{Det} \left\| \begin{matrix} \mathcal{H}_{33}^{(++)} - \varepsilon & \mathcal{H}_{33}^{(+-)} \\ \mathcal{H}_{33}^{(-+)} & \mathcal{H}_{33}^{(--)} - \varepsilon \end{matrix} \right\| = 0 \tag{3.140}$$

Let us now examine the dependence of the matrix elements H_{ij} on the interaction integrals of the EOs $\varphi_j^{(+)}$ and $\varphi_j^{(-)}$. It is obvious that for the matrix elements $H_{ij}^{(++)}$ and $H_{ij}^{(--)}$ this dependence has the same form as in the EO method. It is given by (3.22), where the parameters $\alpha^{(+)}$, $\alpha^{(-)}$, . . . , etc., in turn, are determined in accordance with Eqs. (3.48)–(3.51) or (3.62)–(3.65). For a similar writing of the matrix elements $H_{ij}^{(++)}$ and $H_{ij}^{(--)}$ let us introduce the notations* (see Fig. 3.3)

$$\beta_{\mathrm{A}}^{(+-)} = \langle \varphi'^{(+)} | \hat{H} | \varphi''^{(-)} \rangle; \quad \beta_{\mathrm{A}}^{(-+)} = \langle \varphi'^{(-)} | \hat{H} | \varphi''^{(+)} \rangle \tag{3.141}$$

$$\beta_{\mathrm{B}}^{(+-)} = \langle \varphi^{(+)} | \hat{H} | \varphi'^{(-)} \rangle; \quad \beta_{\mathrm{B}}^{(-+)} = \langle \varphi^{(-)} | \hat{H} | \varphi'^{(+)} \rangle \tag{3.142}$$

$$\beta_t^{(+-)} = \langle \varphi^{(+)} | \hat{H} | \varphi''^{(-)} \rangle; \quad \beta_t^{(-+)} = \langle \varphi^{(-)} | \hat{H} | \varphi''^{(+)} \rangle \tag{3.143}$$

*The matrix elements $\alpha^{(+-)} = \langle\varphi^{(+)}|\hat{H}|\varphi^{(-)}\rangle$, $\alpha^{(-+)} = \langle\varphi^{(-)}|\hat{H}|\varphi^{(+)}\rangle$ obviously become zero since the orbital $\varphi^{(+)}$ is symmetrical and $\varphi^{(-)}$ is antisymmetrical with respect to the middle of the A–B bond. In addition, let us note that for the matrix elements $H_{ij}^{(+-)}$, $H_{ij}^{(-+)}$ the integrals $\beta_{\mathrm{A}}^{(+-)}$ and $\beta_{\mathrm{A}}^{(-+)}$ are different from the integrals $\beta_{\mathrm{B}}^{(+-)}$, $\beta_{\mathrm{B}}^{(-+)}$.

$$\beta_g^{(+-)} = \langle \varphi^{(+)} | \hat{H} | \varphi'''^{(-)} \rangle ; \quad \beta_g^{(-+)} = \langle \varphi^{(-)} | \hat{H} | \varphi'''^{(+)} \rangle \tag{3.144}$$

Then the matrix elements $H_{ij}^{(+-)}, H_{ij}^{(-+)}$ have the form

$$\begin{aligned}
H_{11}^{(+-)} &= -2i\beta_t^{(+-)} \{\sin(q_x+q_y)+\sin(q_x+q_z)+\sin(q_y+q_z)\} \\
H_{22}^{(+-)} &= 2i\beta_t^{(+-)} \{\sin(q_x+q_y)+\sin(q_x-q_z)+\sin(q_y-q_z)\} \\
H_{33}^{(+-)} &= 2i\beta_t^{(+-)} \{\sin(q_x+q_z)+\sin(q_x-q_y)+\sin(q_z-q_y)\} \\
H_{44}^{(+-)} &= 2i\beta_t^{(+-)} \{\sin(q_y+q_z)+\sin(q_y-q_x)+\sin(q_z-q_x)\} \\
H_{12}^{(+-)} &= \beta_B^{(+-)} + \beta_A^{(+-)} e^{i(q_x+q_y)} - \beta_g^{(+-)} \left(e^{iq_y}+e^{iq_x}\right) 2i \sin q_z \\
H_{13}^{(+-)} &= \beta_B^{(+-)} + \beta_A^{(+-)} e^{i(q_x+q_z)} - 2i\beta_g^{(+-)} \left(e^{iq_x}+e^{iq_z}\right) \sin q_y \\
H_{14}^{(+-)} &= \beta_B^{(+-)} + \beta_A^{(+-)} e^{i(q_y+q_z)} - 2i\beta_g^{(+-)} \left(e^{iq_y}+e^{iq_z}\right) \sin q_x \\
H_{23}^{(+-)} &= \beta_B^{(+-)} + \beta_A^{(+-)} e^{i(q_z-q_y)} + 2i\beta_g^{(+-)} \left(e^{iq_z}+e^{-iq_y}\right) \sin q_x \\
H_{24}^{(+-)} &= \beta_B^{(+-)} + \beta_A^{(+-)} e^{i(q_z-q_x)} + 2i\beta_g^{(+-)} \left(e^{iq_z}+e^{-iq_x}\right) \sin q_y \\
H_{34}^{(+-)} &= \beta_B^{(+-)} + \beta_A^{(+-)} e^{i(q_y-q_x)} + 2i\beta_g^{(+-)} \left(e^{iq_y}+e^{-iq_x}\right) \sin q_z \\
H_{12}^{(-+)} &= \beta_B^{(-+)} + \beta_A^{(-+)} e^{i(q_x+q_y)} - 2i\beta_g^{(-+)} \left(e^{iq_y}+e^{iq_x}\right) \sin q_z \\
H_{13}^{(-+)} &= \beta_B^{(-+)} + \beta_A^{(-+)} e^{i(q_x+q_z)} - 2i\beta_g^{(-+)} \left(e^{iq_x}+e^{iq_z}\right) \sin q_y \\
H_{14}^{(-+)} &= \beta_B^{(-+)} + \beta_A^{(-+)} e^{i(q_y+q_z)} - 2i\beta_g^{(-+)} \left(e^{iq_y}+e^{iq_z}\right) \sin q_x \\
H_{23}^{(-+)} &= \beta_B^{(-+)} + \beta_A^{(-+)} e^{i(q_z-q_y)} + 2i\beta_g^{(-+)} \left(e^{iq_z}+e^{-iq_y}\right) \sin q_x \\
H_{24}^{(-+)} &= \beta_B^{(-+)} + \beta_A^{(-+)} e^{i(q_z-q_x)} + 2i\beta_g^{(-+)} \left(e^{iq_z}+e^{-iq_x}\right) \sin q_y \\
H_{34}^{(-+)} &= \beta_B^{(-+)} + \beta_A^{(-+)} e^{i(q_y-q_x)} + 2i\beta_g^{(-+)} \left(e^{iq_y}+e^{-iq_x}\right) \sin q_z
\end{aligned} \tag{3.145}$$

Then, as it is easy to see from Fig. 3.3, the following relations are satisfied:

$$\beta_t^{(+-)} = -\beta_t^{(-+)} = -\beta_t^{(++)} = \beta_t^{(--)} \tag{3.146}$$

$$\beta_g^{(+-)} = -\beta_g^{(-+)} = -\beta_g^{(++)} = \beta_g^{(--)} \tag{3.147}$$

$$\beta_A^{(+-)} = \beta_A^{(-+)} = -\beta_B^{(+-)} = -\beta_B^{(-+)} \tag{3.148}$$

so that the expressions for the matrix elements (3.143), (3.144) in terms of the parameters (3.34)–(3.37) or (3.52), (3.37) are given by Eqs. (3.50), (3.51) or (3.64), (3.65). Using the LCAO expansion for $\varphi^{(+)}$ and $\varphi^{(-)}$, it is easy to find the analogous expression for $\beta_A^{(+-)}$ too:

$$\beta_A^{(+-)} = \beta_A^{(-+)} = -\frac{1}{8}\{\alpha_A^{(s)} - \alpha_A^{(p)}\} \tag{3.149}$$

and, consequently,

$$\beta_B^{(+-)} = \beta_B^{(-+)} = +\frac{1}{8}\{\alpha_A^{(s)} - \alpha_A^{(p)}\} \tag{3.150}$$

Let us note that Eqs. (3.146), (3.147) and (3.149), (3.150) give some criterion of the equivalence of the EO LCAO method and the tight binding method. As follows from Eq. (3.138), these methods would be equivalent if the nondiagonal boxes in the secular determinant (3.138), containing the elements $H_{ij}^{(+-)}$ and $H_{ij}^{(-+)}$, identically became zero submatrices. According to Eq. (3.145), this is possible only for $\beta_A^{(+-)} = \ldots = \beta_g^{(-+)} = 0$. Thus, the difference between the tight binding method and the EO LCAO method is caused, first, by the difference in the energies of the s and p levels of an individual atom and, second, by the interaction of the second (and more remote) neighbor bonds. If the interaction is extracted from the latter, then the necessary condition for the equivalence of the LCAO and EO LCAO methods will be the equality $\alpha_A^{(s)} = \alpha_A^{(p)}$. If $\alpha_A^{(s)} = \alpha_A^{(p)}$, then it can be shown that the "true" EOs in a tetrahedral covalent crystal are exactly the same as the sum of the hybrid tetrahedral sp^3 functions and in this sense all the atoms of the crystal are actually strictly in the sp^3 hybridization states.

In real crystals of group IV elements the energies of the atomic s and p levels are not the same. However, Eqs. (3.149) and (3.150) for the parameters $\beta_A^{(+-)}, \ldots, \beta_B^{(-+)}$ contain the factor 1/8, and the interaction integrals of the second (and more remote) neighbor bonds are small. Therefore even for real crystals with an $\mathscr{E}_{\text{promot}}$ value of the order of several electron volts the EO LCAO method approximates the band structure, calculated by the tight binding method, fairly well.

The structure of the valence band for C and Si, calculated by the tight binding method [163] with the parameters $\alpha_A^{(s)}$, $\alpha_A^{(p)}$, β and β_π, taken from Tables 3 and 4, is shown in Fig. 3.6 (the band structure, calculated by the EO LCAO method, is shown by the dashed lines

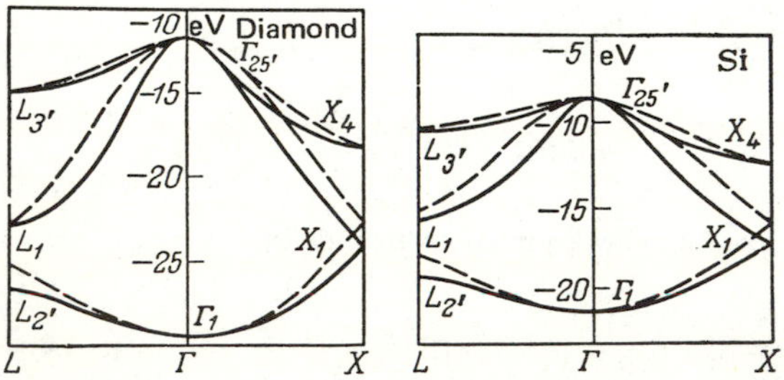

FIG. 3.6. Comparison of the valence band of diamond and silicon in the semiempirical EO LCAO method and in the semiempirical LCAO method.

for comparison). As seen from Fig. 3.6, as one goes from the EO LCAO method to the tight binding approximation, the change in the band structure is quite small. The results of both methods are the same at the center Γ of the band (and also at the edges of the X and L bands for the p level) just as should be expected from symmetry considerations. Thus, a comparison of the results of the application of the tight binding method with experimental data and also with the data of calculations [67, 83–85, 106–108, 162] in Table 6 shows that within the accuracy limits of all these data the difference between the tight binding method and the EO LCAO method is significant.

3.6.3 Nonempirical Calculations of the Band Structure of Covalent Crystals by the EO and LCAO Methods

Finally, in connection with what has been said in the foregoing sections we will discuss the question of a nonempirical approach to the calculation of the band structure of covalent crystals by the EO and LCAO methods. As mentioned, one reason for choosing the semiempirical formulation of the theory of the chemical bond in crystals is the lack, until recently, of reliable nonempirical LCAO calculations of the band structure of A^{IV} and $A^N B^{8-N}$ crystals. Such calculations have been undertaken many times, but have led to unsatisfactory agreement with experiment even for the valence band, at least in a quantitative sense (whereas for the conduction band the qualitative picture is also incorrect). Thus in the well-known papers of Nran'yan [164, 165] values of 36.6 and 33.1 eV were obtained for the width E_v of the valence band in Si and Ge, respectively, whereas experiment gives values of 12–16.5 and 12.3–13 eV. Similarly, in the paper by Cohen et al. [166] a value of 52 eV was obtained for the width of the valence band in diamond, which is 30 eV higher than the actual value. For a somewhat different version of the calculation by Pugh [54, 167] the value of E_v was reduced to 34 eV, but at the same time the qualitative picture of the valence band structure was drastically degraded. A value of $E_g = 13.5$ eV, in contrast to an experimental value of $E_g = 5.4$ eV, was obtained in [54] for the forbidden band gap; and the minimum of the conduction band was erroneously found at the point Γ. Satisfactory results for the tetrahedral A^{IV} and $A^N B^{8-N}$ semiconductors have been obtained by a nonempirical method only for diamond up to now (and also partial results for silicon) in work by Chaney [55] and Painter et al. [52]; as seen from Table 6, they agree with the data of the semiempirical EO LCAO method.

3.6.4 Structure of Bands in Diamond and Silicon and the Question of Silicon Analogs of Conjugated Systems

In concluding the investigation of the valence band of diamond-type crystals, let us again discuss the relationship between the band structure of a crystal and the nature of the bond in molecules. Until now we have been using information about the electron structure of molecules for investigating the band structure of crystals. One can, however, take the opposite approach: Use experimental data on the band structure of crystals for investigating the chemical bond in molecules. Here we shall illustrate such an approach for the example of the multiple bonds in the chemistry of silicon [168]. Another similar example is considered in Sec. 4.5.3.

It is known that despite the presence of a rather large number of Si analogs of the saturated hydrocarbons there are apparently no analogs of organic compounds with multiple or conjugated bonds for silicon.* There have been reports on the synthesis of three such compounds

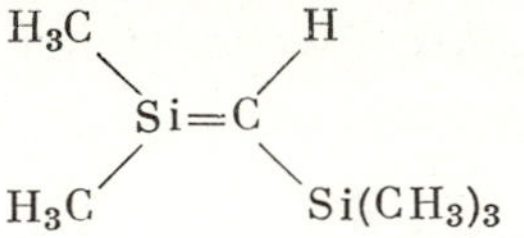

(I) [169]

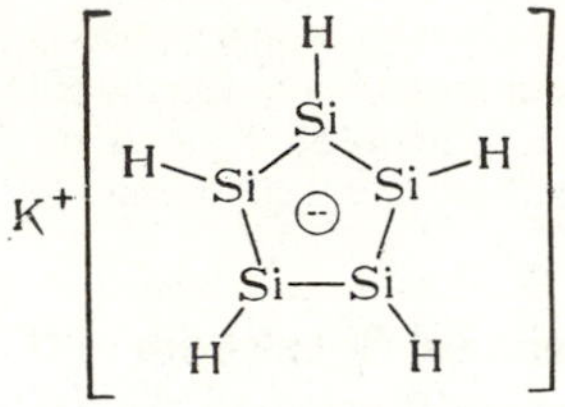

(II) [170]

as well as a silicon analog of hexachlorobenzene Si_6Cl_6 (III) [171]. It was found, however, that the compound (I) does not exist, and its molecular formula belongs to a saturated compound of the type (IV):

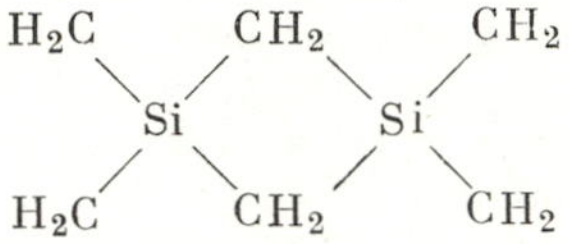

(IV)

The arguments in favor of the existence of the compound (II) are subject to doubt, and the existence of the compound (III) has not been confirmed [172].

The absence of double and multiple bonds between Si atoms because of the $3p\pi$–$3p\pi$ interaction appears strange if the strength of the $p\pi$–$p\pi$ bonds is compared for carbon and silicon, employing methods used in quantum chemistry. Thus an estimate of the ratio $\beta_\pi(\mathrm{Si}):\beta_\pi(\mathrm{C})$ by means of the overlap integrals with the Slater functions for interatomic distances corresponding to the ordinary C–C (1.54 Å) and Si–Si (2.32 Å in H_3Si–SiH_3) bonds

$$\beta_\pi(\mathrm{Si}):\beta_\pi(\mathrm{C}) \approx S_\pi(\mathrm{Si}):S_\pi(\mathrm{C}) \tag{3.151}$$

gives $\beta_\pi(\mathrm{Si}):\beta_\pi(\mathrm{C}) = 1$. A similar result is obtained for an estimate by the Wolfsberg-Helmholz method:

$$\beta_\pi(\mathrm{Si}):\beta_\pi(\mathrm{C}) \approx S_\pi(\mathrm{Si})\,\alpha^{(p)}_{\mathrm{Si}}:S_\pi(\mathrm{C})\,\alpha^{(p)}_{\mathrm{C}} \tag{3.152}$$

*The Si atoms may form bonds of increased multiplicity in the silicates, siloxanes, and such molecules as H_3Si–O–SiH_3 or H_3SiNCS [126, 173], however, because of the partial acceptance of an electron in the $3d$ orbital by the Si atom. Such bonds are the $p\pi$–$d\pi$ bonds, but not the $p\pi$–$p\pi$ bonds, and they can play a marked role apparently only in compounds of silicon with donor atoms such as O or N.

In this case this ratio is also close to one and amount to 1:1.25 if the usual ionization potentials are used in (3.152) and 1:1.1 if the energies to remove an electron from the $p\pi$ orbital into the valence state $trtrtr\pi$ are taken for $\alpha_A^{(p)}$ (11.16 eV for C and 9.17 eV for Si [174]).

The estimates (3.151), (3.152) can be shown to be incorrect since the resonance integrals for the π bonds are comparable for distances corresponding to the ordinary C–C and Si–Si bonds, not double bonds. It is hard, of course, to say what Si–Si distance should correspond to a hypothetical Si=Si double bond because of the $3p\pi$-$3p\pi$ interaction. The simplest assumption is that the same degree of double bondedness corresponds to the same relative decrease in the interatomic distance, i.e., that to each distance R_C between the carbon atoms there corresponds a distance between the silicon atoms $R_{Si} = (2.32/1.54)R_C$ Å. Graphs of the overlap integrals S_π(Si) and S_π(C) at these "matched" distances are compared in Fig. 3.7, from which it is seen that within the entire range of variation of R_C and R_{Si} at the "matched" distances the ratio S_π(Si):S_π(C) remains practically unchanged and close to one. Thus for the matched distances R_C = 1.34 Å (the length of a double bond in ethylene) and R_{Si} = 2.02 Å this ratio is 1:1.09.

Let us point out that within reasonable limits the result is no different if the method of determining the "matched" distances is changed. If it is assumed, for example, that the Si–Si distance changes faster with an increase in the order of the bond than the C–C distance, then the ratio S_π(Si):S_π(C) will approach one. On the other hand if it is assumed that the distance R_{Si} is shortened more slowly than the distance R_C with an increase in the order of the bond, so that, for example, the distance R_C = 1.34 Å in ethylene corresponds to the distance R_{Si} = 2.10 Å, corresponding to the "sesqui" bond in benzene in Fig. 3.7, then in this case too the ratio β_π(Si):β_π(C) varies insignificantly and amounts to ~1:1.2. Thus from the viewpoint of the usual methods of estimating the bond strength the $p\pi$ bonds for the Si atoms should be nearly as strong (and compared with the σ bonds, even stronger) as the π bonds for the C atoms. It is hard to reconcile such a conclusion with the fact that there are no $p\pi$-$p\pi$ bonds for Si atoms, but it can hardly be proved by the usual data of "molecular" inorganic chemistry.

However, the experimental data on the band structure of crystals make it possible to estimate the value of β_π for silicon by a direct method. As was

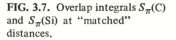

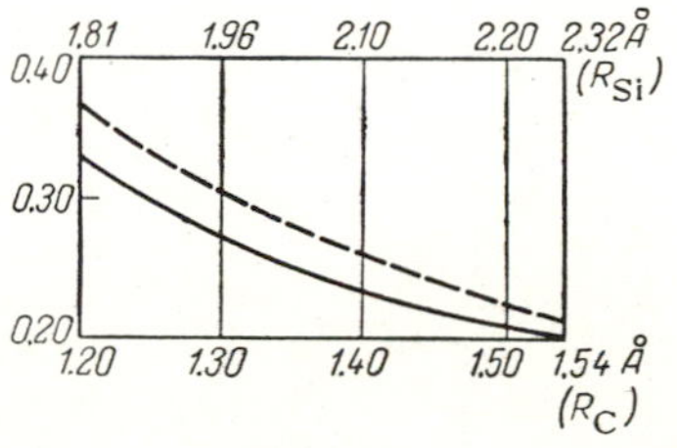

FIG. 3.7. Overlap integrals S_π(C) and S_π(Si) at "matched" distances.

indicated, the function $\varepsilon(\mathbf{k})$ for the p level in crystals with the diamond lattice is determined by the integral β_π, and the experimental estimates of β_π can be obtained from test data for diamond and silicon (see Sec. 3.5.2). According to these estimates, the ratio $\beta_\pi(\mathrm{Si}):\beta_\pi(\mathrm{C})$ is equal to 1/2–1/3, and the $p\pi$–$p\pi$ bonds for silicon should be about 2-3 times weaker than for carbon. This result agrees with the fact that silicon compounds with double and triple $p\pi$–$p\pi$ bonds do not exist, and it shows that band structure data in many cases can be useful for interpreting molecular structure data.

4

Chemical Bond and Band Structure of Covalent Crystals. Conduction Band and Complete Band Structure

Having investigated the valence band, we can now turn to an investigation of the conduction band and the complete band structure of crystals with a diamond lattice. In this situation, as we shall see, it will be necessary to switch from the "simplified" to the "complete" version of the theory because of the unique properties of the conduction band. A more careful investigation of the nature of the chemical bond in compounds of group IV elements will also be necessary.

4.1 STRUCTURE OF THE CONDUCTION BAND AS AN INDIVIDUAL PROPERTY OF A GROUP IV ELEMENT

4.1.1 Specific Properties of Conduction Band

Let us first consider what problems arise in the changeover from the valence band to the conduction band structure (within the framework of the EO LCAO method and the tight binding approximation).

It was already stated in Sec. 3.2.3 that Eqs. (3.23)-(3.26) for the dispersion laws can be used to study both the conduction band and the valence band, but in place of the parameters $\alpha^{(+)}$, $\beta_A^{(+)}$, . . . , etc., in the function $\varepsilon(\mathbf{k})$ it is now necessary to substitute the parameters $\alpha^{(-)}$, $\beta_A^{(-)}$, ... As follows from

Eqs. (3.48)-(3.51), the difference between the parameters designated by the superscript (+) and the parameters with the superscript (−) is in the signs in front of the resonance integrals. Therefore each formula for the level difference in the valence band corresponds to a similar formula for the conduction band, differing, however, from the former in the sign of the "resonance" terms. Thus, the formula

$$\varepsilon(\Gamma^v_{25'}) - \varepsilon(\Gamma^v_1) = \{\alpha^{(p)}_A - \alpha^{(s)}_A\} - 4\left\{\beta_{ss} - \frac{1}{3}\beta_{pp} + \frac{2}{3}\beta_\pi\right\} \tag{4.1}$$

for the difference in the bonding p and s levels $\Gamma^v_{25'}$, Γ^v_1 in the valence band corresponds to the analogous formula for the distance between the antibonding p and s levels in the conduction band:

$$\varepsilon(\Gamma^c_{15}) - \varepsilon(\Gamma^c_{2'}) = \{\alpha^{(p)}_A - \alpha^{(s)}_A\} + 4\left\{\beta_{ss} - \frac{1}{3}\beta_{pp} + \frac{2}{3}\beta_\pi\right\} \tag{4.2}$$

Similarly, the distance between the bonding s level Γ^v_1 and the bonding s–p level X^v_1

$$\varepsilon(X^v_1) - \varepsilon(\Gamma^v_1) = \frac{1}{2}\{\alpha^{(p)}_A - \alpha^{(s)}_A\} - 4\left\{\beta_{ss} - \frac{1}{\sqrt{3}}\beta_{sp}\right\} \tag{4.3}$$

corresponds to the distance between the antibonding s level $\Gamma^c_{2'}$ and the antibonding s–p level X^c_1

$$\varepsilon(X^c_1) - \varepsilon(\Gamma^c_{2'}) = \frac{1}{2}\{\alpha^{(p)}_A - \alpha^{(s)}_A\} + 4\left\{\beta_{ss} - \frac{1}{\sqrt{3}}\beta_{sp}\right\} \tag{4.4}$$

etc.

The above-stated difference in the signs of the resonance terms is important in two respects. It is not hard to see that in formulas such as (4.1) and (4.3) the right sides are the sums of two quantities with identical signs (the resonance integrals are negative). Therefore the scheme of level arrangement in the valence band is practically independent, in its important aspects, of the numerical values of the corresponding matrix elements and is preserved in practically any approximation. On the other hand, the structure of the conduction band, as is seen, for example, from Eqs. (4.2) and (4.4), is determined by the differences of two quantities (moreover, they are of the same order of magnitude; see below) and, consequently, is much more sensitive to the choice of approximation. Therefore in describing the conduction band in the complete band structure the "simplified" version of the theory (Sec. 3.3.2) must be discarded and all three types of resonance integrals (β_{ss}, β_{sp}, and β_{pp}) must be introduced into the discussion.

An even more important consequence of the sensitivity of the conduction band with respect to possible variations of the corresponding matrix elements is the fact that the individual properties of the different group IV elements should have a much stronger effect on the structure of the conduction band than on the structure of the valence band. On the basis of formulas of the type (4.2) and (4.4) it is not hard to prove that even small variations in the magnitude of certain parameters, having practically no effect on the valence band, can alter considerably the scheme of level arrangement in the conduction band, even to the extent of changing their order.

4.1.2 Individual Properties of the Band Structure in Crystals of Group IV Elements

The currently available experimental data on band-to-band transitions [146, 147] confirm that the form of the conduction band is in fact a particularly individualistic property of the crystal. To do this it is sufficient to examine Fig. 3.2. Thus the antibonding s level $\Gamma_{2'}^c$, which lies below the antibonding p level Γ_{15}^c for germanium and gray tin crystals (as in the valence band), is located above this level for diamond and silicon crystals. A similar situation occurs for the pair of levels $\Gamma_{2'}^c$ and X_1^c. Whereas for diamond and silicon the antibonding s-p level X_1^c lies below the antibonding s level $\Gamma_{2'}^c$, these levels X_1^c and $\Gamma_{2'}^c$ are arranged in the opposite order for germanium and gray tin crystals.

It is significant that the change in the conduction band structure, as well as the complete band structure along with it, has a completely regular behavior in the series C-Si-Ge-α-Sn. This "evolution" of the band structure is clearly evident from Fig. 3.2 and can be described as follows:

(a) As one goes from C to α-Sn, the valence band and the conduction band approach each other monotonically on the "average," i.e., the average width of the energy gap separating one band from the other decreases.

(b) The antibonding s level $\Gamma_{2'}^c$ of the conduction band is gradually shifted with respect to the antibonding p level Γ_{15}^c in the series of group IV elements. Thus for diamond and silicon it lies above; for germanium and gray tin, it lies below the Γ_{15}^c level; and the level difference $\varepsilon(\Gamma_{15}^c) - \varepsilon(\Gamma_{2'}^c)$ increases monotonically.

(c) As one goes from C to α-Sn, the principal minimum of the conduction band is shifted from the point Δ, close to the point X, where it is for diamond and Si, to the center Γ of the band, where this minimum is located for α-Sn.

(d) For germanium, which lies between silicon and tin in the periodic table, the conduction band assumes a complicated structure with three minimums—at X, Γ, and L, the deepest of which is located at L.

(e) The forbidden band gap E_g decreases monotonically in the series C-Si-Ge-α-Sn.

Below we shall consider the dependence of this regular change in band structure on the nature of the chemical bond in the series of group IV elements.

4.1.3 Common Convergence of Conduction Band with Valence Band

Let us first consider the question of the common convergence of the conduction band with the valence band since no detailed numerical estimation of parameters is required to do this.

As a reasonable measure for the average distance between both bands let us take the quantity $\langle\Delta\varepsilon\rangle$, equal to the distance between the average levels of both bands $\langle\varepsilon^v\rangle$ and $\langle\varepsilon^c\rangle$: $\langle\Delta\varepsilon\rangle = \langle\varepsilon^c\rangle - \langle\varepsilon^v\rangle$. (Let us recall that the energy of the average level is defined as the arithmetic average of the energies of all the individual levels of a given band.) According to Eqs. (2.65) and (2.66) the energy values of the average levels $\langle\varepsilon^v\rangle$, $\langle\varepsilon^c\rangle$ are equal to the average values of the Hamiltonian $\hat{H}$ of the crystal in the corresponding equivalent orbitals. Hence for diamond-type crystals we have

$$\langle \varepsilon^v \rangle = \langle \varphi^{(+)} | \hat{H} | \varphi^{(+)} \rangle = \alpha_A + \beta \tag{4.5}$$

$$\langle \varepsilon^c \rangle = \langle \varphi^{(-)} | \hat{H} | \varphi^{(-)} \rangle = \alpha_A - \beta \tag{4.6}$$

so that in this case the distance between the average levels is simply equal to twice the value of the interaction integral between the tetrahedral sp^3 orbitals of the valence bound atoms:

$$\langle \Delta\varepsilon \rangle = \langle \varepsilon^c \rangle - \langle \varepsilon^v \rangle = -2\beta \tag{4.7}$$

As we have already seen in the previous chapter, the values of β change approximately proportionally to the heats of atomization $-\Delta H_{at}$, which decrease monotonically in the series C(diamond)–Si–Ge–α-Sn. Accordingly, Eq. (4.7) shows that the average distance between the valence band and the conduction band should behave in this same manner:

$$\langle \Delta\varepsilon \rangle \approx \text{const} \, (-\Delta H_{aT}) \tag{4.8}$$

This also explains the first unique property of the band structure in qualitative form.

Let us note that the relation (4.7) allows an even more detailed comparison with experiment by means of the density of states* $G(\varepsilon)$.

The density of states can be found from X-ray spectra studies or formulated from the band structure, calculated, for example, by the EP or EC OPW method. Such a curve was recently calculated for germanium by Herman et al. [103] on the basis of careful

*The density of states is the number of electron states in a unit interval of energy; thus the number of electron states with an energy in the interval between ε and $\varepsilon + d\varepsilon$ is equal to $G(\varepsilon)d\varepsilon$.

calculations by the EC OPW method. According to the data of [103], in this case $\langle \varepsilon^c \rangle - \langle \varepsilon^v \rangle \approx$ 8–10 eV, which agrees satisfactorily with the value of -2β = 6–7.6 eV, where β is taken from Table 4.

There is another method of estimating the value of $\langle \varepsilon^c \rangle - \langle \varepsilon^v \rangle$ from test data. At the center of the first Brillouin zone we take the average $(1/4)\{3\varepsilon(\Gamma_{25'}^v) + \varepsilon(\Gamma_1^v)\}$ of the energies of the triply degenerate level $\Gamma_{25'}^v$ and the single level Γ_1^v of the valence band and this same average $(1/4)\{3\varepsilon(\Gamma_{15}^c) + \varepsilon(\Gamma_{2'}^c)\}$ for the corresponding levels Γ_{15}^c, $\Gamma_{2'}^c$ in the conduction band, after which we subtract the first quantity from the second. On the basis of Eqs. (3.23)–(3.25) it is easy to see that the resulting difference $\langle \Delta\varepsilon(\Gamma) \rangle$ will be equal to

$$\langle \Delta\varepsilon(\Gamma) \rangle = \frac{1}{4}\{3[\varepsilon(\Gamma_{15}^c) - \varepsilon(\Gamma_{25'}^v)] + \varepsilon(\Gamma_{2'}^c) - \varepsilon(\Gamma_1^v)\} = -2\beta - 12\beta_t^{(+)} \quad (4.9)$$

or, measuring the levels from the top of the valence band, as is equal,

$$\langle \Delta\varepsilon(\Gamma) \rangle = \frac{1}{4}\{3\varepsilon(\Gamma_{15}^c) + \varepsilon(\Gamma_{2'}^c) - \varepsilon(\Gamma_1^v)\} = -2\beta - 12\beta_t^{(+)} \quad (4.10)$$

hence

$$\langle \Delta\varepsilon \rangle = \frac{1}{4}\{3\varepsilon(\Gamma_{15}^c) + \varepsilon(\Gamma_{2'}^c) - \varepsilon(\Gamma_1^v)\} + 12\beta_t^{(+)} \quad (4.11)$$

The value of $12\beta_t^{(+)}$ in Eq. (4.11) is easily estimated approximately from test data. As follows from Eqs. (3.51), $12\beta_t^{(+)} = 12\beta_{t\sigma} - 4\beta_\pi$, where the value of $12\beta_{t\sigma}$ does not exceed about −2 eV for diamond and about −1 eV for Si, Ge, and α-Sn; these numerical estimates follow from Table 4. At the same time it can be assumed that the value of β_π is estimated sufficiently accurately from spectroscopic data (see Sec. 3.5.2). Then, using the fundamental spectral data from [146, 147] for estimating $\varepsilon(\Gamma_{15}^c)$, $\varepsilon(\Gamma_{2'}^c)$ and X-ray spectral data for estimating $\varepsilon(\Gamma_1^v)$, we obtain the following experimental estimates of $\langle \Delta\varepsilon \rangle$ from Eq. (4.11):

$$\langle \Delta\varepsilon(\text{diamond}) \rangle \approx 15 \text{ eV}; \quad \langle \Delta\varepsilon(\text{Si}) \rangle \approx 8 \text{ eV}; \quad \langle \Delta\varepsilon(\text{Ge}) \rangle \approx 7 \text{ eV} \quad (4.12)$$

whereas the corresponding "theoretical" values of $\langle \Delta\varepsilon \rangle = -2\beta$ with β taken from Table 4 amount to

$$\langle \Delta\varepsilon(\text{diamond}) \rangle \approx 13 \div 16.5 \text{ eV}; \quad \langle \Delta\varepsilon(\text{Si}) \rangle \approx 8 \div 10 \text{ eV}; \quad \langle \Delta\varepsilon(\text{Ge}) \rangle \approx 6 \div 7.5 \text{ eV} \quad (4.13)$$

The distance between two analogous levels of different bands, for example, between the bonding and antibonding p levels $\Gamma_{25'}^v$ and Γ_{15}^c or between the bonding and antibonding s levels Γ_1^v and $\Gamma_{2'}^c$, can serve as another measure of the nearness of the valence band and conduction band and is more convenient for comparing with experiment. As follows from Eqs. (3.23)–(3.26) and (3.48)–(3.51), for $\mathbf{k} = 0$ we have

$$\varepsilon(\Gamma_{15}^c) - \varepsilon(\Gamma_{25'}^v) = -\frac{8}{3}\{\beta_{pp} - 2\beta_\pi\} \quad (4.14)$$

A direct calculation of the difference (4.14) requires a numerical estimate of the resonance integrals β_{pp} and β_π for each of the group IV elements (see below). One can, however, indicate how the difference $\varepsilon(\Gamma_{15}^c) - \varepsilon(\Gamma_{25'}^v)$ will vary in the series C–Si–Ge–α-Sn without such a detailed estimate. If it is assumed, as before, that the resonance integrals are proportional to the heats of atomization $-\Delta H_{at}$, then we again find that the quantity $\varepsilon(\Gamma_{15}^c) - \varepsilon(\Gamma_{25'}^v)$ should decrease in the series C–Si–Ge–α-Sn. In this case one can write

$$\Delta\varepsilon\,(\mathrm{C}) : \Delta\varepsilon\,(\mathrm{Si}) : \Delta\varepsilon\,(\mathrm{Ge}) : \Delta\varepsilon\,(\alpha\text{-Sn}) \approx \Delta H_{\mathrm{aT}}\,(\mathrm{C}) : \Delta H_{\mathrm{aT}}\,(\mathrm{Si}) : \Delta H_{\mathrm{aT}}\,(\mathrm{Ge}) : \Delta H_{\mathrm{aT}}\,(\mathrm{Sn}) \tag{4.15}$$

for the succession of ratios $\Delta\varepsilon(\mathrm{C}) : \Delta\varepsilon(\mathrm{Si}) : \Delta\varepsilon(\mathrm{Ge}) : \Delta\varepsilon(\alpha\text{-Sn})$, where $\Delta\varepsilon = \varepsilon(\Gamma_{15}^c) - \varepsilon(\Gamma_{25'}^v)$. Then

$$\Delta\varepsilon\,(\mathrm{C}) : \Delta\varepsilon\,(\mathrm{Si}) : \Delta\varepsilon\,(\mathrm{Ge}) : \Delta\varepsilon\,(\alpha\text{-Sn}) \approx 1 : 0.6 : 0.45 : 0.4 \tag{4.16}$$

Moreover, the level differences $\Delta\varepsilon = \varepsilon(\Gamma_{15}^c) - \varepsilon(\Gamma_{25'}^v)$ are equal to the energies of the band-to-band transitions $\Gamma_{25'}^v \to \Gamma_{15}^c$, which are well known from the optical spectra [146, 147]. Therefore the relation (4.16) is easily subjected to experimental verification. According to spectroscopic data*

$$\Delta\varepsilon\,(\mathrm{C}) : \Delta\varepsilon\,(\mathrm{Si}) : \Delta\varepsilon\,(\mathrm{Ge}) : \Delta\varepsilon\,(\alpha\text{-Sn}) = 1 : 0.5 : 0.45 : 0.4 \tag{4.17}$$

which agrees well with the "theoretical" relation (4.16).

Let us note that Eq. (4.15), like (3.104), can be expressed in the form of the linear dependence $\Delta\varepsilon \approx \mathrm{const} \cdot Q_{at}$, where $Q_{at} = (-1/2)\Delta H_{at}$. In this form it resembles Manca's [255] well-known empirical relation between the thermochemical binding energy Q_{at} and the forbidden band gap E_g : $E_g(\mathrm{IV}) = 2.17$ ($Q_{at} = 1.34$). Unlike the latter, however, Eqs. (3.104) and (4.16) have a clearer physical meaning since E_g corresponds to transitions between different levels for different crystals with the diamond lattice.

It is not hard to see that the distance between the bonding and antibonding s level Γ_1^v and $\Gamma_{2'}^c$ should also decrease with a decrease in the heat of atomization. It follows from Eqs. (3.23), (3.25), and (3.48)–(3.51) that

$$\varepsilon\,(\Gamma_{2'}^c) - \varepsilon\,(\Gamma_1^v) = -8\beta_{ss} \tag{4.18}$$

In this case the experimental verification of Eq. (4.18) is more complicated since the

*Here the universally accepted value of ≈3.5 eV [146, 147] is used for the $\Gamma_{25'}^v \to \Gamma_{15}^c$ transition in Si. Recently, Herman et al. [85, 103] have indicated a value of 2.9 eV for the energy of this transition, which is practically the same as for Ge. However, doubt was cast upon Herman's interpretation by Saravia and Brust [175], who prefer the former value of 3.4 eV for the energy of the $\Gamma_{25'}^v \to \Gamma_{15}^c$ transition in Si.

transition from the bottom of the valence band Γ_1^v into the s state of the conduction band $\Gamma_{2'}^c$ is not observable. Here, however, one can make use of the obvious relation

$$\varepsilon(\Gamma_{2'}^c) - \varepsilon(\Gamma_1^v) = E_v + \varepsilon(\Gamma_{25'}^v \to \Gamma_{2'}^c) \tag{4.19}$$

after taking the energies of the $\Gamma_{25'}^v \to \Gamma_{2'}^c$ transitions from spectroscopic data or from EP and EC OPW calculations and E_v from X-ray spectral estimates or from these same calculations (where experimental estimates are lacking). According to Eq. (4.19) we then obtain: $\Delta\varepsilon$(diamond) $\approx$ 35 eV; $\Delta\varepsilon$(Si) $\approx$ 17–20 eV; $\Delta\varepsilon$(Ge) $\approx$ 15–17 eV; and $\Delta\varepsilon$(α-Sn) $\approx$ 12 eV, where $\Delta\varepsilon = \varepsilon(\Gamma_{2'}^c) - \varepsilon(\Gamma_1^v)$. Thus for the difference in the levels $\Gamma_{2'}^c$ and Γ_1^v we obtain

$$\Delta\varepsilon\,(\mathrm{C}) : \Delta\varepsilon\,(\mathrm{Si}) : \Delta\varepsilon\,(\mathrm{Ge}) : \Delta\varepsilon\,(\alpha\text{-Sn}) = 1 : 0.55 : 0.45 : 0.35 \tag{4.20}$$

which is also found to be in good agreement with the ratio of the heats of atomization (1: 0.6: 0.45: 0.4).

4.2 STRUCTURE OF HYDROCARBONS AND REFINEMENT OF THE RESONANCE PARAMETERS FOR DIAMOND

4.2.1 Features of Conduction Band of Diamond and Silicon and Pauling's "Rule of Angular Parts"

A more detailed study of the conduction band already assumes a detailed investigation of the system of resonance parameters. As we shall see, such an investigation leads to a review of the "simplified" version of the theory (which was adopted in the previous chapter solely because of its simplicity).

It is not hard to show that the description of the conduction band, at least for diamond and silicon, is in fact incompatible with the "simplified" version of the theory.

It was stated in Sec. 3.2.4 that the inequality (3.21) is true for the conduction band of C and Si [127]. This inequality, as seen from Eqs. (3.50) and (3.51), can be written in the form

$$2\beta_{g\sigma}^{(-)} + \frac{1}{6}\beta_\pi > 0, \qquad \beta_{g\sigma} = -\frac{1}{8}\left\{\beta_{ss} - \frac{2}{\sqrt{3}}\beta_{sp} + \frac{1}{3}\beta_{pp}\right\} \tag{4.21}$$

where β_π is certainly less than zero. Thus for C and Si $\beta_{g\sigma}^{(-)}$ is greater than zero in every case. Such an inequality, however, is incompatible with the simplified version since the values of the resonance parameters, determined from Eqs. (3.58)–(3.61), always lead to an exact equality of $\beta_{g\sigma}^{(-)} = 0$. Let us note that the inequality (4.21) negates certain other estimates besides the estimates (3.58)–(3.61). Thus if the resonance integrals themselves rather than the overlap integrals in (3.58)–(3.61) are estimated from "Pauling's rule," then in this case again the equality $\beta_{g\sigma}^{(-)} = 0$ is obtained. This contradicts the inequality (4.21).

4.2.2 Nature of Bond between Carbon Atoms and the Resonance Integrals β_{ss}(C) and β_{pp}(C)

Let us now turn to the evaluation of the parameters β_{ss}, β_{sp}, and β_{pp}. We shall start with carbon, as in the previous chapter, and the empirical data of [176].

Let us take a C atom, which is in a state of arbitrary $sp^{\lambda 2}$ hybridization and is bound to some other atom X. Let us examine the interaction integral between the $sp^{\lambda 2}$ orbital of the C atom and some q orbital of the X atom:

$$\beta_{C-X}(\lambda) = \frac{1}{\sqrt{1+\lambda^2}}\{\beta_{sq} + \lambda\beta_{pq}\} \tag{4.22}$$

In the most important special case when X is another C atom also in the state of $sp^{\lambda 2}$ hybridization and its q orbital is another $sp^{\lambda 2}$ orbital, Eq. (4.22) becomes

$$\beta(\lambda) = \beta_{C-C}(\lambda) = \beta_{C-C}(sp^{\lambda} - sp^{\lambda^2}) = \frac{1}{1+\lambda^2}\{\beta_{ss} + 2\lambda\beta_{sp} + \lambda^2\beta_{pp}\} \tag{4.23}$$

Then the behavior of the integral β_{C-X} or β_{C-C} as a function of λ is determined by the relative magnitude of the coefficients for the different powers of λ, i.e., by the magnitude of the resonance integrals β_{sq}, β_{pq} or β_{ss}, β_{sp}, and β_{pp}. As usual, we shall assume that the strength of the C–X or C–C bond is determined by the resonance integral β_{C-X} or β_{C-C}* and that the strength of the C–X and C–C bonds increases with an increase in λ (i.e., with an increase in the p character of the bonds of the C atom). As seen from Eqs. (4.22) and (4.23), this is possible when β_{pq} is sufficiently large compared with β_{sq} or, correspondingly, when β_{pp} or β_{sp} are sufficiently large compared with β_{ss}. Thus, an analysis of experimental data on the strength of the bonds formed by the C atom in different hybridization states can give information on the values of the unknown resonance parameters.

Pauling's "rule of angular parts" will be used for estimating the resonance integrals β_{sq}, β_{pq} or β_{ss}, β_{sp}, β_{pp}. According to this rule $\beta_{pq}:\beta_{sq} = \sqrt{3} : 1$, and $\beta_{pp}:\beta_{sp}:\beta_{ss} = 3 : \sqrt{3} : 1$. It is not hard to see that in this case the strongest bonds would be the ordinary bonds of the type C(sp^3)–X and C(sp^3)–C(sp^3); the weakest would be bonds of the type C(sp)–X and C(sp)–C(sp). A similar result is also obtained in the "simplified" version of the theory (see Sec. 3.2.2). In this case we have $\beta_{pp}:\beta_{sp}:\beta_{ss} \approx 1.65 : 1.35 : 1$ for the ordinary C–C bonds according to Eqs. (3.58) and (3.61). In the "simplified"

*The assumption of a proportionality between the bond strength and the magnitude of the interaction of the atomic functions was used successfully by Mulliken [177] in a study of the bonds in simple compounds of the elements of the short periods of the periodic table although Mulliken used the overlap integrals as a measure of the AO interaction (as was often done then) rather than the resonance integrals.

version, as it is easy to see, it was observed that all types of bonds have approximately equal strength [the strength is possibly slightly greater for the $C(sp^2)$–$C(sp^2)$ bonds].

As numerous experimental data from organic chemistry show, the strength of the bonds of C atoms, conversely, decreases monotonically with an increase in the p character of the bonds. In the series of bonds $C(sp^3)$–$C(sp^3)$, $C(sp^2)$–$C(sp^2)$, $C(sp)$–$C(sp)$ the maximum strength is reached for the $C(sp)$–$C(sp)$ bonds. Thus it is known that in the series ethane $H_3C(sp^3)$–$H_3C(sp^3)$, ethylene $H_2C(sp^2){=}C(sp^2)H$, acetylene $HC(sp){\equiv}C(sp)H$ the lengths of the C–H bonds decrease, and the thermochemical energies and force constants of these bonds systematically increase [178] (Table 7).

As shown in [179], the C–Hal bonds possess similar properties. Finally, the same characteristic increase in the strength as one goes from sp^3 to sp hybridization of the C atoms occurs also for the so-called "purely ordinary" C–C bonds [26, 180] (Table 7). This characteristic is retained even after the energy of the bonds is reduced to the standard $C(sp^3)$–$C(sp^3)$ separation, equal to 1.54 Å.

Thus experimental data show that the maximum strength of the bonds formed by the C atoms is reached for a hybridization close to sp. It is not hard to see that this leads to the estimate [176]

$$\beta_{ss} \approx \beta_{pp} \, (\approx \beta_0) \tag{4.24}$$

where β_0 denotes the average value of the integrals β_{ss} and β_{pp}: $\beta_0 = (1/2)\{\beta_{ss} + \beta_{pp}\}$.

TABLE 7. The Separation, Energy, and Force Constants for Ordinary C Atom Bonds

Type of Bond[a]	Scheme	Length (Å)	Energy (kcal/mol)	Force Constant (10^5 dynes/cm)
Methane $C(sp^3)$–H	—	1.09	103 —	5.387
Ethylene $C(sp^2)$–H	—	1.07	106 —	6.126
Acetylene $C(sp)$–H	—	1.06	121 —	6.397
$C(sp^3)$—$C(sp^3)$	Dewar	1.54	85 (85)	—
	Lorquet	1.54	85 (85)	
$C(sp^2)$—$C(sp^2)$	Dewar	1.49	95 —	—
	Lorquet	1.52	92 —	—
$C(sp)$—$C(sp)$	Dewar	1.38	120 (89)	—
	Lorquet	1.46	103 (88)	—

[a]The data for the C–H bonds are taken from [178]. The energy values in parentheses are calculated for a separation of 1.54 Å.

Actually, equating the derivative $\partial\beta/\partial\lambda$ of the function $\beta(\lambda)$, defined by Eq. (4.23), to zero, we have

$$(1-\lambda^2)\,\beta_{sp}+\lambda\,(\beta_{pp}-\beta_{ss})\approx 0 \tag{4.25}$$

The estimate (4.24) is also obtained from this for $\lambda = 1$ [the minimum in $\beta(\lambda)$ is reached for *sp* hybridization].

One can arrive at this same estimate by another method by using the data on the strength of "purely ordinary" C–C bonds from Table 7 instead of the maximum condition (see below for more details about these data). As seen from Table 7, the values of the function $\beta(\lambda)$ for the C–C distance of 1.54 Å are generally only slightly dependent on λ, so that the derivative $\partial\beta/\partial\lambda$ is small* for all λ values. Since the derivative $\partial\beta/\partial\lambda$ then assumes the form

$$\left\{\frac{\partial\beta\,(\lambda)}{\partial\lambda}\right\}_{\lambda=1}=\beta_{pp}-\beta_{ss} \tag{4.26}$$

for $\lambda = 1$, we again arrive at the estimate (4.24).

Let us point out again it was essentially this estimate that was adopted by Mulliken [177] during his discussion of a similar problem, except that Mulliken simply operated with the overlap integrals (for the Slater AOs), comparing their values for different hybridization states.

4.2.3 Resonance Integral β_{sp}

Now let us examine the question of the resonance integral β_{sp}. As follows from Eqs. (4.25) and (4.26), the value of β_{sp} cannot be determined from the maximum condition $\{\partial\beta(\lambda)/\partial\lambda\} = 0$ or from the condition that the derivative $\partial\beta/\partial\lambda$ is small. Therefore to find β_{sp} additional data are needed. One can attempt to find these data by comparing the energies of the "purely ordinary" C–C bonds with C atoms in different hybridization states.

As before, we shall assume that the thermochemical strength $Q_{at}(sp^{\lambda 2} - sp^{\lambda 2})$ of the $C(sp^{\lambda 2})$–$C(sp^{\lambda 2})$ bonds is proportional to the corresponding integral (4.23), and we shall denote the ratio β_{sp}/β_0 by κ and the ratio $Q_{at}(sp^3 - sp^3)$: $Q_{at}(sp - sp)$ by K. Then we have from Eq. (4.23)

*We assume that

$$(1/\beta)\,(\partial\beta/\partial\lambda)\approx(1/\beta)\,(\Delta\beta/\Delta\lambda)\approx\frac{Q_{at}\,(sp^3-sp^3)_{1.54\text{Å}}-Q_{at}\,(sp-sp)_{1.54\text{Å}}}{Q_{at}\cdot(\sqrt{3}-1)}$$

where Q_{at} is any of the energies $Q_{at}(sp^{\lambda 2} - sp^{\lambda 2})$ in the interval $1 \leqslant \lambda \leqslant \sqrt{3}$. Then on the basis of Table 7 we have the estimate $(1/\beta)(\partial\beta/\partial\lambda) \approx 0.06$.

$$K=\frac{Q_{at}\,(sp^3-sp^3)}{Q_{at}\,(sp-sp)}=\frac{\beta\,(sp^3-sp^3)}{\beta\,(sp-sp)}=\frac{\beta_0+\sqrt{\frac{3}{2}}\,\beta_{sp}}{\beta_0+\beta_{sp}}=\frac{1+\varkappa\sqrt{\frac{3}{2}}}{1+\varkappa} \tag{4.27}$$

from which we obtain for $\kappa = \beta_{sp}/\beta_0$

$$\varkappa=\frac{\beta_{sp}}{\beta_0}=\frac{1-K}{K-\sqrt{\frac{3}{2}}} \tag{4.28}$$

For a numerical estimation of the ratio β_{sp}/β_0 it is necessary to know the dependence of the thermochemical energy $Q_{\text{at}}(sp^{\lambda 2}-sp^{\lambda 2})$ of purely ordinary C–C bonds on λ. Recently two schemes (Dewar [26] and Lorquet [180]) have been developed in considerable detail that provide, on the basis of test data, the dependence of the length and strength $Q_{\text{at}}(sp^{\lambda 2}-sp^{\lambda 2})$ of "purely ordinary" C–C bonds on the hybridization state of both carbon atoms. We shall make use of these schemes after reducing, of course, all the energies to the same C–C distance (1.54 Å).

As seen from Table 7, the length of the purely ordinary *sp–sp* bond (1.38 Å) in the Dewar scheme is practically the same as the average C–C separation in aromatic hydrocarbons (1.40 Å).* If it is again assumed that the binding energies are proportional to the corresponding resonance integrals, then in this case we obtain the estimate

$$Q_{at}\,(sp-sp)_{1.54\text{Å}}\approx Q_{at}\,(sp-sp)_{1.38\text{Å}}\cdot\frac{\beta_\pi\,(1.54\text{Å})}{\beta_\pi\,(\text{aromatic})}\approx 89\,\frac{\text{kcal}}{\text{mol}} \tag{4.29}$$

for $Q_{\text{at}}(sp-sp)_{1.54\,\text{Å}}$, from which

$$\varkappa=\frac{\beta_{sp}}{\beta_0}\approx 0.5 \tag{4.30}$$

The Lorquet scheme leads to a similar estimate. Since the length of a purely ordinary C(*sp*)–C(*sp*) bond in this scheme lies between 1.40 and 1.54 Å, then, assuming an intermediate value of ≈–2 eV for β_π, we again obtain $Q_{\text{at}}(sp-sp)_{1.54\,\text{Å}}\approx 88$ kcal/mol and $\kappa\approx 0.4$.

Other estimates, based on the schemes in [26, 180], also agree with (4.30).

If, for example, a value of –2.45 eV is assumed for $\beta_\pi(1.40\ \text{Å})$ instead of a value of –2.3 eV as in the aromatic hydrocarbons, i.e., the average of the values for the aromatic

*And in the polyenes. However, the separations in the polyenes, especially in the first members of the series, exhibit too much variation: ~0.07–0.08 Å about the mean value.

hydrocarbons and the polyenes, or in general any value falling within the interval from −2.3 to −2.6 eV is assumed, then the ratio κ becomes even smaller than 0.5–0.4. A similar result is also obtained for a more precise estimate that takes account of the difference between the C(*sp*)–C(*sp*) separation in the Dewar scheme (1.38 Å) and the C–C separation in the aromatic hydrocarbons (1.40 Å) or the deviation of this separation in the Lorquet scheme (1.46 Å) from the average of the 1.40 and 1.54 Å separations. Besides, such precision is hardly in keeping with the nature of the assumptions made or the actual accuracy of the schemes [26, 180]. This is especially true since modern studies have shown that the hybridization indices of the C atoms in hydrocarbons are not actually integers but only close to them.

4.3 STRUCTURE OF CONDUCTION BAND AND THE COMPLETE BAND STRUCTURE OF DIAMOND AND SILICON

4.3.1 Organic Chemistry and Characteristic Features of the Band Structure of Diamond

Let us first consider how the relations $\beta_{ss} \approx \beta_{pp} \approx \beta_0$ and $\beta_{sp} \lesssim 0.5\beta_0$ found from organic chemistry without a numerical estimate of the parameters describe the characteristic features of the band structure of diamond [176].

As seen from Fig. 3.2, these features include the fact that in the conduction band of the diamond crystal the *s*-level $\Gamma^c_{2'}$ lies above the triple *p*-level Γ^c_{15}, and the absolute minimum of the conduction band is located on the Δ line near the edge X of the first zone, with the value of the energy at the minimum being close to the value of the energy of the X^c_1 level.

Let us now take any crystal with a diamond lattice. Then, as follows from Eqs. (3.24) and (3.26), all the doubly degenerate *p* states, described by the $\varepsilon = \varepsilon^{(-)}_{3,4}(\mathbf{k})$ branch of the dispersion law, for the conduction band are always located above the triply degenerate *p* state Γ^c_{15}. Therefore the X^c_3 and L^c_3 levels must lie above the Γ^c_{15} level. For a qualitative description of the conduction band structure in a diamond-type crystal it only remains to compare the energies of the Γ^c_{15}, $\Gamma^c_{2'}$ and X^c_1 levels, as well as the $\bar{L}^c$ and $\bar{\bar{L}}^c$ levels at the edge L of the first zone. Here and below we will use $\bar{L}^c$ and $\bar{\bar{L}}^c$ as a convenient way to denote the levels belonging to that branch of the dispersion law that becomes, at the center Γ of the zone, the $\Gamma^c_{2'}$ and Γ^c_{15} states, respectively. Thus, in the conventional notations $\bar{L}^c = L^c_{2'}$, $\bar{\bar{L}} = L_1$ for diamond and silicon, and, conversely, $\bar{L}^c = L^c_1, \bar{\bar{L}}^c = L^c_{2'}$ for germanium and gray tin.

To investigate the relative arrangement of the levels Γ^c_{15}, $\Gamma^c_{2'}$, X^c_1, $\bar{L}^c$, and $\bar{\bar{L}}^c$ it is convenient to start from the expressions describing the position of the $\Gamma^c_{2'}$ and X^c_1 levels with respect to the Γ^c_{15} level as well as the position of the $\bar{L}^c$ and $\bar{\bar{L}}^c$ levels with respect to the X^c_1 level. Let us write the corresponding expressions in general form. It follows from Eqs. (3.23)–(3.26) for the dispersion law with Eqs. (3.48)–(3.51) taken into consideration:

$$\varepsilon(\Gamma^c_{15}) - \varepsilon(\Gamma^c_{2'}) = \mathscr{E}_{\text{promot}} + 4\left\{\beta_{ss} - \frac{1}{3}\beta_{pp}\right\} + \frac{8}{3}\beta_\pi \tag{4.31}$$

$$\varepsilon(\Gamma_{15}^c) - \varepsilon(X_1^c) = \frac{1}{2}\mathscr{E}_{\text{promot}} - \frac{4}{3}\{\beta_{pp} - \sqrt{3}\,\beta_{sp}\} + \frac{8}{3}\beta_\pi \tag{4.32}$$

$$\varepsilon(X_1^c) - \varepsilon(\overline{L}^c) = \frac{1}{4}\mathscr{E}_{\text{promot}} + \frac{3}{2}\left\{\beta_{ss} - \frac{1}{9}\beta_{pp} - \frac{2}{3\sqrt{3}}\beta_{sp}\right\} - \frac{2}{3}\beta_\pi \tag{4.33}$$

$$\varepsilon(X_1^c) - \varepsilon(\overline{\overline{L}}^c) = -\frac{1}{4}\mathscr{E}_{\text{promot}} - \frac{1}{2}\{\beta_{ss} - \beta_{pp}\} - \frac{1}{\sqrt{3}}\beta_{sp} + 2\beta_\pi \tag{4.34}$$

where the numerical values of $\mathscr{E}_{\text{promot}}(\mathrm{C}) = \alpha_{\mathrm{C}}^{(p)} - \alpha_{\mathrm{C}}^{(s)}$ and of the resonance integral β_π can be assumed to be known quite accurately (see Sec. 3.4.1 and 3.4.3).

Using the values from Tables 4 and 5 for these parameters and Eq. (4.24) for the integral β_{ss} and β_{pp}, we have (in eV)

$$\varepsilon(\Gamma_{15}^c) - \varepsilon(\Gamma_{2'}^c) \approx 4.3 + \frac{8}{3}\beta_0 \tag{4.35}$$

$$\varepsilon(\Gamma_{15}^c) - \varepsilon(X_1^c) \approx -\frac{4}{3}\beta_0(1 - \varkappa\sqrt{3}) \tag{4.36}$$

$$\varepsilon(X_1^c) - \varepsilon(\overline{L}^c) \approx 3.3 + \frac{4}{3}\beta_0\left(1 - \frac{\sqrt{3}}{4}\varkappa\right) \tag{4.37}$$

$$\varepsilon(X_1^c) - \varepsilon(\overline{\overline{L}}^c) \approx -5.6 - \frac{1}{\sqrt{3}}\beta_0\varkappa \tag{4.38}$$

Now let us point out that regardless of the exact value of β_0 the inequality

$$|\beta_0| > |\beta_\pi| \approx 1.7 \text{ eV}$$

will necessarily be satisfied in any case.*

It is then seen from Eqs. (4.35)-(4.38) that for $\kappa \lesssim 0.5$ the right side of Eq. (4.36) is positive, and the right sides in Eqs. (4.35), (4.37), and (4.38) are negative. [The last of these differences can, of course, be positive too for $|\beta|_0 >$ 20 eV; however, a value of $|\beta_i| \gtrsim 20$ eV is clearly unrealistic (see below).] Thus, for practically any values of the parameters that have a realistic meaning the estimates $\beta_{ss} \approx \beta_{pp}$ and $(\beta_{sp}/\beta_0) = \kappa \lesssim 0.5$ will lead to the qualitatively correct

*For C atoms, as for other atoms of the second period, the $p\pi-p\pi$ bonds are comparable in energy to the $p\sigma-p\sigma$ and $s\sigma-s\sigma$ bonds only at separations of the order of the separations in diatomic molecules (at a separation of 1.3 Å for C), which is observed in the C_2 molecule. At separations greater by only 0.1-0.2 Å the $p\pi-p\pi$ bonds are considerably weaker than the σ bonds.

conduction band structure of diamond:

$$\varepsilon(\Gamma^c_{2'}) > \varepsilon(\Gamma^c_{15}); \quad \varepsilon(\Gamma^c_{15}) > \varepsilon(X^c_1); \\ \varepsilon(\bar{L}) > \varepsilon(X^c_1); \quad \varepsilon(\bar{\bar{L}}) > \varepsilon(X^c_1) \tag{4.39}$$

4.3.2 Numerical Estimates of the Structure of the Conduction Band, Band-to-Band Transitions, and the Forbidden Band Gap in Diamond

Let us now proceed to a quantitative study of the complete band structure of diamond. For a numerical illustration of the inequalities (4.39) let us recall that in the case of diamond (as is evident from the example of the parameters β and β_π; see Sec. 3.4.2 and 3.4.3) the resonance integrals are fairly well approximated by the expressions $S\alpha_A$ and $S_\pi\alpha_A^{(p)}$, where α_A and $\alpha_A^{(p)}$ are the Coulomb integrals and S and S_π are the corresponding overlap integrals, calculated for the Slater orbitals. Therefore it is reasonable to take the average

$$\beta_0 \approx \frac{1}{2}\{S_{ss}\alpha_C^{(s)} + S_{pp};\ \alpha_C^{(p)}\} \approx 5.1 \text{ eV} \tag{4.40}$$

for the average value β_0 of the two resonance integrals β_{ss} and β_{pp}.

Accordingly, for the integral β_{sp} we obtain the estimate*

$$\beta_{sp} \approx \beta_0 \cdot 0.5 \approx 2.55 \text{ eV} \tag{4.41}$$

Then, according to Eqs. (4.31)-(4.34) for diamond we have $\varepsilon(\Gamma^c_{2'}) - \varepsilon(\Gamma^c_{15}) \approx 9$ eV; $\varepsilon(\Gamma^c_{15}) - \varepsilon(X^c_1) = 0.8$ eV; and $\varepsilon(L^c)_{\min} - \varepsilon(X^c_1) = 2$ eV. These results agree satisfactorily, for example, with the recent EC OPW calculations of Herman and co-workers [83-85], according to which $\varepsilon(\Gamma^c_{2'}) - \varepsilon(\Gamma^c_{15}) = 5.8 \pm 1.0$ eV; $\varepsilon(\Gamma^c_{15}) - \varepsilon(X^c_1) = 1.2 \pm 0.3$ eV; and $\varepsilon(L^c)_{\min} - \varepsilon(X^c_1) = 2.7 \pm 0.4$ eV.

Let us now turn to an investigation of the relative arrangement of the valence band and conduction band, after estimating the energies of the direct transitions $\Gamma^v_{25'} \to \Gamma^c_{15}$ and $\Gamma^v_{25'} \to \Gamma^c_{2'}$ and the forbidden band gap E_g. Here the expression for $\Gamma^v_{25'} \to \Gamma^c_{15}$ is given by Eq. (4.14), the energy of the $\Gamma^v_{25'} \to \Gamma^c_{2'}$ transition

$$\varepsilon(\Gamma^v_{25'} \to \Gamma^c_{2'}) = -\{\alpha_A^{(p)} - \alpha_A^{(s)}\} - 4\left\{\beta_{ss} + \frac{1}{3}\beta_{pp}\right\} + \frac{8}{3}\beta_\pi \tag{4.42}$$

is easy to obtain by combining Eqs. (4.14) and (4.31), and the forbidden band

*Note that when the estimate for β_{ss}, β_{pp}, and β_{sp} is refined, the value of the parameter β changes hardly at all.

gap can be approximately estimated* as the energy of the indirect transition $\Gamma^v_{25'} \to X^c_1$:

$$E_g \approx \varepsilon(\Gamma^v_{25'} \to X^c_1) = -\frac{1}{2}\{\alpha^{(p)}_A - \alpha^{(s)}_A\} - \frac{4}{3}\{\beta_{pp} + \sqrt{3}\,\beta_{sp}\} + \frac{8}{3}\beta_\pi \quad (4.43)$$

Then for β_0, as given by the estimate (4.40), the value of $\kappa \approx 0.5$, and the value of $\beta_\pi \approx -1.7$ eV we obtain from Eqs. (4.41), (4.42), and (4.43): $\varepsilon(\Gamma^v_{25'} \to \Gamma^c_{15})_{\text{diamond}} \approx 4.5$ eV, $\varepsilon(\Gamma^v_{25'} \to \Gamma^v_{2'})_{\text{diamond}} \approx 13.5$ eV, $(E_g)_{\text{diamond}} \approx 3.7$ eV. Approximately these same estimates are obtained if a value of $\beta_\pi \approx -1.6$ is taken for β_π (obtained by an extrapolation from the spectroscopic data for only the aromatic hydrocarbons, which, perhaps, are more reliable in view of the alternation of the lengths of the C–C bonds in the polyenes). In this case $\varepsilon(\Gamma^v_{25'} \to \Gamma^c_{15})_{\text{diamond}} \approx 5.1$ eV, $\varepsilon(\Gamma^v_{25'} \to \Gamma^c_{2'})_{\text{diamond}} \approx 13.8$ eV, $(E_g)_{\text{diamond}} \approx 4.1$ eV.

The estimates obtained in this way can be compared with experimental data or with the results of EC OPW and EP calculations [83-85, 106, 108]. Usually the value of the energy of the $\Gamma^v_{25'} \to \Gamma^c_{15}$ transition for diamond, adopted on the basis of spectroscopic data, amounts to ≈7 [147] or 7.3 eV [106, 145] (although other estimates, such as 8.7 eV [146], are encountered). Similar data lead to a value of 5.2-5.5 eV [148] for the forbidden band gap of diamond. In the case of diamond there are no direct experimental data for the $\Gamma^v_{25'} \to \Gamma^c_{2'}$ transition. However, EP and EC OPW calculations give a value of 12.5 or 12.9 eV for the energy of this transition. The final data on the band structure of diamond, obtained recently in several experimental and theoretical papers, are listed in Table 8. They show that the agreement between the experimental data and the data of detailed calculations [67, 85, 106], on the one hand, and computer estimates, on the other, is fully satisfactory, especially if the indirect

*In fact, as already stated, the minimum of the conduction band of diamond is not at the point X itself, but on the line Δ near this point. However, the value of the energy at the minimum $\varepsilon(\Delta)_{\min}$ is close to the value $\varepsilon(X^c_1)$.

TABLE 8. Conduction Band Structure of Diamond According to Various Data[a]

Levels	Experiment	Semiempirical EO LCAO	APW [67]	OPW [162]	SC OPW [85]	EC OPW [85]	EP [106]
Γ^c_{15}	7—7.3	5.1	5.8	5.1	5.6	7.1	7.3
$\Gamma^c_{2'}$	—	13.8	—	13.7	11.2	12.9	12.5
X^c_1	5.5	4.1	5.9	5	4.4	5.9	5.6
L^c_1	—	7	7.9	5.3	6.6	8.7	9.8
Γ^v_{25}	0.0	0.0	0.0	0.0	0.0	0.0	0.0

[a]All transition energy values are given in eV.

nature of the estimate of the numerical values of the parameters β_{ss}, β_{pp}, and β_{sp} is taken into consideration.

4.3.3 Saravia-Brust Model and the Question of the Band Structure of Diamond

Progress in the uv spectroscopy of saturated hydrocarbons will probably make it possible to improve these estimates; however, even in their present form they can be used in band theory, in particular, as a method of testing the validity of alternative models. Let us illustrate this [181] on the example of the Saravia-Brust model [182], suggested not too long ago for the band structure of diamond. As pointed out by Savaria and Brust [182], the spectral dependence $\varepsilon_2(\omega)$ for diamond agrees somewhat better with the model (II) in which the band structure for diamond resembles the band structure for germanium than with the usual model of its band structure (I) (Fig. 4.1).

Let us examine to see which of these models agrees better with the data presented above concerning the chemical bond between the carbon atoms.

In the usual model (I) a value of 12.6-12.7 eV or 12.2 eV is assumed for the energy of the $X_4^v \to X_1^c$ transition on the basis of spectroscopic data; E_g is assigned a value of $\approx$5.5 eV, and the difference $\varepsilon(X_1^c) - \varepsilon(\Delta^c)_{\min}$ is usually estimated to be $\approx$0.5 eV [83-85], from which $\varepsilon(\Gamma_{25'}^v) - \varepsilon(X_4^v) = 4|\beta_\pi| = 6.5$ eV and $\beta_\pi = -1.6$ eV. Let us now take the transition $\Gamma_{25'}^v \to \Gamma_{15}^c$, the expression for which is given by Eq. (4.14). Substituting $\beta_\pi = -1.6$ eV into (4.14), we find that $\beta_{pp} = -5.9$ eV, so that with a probable value of $\varepsilon(\Gamma_{2'}^c) - \varepsilon(\Gamma_{15}^c) \approx$ 5.2-5.8 eV [83-85] and with $\mathcal{E}_{\text{promot}} = 8.8$ eV (see Table 3) we obtain from Eqs. (4.42) and (4.32): $\beta_{ss} \approx -4.3$ eV and $\beta_{sp} \approx -2.6$ eV. [Let us note that all the formulas, with the exception of Eq. (4.32), have one and the same form in the EO LCAO method and also in the tight binding approximation; see below. In the tight binding method the difference $\varepsilon(\Gamma_{15}^c) - \varepsilon(X_1^c)$ is given not by Eq. (4.43), but by Eq. (4.50); however, this fact has practically no effect on the values of the parameter β_{sp}.]

Table 9 lists the values of the resonance parameters, obtained by the method described above, for the usual model and for the Saravia-Brust model, as well as the estimates (4.40)

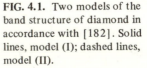

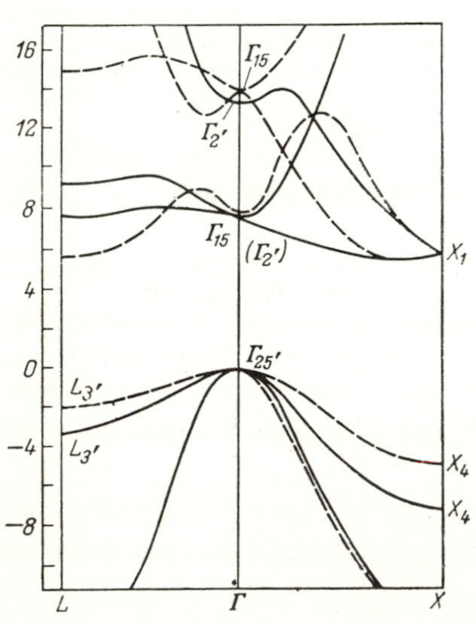

FIG. 4.1. Two models of the band structure of diamond in accordance with [182]. Solid lines, model (I); dashed lines, model (II).

TABLE 9. Value of Resonance Integrals (in eV), Obtained from Two Models of the Band Structure of Diamond and from Data on Hydrocarbons

Source of Data	$-\beta_{ss}$	$-\beta_{pp}$	$-\beta_{sp}$	$-\beta_{\pi}$
Model II	2.1	7.3	1.5	1
Model I	4.3	5.9	2.6	1.6
Hydrocarbons	5.1	5.1	2.55	1.6—1.7

and (4.41). As seen from this table, the values of the parameters, obtained from model (I), agree fairly well with the estimates obtained by an analysis of data for hydrocarbon molecules, but do not agree with the values obtained from the Savaria-Brust model; thus the usual model (I) of the band structure of diamond corresponds much better to the data on the nature of the bond between the carbon atoms.

4.3.4 Band Structure and Chemistry of Silicon

Let us now examine the question of the band structure of crystalline silicon [176]. In this case it is impossible to estimate the resonance integrals β_{ss}, β_{pp}, and β_{sp} in the same manner as for diamond since no compounds have been found for silicon with multiple or with conjugated $p\pi$-$p\pi$ bonds in which the σ bonds of the atoms would be formed by hybrid sp^2 or sp orbitals. Therefore for bonds formed by Si atoms no scheme exists, similar to the Dewar [26] or Lorquet [180] schemes. At the same time data on the uv spectra of the silanes Si_nH_{2n+2}, as well as similar data for all the saturated compounds of the rest of the group IV elements, are nonexistent. Thus for estimating the parameters $\beta_{ss}(Si)$, $\beta_{pp}(Si)$, and $\beta_{sp}(Si)$ we can thus far only be on safe ground with a general analogy between the chemical behavior of carbon and silicon, using the estimates for the interaction integrals $\beta_{ss}(C)$, $\beta_{pp}(C)$, and $\beta_{sp}(C)$ already established in the previous section.

It is known that the minimum valence of silicon in stable compounds is equal to four,* just as in carbon, unlike in germanium, tin, and lead, the atoms of which also form stable compounds in the divalent state.

Moreover, it is obvious that for a group IV atom (which in the ground state has an ns^2np^2 electron configuration) a valence of four is possible only in the case when its valence ns orbital takes part in the bonds. Therefore one can assume that the valence 3s orbital of the Si atom has about the same tendency to take part in a bond as the 2s orbital of the C atom; consequently, one can, with a certain justification, retain for silicon the estimate $\beta_{ss} \approx \beta_{pp}$, previously obtained for carbon. Also retaining the estimate (4.30) for the integral β_{sp} and,

*Compounds of divalent C and Si, such as the dihalides $CHal_2$ and $SiHal_2$, exist only under special conditions.

as always, assuming that the resonance integrals vary proportionally to the binding energy, we have

$$\beta_{ss}\,(\mathrm{Si}) \approx \beta_{pp}\,(\mathrm{Si}) \approx \beta_0\,(\mathrm{Si});\; \beta_0\,(\mathrm{Si}) \approx \beta_0\,(\mathrm{C})\,\frac{\Delta H_{\mathrm{aT}}\,(\mathrm{Si})}{\Delta H_{\mathrm{aT}}\,(\mathrm{C})} \approx -3.1\ \mathrm{eV} \quad (4.44)$$

and

$$\beta_{sp}\,(\mathrm{Si}) \approx 0.5\beta_0\,(\mathrm{Si}) \approx -1.55\ \mathrm{eV} \quad (4.45)$$

Let us now use the estimates (4.44) and (4.45) to investigate the conduction band structure and the band-to-band transitions for silicon. Then, as follows from Eqs. (4.31)-(4.33), the inequalities (4.39) will be satisfied for silicon just as for diamond, and this means the estimates (4.44) and (4.45) lead to a qualitatively correct conduction band structure in silicon. A quantitative estimate of the energy of the direct transition $\Gamma^v_{25'} \to \Gamma^c_{15}$ from the top of the valence band into the lower part of the conduction band and the forbidden band gap E_g gives these values for silicon: $\varepsilon(\Gamma^v_{25'} \to \Gamma^c_{15}) \approx 2.8$ eV; $E_g = \varepsilon(X^c_1) - \varepsilon(\Gamma^v_{25'}) \approx 1.35$ eV, which also agree fairly well with experimental data: $\varepsilon(\Gamma^v_{25'} \to \Gamma^c_{15}) \approx 3.5$ eV [146, 147, 175] or 3 eV [83-85] and $E_g = 1.2$ eV.

4.3.5 Refinement of the Scale of the Resonance Parameters and the Valence Band Structure of Covalent Crystals

We have seen that to describe the conduction band and the band-to-band transitions in diamond and silicon it was necessary to switch from the "simplified" to the "complete" version of the theory and also to refine the scale of the resonance parameters. As a result of this a legitimate question arises with regard to the extent of the effect of these circumstances on the results obtained in the previous chapter. Let us examine here what changes this introduces into the structure of the valence band of diamond and silicon as well as two other covalent crystals—germanium and gray tin [183].

It is obvious that all the qualitative conclusions concerning both the valence band of individual crystals as well as the behavior of its various parameters in the series of group IV elements remain valid in the new version. (The valence band is insensitive to the form of the approximation and the choice of the resonance integrals, and the proportionality between the latter and the heats of atomization in the series C–Si–Ge–α-Sn also remains in the "complete" version of the theory.)

Returning to the numerical estimates, let us first discuss the question of the p band $L^v_{3'}$–$\Gamma^v_{25'}$–X^v_4 in the valence band. As follows from Eqs. (3.64) and (3.65) β_t and β_g in the "simplified" version are expressed only in terms of the resonance integral β_π; this is no longer true for the "complete" version of the theory since other terms appear in the expressions for β_t and β_g besides β_π. It is not hard to see, however, that the appearance of the nonzero σ component $\beta_{t\sigma}$, $\beta_{g\sigma}$ in the parameters β_t, β_g does not lead to a change in the nature of the dependence of $\varepsilon_{3,4}$ on **k** since $\beta_{t\sigma} = \beta_{g\sigma}$ [see Eqs. (3.50) and (3.51)], and the function $\varepsilon_{3,4}(\mathbf{k})$ is determined by the difference $\beta_t - \beta_g$. For this same reason the width of the p band ΔE_g and the level difference δE_p in the complete version of the theory remain unchanged; consequently, the agreement of the values of β_π and $(\beta_\pi)_{\text{exp}}$ remains true (see Sec. 2.4.2).

A comparison of the new estimates for the width E_v of the valence band with the estimates previously obtained in the "simplified" version of the theory and also with experiment and the data of EC OPW and EP calculations shows that the new values of E_v for C and Si are somewhat larger than the old and approach the upper limit of the experimental estimates rather than the lower limit as before. Let us note that the value of E_v for diamond falls between the values obtained from the EC OPW method [83-85] and from the EP method [106].

Here we will not spend time discussing the other parameters of the valence band structure considered in the previous chapter. However, it is not hard to see from Eqs. (3.88)-(3.96) and (3.48)-(3.51) that they change by an insignificant amount as one goes to the "complete" version of the theory and more refined estimates for the resonance parameters.

4.3.6 Chemical Bond and the Conduction Band of Diamond and Silicon in the Semiempirical LCAO Method

In conclusion, let us examine the structure of the conduction band in C and Si crystals by means of the LCAO method, and let us show that the proper empirical choice of parameters on the basis of organic chemistry data makes it possible to describe the characteristic features of the conduction band within the framework of this method also [184].

Let us point out first of all that the inequality $\varepsilon(\Gamma^c_{2'}) > \varepsilon(\Gamma^c_{15})$ for C and Si does not require a separate proof in the LCAO method. As seen from Eqs. (3.145) and (3.148), for the matrix elements $H^{(+-)}_{ij}$, $H^{(-+)}_{ij}$ of the secular determinant (3.139) these elements become zero for $\mathbf{k} = 0$. Therefore the position of the levels $\Gamma^c_{2'}$ and Γ^c_{15} at the center Γ of the first Brillouin zone generally does not change as one goes from the EO LCAO method to the LCAO method. Similarly, there is no change in the position of the doubly degenerate p levels X^c_5 and L^c_3 at the edges X and L of the first zone. Consequently, the inequalities $\varepsilon(X^c_5) > \varepsilon(\Gamma^c_{15})$ and $\varepsilon(L^c_3) > \varepsilon(\Gamma^c_{15})$ remain in force. This is easy to understand from the secular equation (3.140) since the matrix elements $H^{(+-)}_{33}$, $H^{(-+)}_{33}$ become zero at X and L.

At the same time it is easy to show that the single levels $\bar{L}^c, \bar{\bar{L}}^c$ at the edge L of the first zone can only be raised with the changeover to the LCAO approximation, so that the inequalities $\varepsilon(\bar{L}^c) > \varepsilon(\Gamma^c_{15})$ and $\varepsilon(\bar{\bar{L}}^c) > \varepsilon(\Gamma^c_{15})$ for C and Si are satisfied even better in the tight binding approximation. In fact, let us represent the matrix of the secular determinant $\mathscr{H}_{\text{LCAO}}$ (3.139) in the form of the following sum:

$$\mathscr{H}_{\text{LCAO}} = \mathscr{H}_{\text{EO}} + \{\mathscr{H}_{\text{LCAO}} - \mathscr{H}_{\text{EO}}\} \tag{4.46}$$

$$\mathscr{H}_{\text{EO}} = \left\| \begin{matrix} \mathscr{H}^{(++)}_{11} & \mathscr{H}^{(++)}_{12} & 0 & 0 \\ \mathscr{H}^{(++)}_{21} & \mathscr{H}^{(++)}_{22} & 0 & 0 \\ 0 & 0 & \mathscr{H}^{(--)}_{11} & \mathscr{H}^{(--)}_{12} \\ 0 & 0 & \mathscr{H}^{(--)}_{21} & \mathscr{H}^{(--)}_{22} \end{matrix} \right\|$$

$$\mathcal{H}_{\text{LCAO}} - \mathcal{H}_{\text{EO}} = \begin{Vmatrix} 0 & 0 & \mathcal{H}_{11}^{(+-)} & \mathcal{H}_{12}^{(+-)} \\ 0 & 0 & \mathcal{H}_{21}^{(+-)} & \mathcal{H}_{22}^{(+-)} \\ \mathcal{H}_{11}^{(-+)} & \mathcal{H}_{12}^{(-+)} & 0 & 0 \\ \mathcal{H}_{21}^{(-+)} & \mathcal{H}_{22}^{(-+)} & 0 & 0 \end{Vmatrix} \qquad (4.46)\ (continued)$$

The first term here is the matrix of the Hamiltonian $\hat{H}$ of the system in the EO method; the second term will be considered as a perturbation. Let us now take the direction Λ and choose as the functions of the zero approximation the eigenfunctions $\Psi_{1,0}^{(+)}$, $\Psi_{2,0}^{(+)}$, $\Psi_{1,0}^{(-)}$, $\Psi_{2,0}^{(-)}$ of the matrix $\mathcal{H}_{\text{EO}}$, corresponding to the branches $\varepsilon_1^{(+)}$, $\varepsilon_2^{(+)}$, $\varepsilon_1^{(-)}$, $\varepsilon_2^{(-)}$ of the dispersion law [see Eq. (3.25)]. The functions $\Psi_1^{(+)}$, $\Psi_2^{(+)}$ are linear combinations of only the functions $\Psi_1^{(+)}$, $\Psi_2^{(+)}$, while the functions $\Psi_{1,0}^{(-)}$, $\Psi_{2,0}^{(-)}$ are linear combinations of only the functions $\Psi_1^{(-)}$, $\Psi_2^{(-)}$, where the functions $\Psi_1^{(+)}, \ldots, \Psi_2^{(-)}$ are defined by Eq. (3.8).

Then it is not hard to see that in the basis of the eigenfunctions $\Psi_{1,0}^{(+)}, \ldots, \Psi_{2,0}^{(-)}$ the perturbation matrix $\mathcal{H}_{\text{LCAO}}$-$\mathcal{H}_{\text{EO}}$ retains its diagonal-box form since the perturbation $\mathcal{H}_{\text{LCAO}}$-$\mathcal{H}_{\text{EO}}$ corresponds to taking account of the interaction of the states $\Psi_1^{(+)}$, $\Psi_2^{(+)}$ with only the states $\Psi_1^{(-)}$, $\Psi_2^{(-)}$. Let us assume this matrix in the basis $\Psi_{1,0}^{(+)}, \ldots, \Psi_{2,0}^{(-)}$ has the form

$$\mathcal{H}_{\text{LCAO}} - \mathcal{H}_{\text{EO}} = \begin{Vmatrix} 0 & 0 & a & b \\ 0 & 0 & c & d \\ a^* & c^* & 0 & 0 \\ b^* & d^* & 0 & 0 \end{Vmatrix} \qquad (4.47)$$

where $a, \ldots, d$ are some matrix elements. Then in the first order of the theory of perturbations the corrections to the energies $\varepsilon_1^{(+)}, \ldots, \varepsilon_2^{(-)}$ become zero. Considering the second-order corrections to the energies $\varepsilon_1^{(-)}$ and $\varepsilon_2^{(-)}$, we obtain

$$\begin{aligned} \varepsilon_{1,\,\text{LCAO}}^{(-)} &= \varepsilon_{1,\,\text{EO}}^{(-)} + \frac{|a|^2}{\varepsilon_{1,\,\text{EO}}^{(-)} - \varepsilon_{1,\,\text{EO}}^{(+)}} + \frac{|b^2|}{\varepsilon_{1,\,\text{EO}}^{(-)} - \varepsilon_{2,\,\text{EO}}^{(+)}} \\ \varepsilon_{2,\,\text{LCAO}}^{-} &= \varepsilon_{2,\,\text{EO}}^{(-)} + \frac{|c|^2}{\varepsilon_{2,\,\text{EO}}^{(-)} - \varepsilon_{1,\,\text{EO}}^{(+)}} + \frac{|d|^2}{\varepsilon_{2,\,\text{EO}}^{(-)} - \varepsilon_{2,\,\text{EO}}^{(+)}} \end{aligned} \qquad (4.48)$$

where, obviously, $\varepsilon_{1,\text{EO}}^{(-)} > [\varepsilon_{1,\text{EO}}^{(+)}, \varepsilon_{2,\text{EO}}^{(+)}]$ and $\varepsilon_{2,\text{EO}}^{(-)} > [\varepsilon_{1,\text{EO}}^{(+)}, \varepsilon_{2,\text{EO}}^{(+)}]$. It is seen from this that for the entire direction Λ and, in particular, for the point L of the first zone the second-order corrections have a plus sign. As a result the levels $\overline{L}^c$, $\overline{\overline{L}}^c$ can only be raised as one goes from the EO method to the LCAO method.

Thus of all the inequalities (4.39) for the conduction band of diamond and silicon it remained to be proved whether the inequality $\varepsilon(X_1^c) < \varepsilon(\Gamma_{15}^c)$ will be satisfied in the LCAO method. From the secular determinant (3.139) with the form of $\mathcal{H}_{ij}$ taken into consideration we have

$$\varepsilon_{1,\,2,\,\text{LCAO}}^{(\pm)}(X_1) = \frac{1}{2}\left\{\alpha_A^{(s)} - \alpha_A^{(p)}\right\} \pm \sqrt{\frac{1}{4}\left\{\alpha_A^{(s)} - \alpha_A^{(p)}\right\}^2 + \frac{16}{3}\beta_{sp}^2} \qquad (4.49)$$

from which we obtain for the difference $\varepsilon(\Gamma_{15}^c) - \varepsilon(X_1^c)$

$$\varepsilon(\Gamma_{15}^{c})-\varepsilon(X_{1}^{c})=\frac{1}{2}\{\alpha_{A}^{(p)}-\alpha_{A}^{(s)}\}-\frac{4}{3}\beta_{pp}+\frac{8}{3}\beta_{\pi}-\sqrt{\frac{1}{4}\{\alpha_{A}^{(s)}-\alpha_{A}^{(p)}\}^{2}+\frac{16}{3}\beta_{sp}^{2}} \tag{4.50}$$

Using Eq. (4.40), it is not hard to see that the difference $\varepsilon(\Gamma_{15}^c) - \varepsilon(X_1^c)$ will be greater than zero within the entire interval $|\beta_{sp}| < 0.43|\beta_0|$, so that the estimates of the parameters, made on the basis of organic chemistry data and in the LCAO approximation, correctly describe the features of the conduction band in the diamond crystal. A similar conclusion is valid for silicon too. Using the estimate (4.44) for silicon, it is easy to show that the difference $\varepsilon(\Gamma_{15}^c) - \varepsilon(X_1^c) > 0$ for all $|\beta_{sp}| < 0.5\ |\beta_0|$, where β_0 (Si) ≈ -3.1 eV.

Let us mention one more fundamental point here. According to the opinion prevalent until now, the tight binding and EO methods are fundamentally incapable of giving even a qualitatively correct picture of the conduction band since these methods do not use plane waves.* This point of view was partially motivated by the fact that the wave functions of an electron in the conduction band, it would appear, should be quite close to the wave functions for a free electron. At the same time it is not contradicted by known calculations since the correct conduction band structure, until recently, was used only in calculations with the OPW, APW, etc., methods, whereas in calculations with the EO and LCAO methods the structure of the conduction band was either completely ignored [142, 164, 165], or it was incorrect† [54, 166, 167]. At present it is hard to say what causes the poor results obtained in the nonempirical EO and LCAO calculations: the lack of a basis of plane waves among the functions or an incorrect form for the atomic functions and potential used (the satisfactory results obtained in the nonempirical calculations in [52, 55] may favor the latter assumption). However, starting from the experience of molecular quantum chemistry, one can expect that the successive semiempirical approach in both cases will make it possible to compensate the imprecision of the model and to give a correct description of the structure of the energy levels, as is obviously confirmed by the investigation of the band structure of diamond and silicon presented above.

4.4 CHEMICAL BOND AND BAND STRUCTURE IN GERMANIUM AND GRAY TIN CRYSTALS

4.4.1 Nature of Chemical Bond in Ge and Sn Compounds

Diamond and silicon crystals were discussed in the previous sections of this chapter. Let us now examine the relationship between the nature of the bond and the structure of the energy bands for two other covalent crystals—germanium and gray tin [186-188]. As will be explained below, it is impossible in all probability to extend the estimate (4.24) for the resonance integrals in diamond and silicon to germanium and gray tin crystals. Moreover, it is impossible to extract the necessary estimates of the resonance integrals from

*It was usually assumed that the LCAO method in the best case can be used for interpolation of the results obtained by other methods (for example, by the OPW method), as was suggested in the paper by Slater and Koster [185].

†As a typical example of such unsuccessful calculations let us cite the recent nonempirical calculation in [54]. In this calculation a single minimum at the center Γ of the zone was obtained for the conduction band of the diamond crystal and the forbidden band gap was estimated to be 13.5 eV.

molecular data since information on the physical and chemical properties of the saturated compounds Ge_nX_{2n+2} and Sn_nX_{2n+2} is even scarcer than information on the silanes. Therefore we shall limit our discussion to the qualitative aspect of the problem, based on data on the general nature of the chemical properties of germanium and tin.

It was already mentioned that a characteristic feature of Ge and Sn atoms, unlike for C and Si atoms, is the inertness of the valence *ns* electrons (the so-called "inert Sedgwick pair").

In this regard let us note that the inertness of the valence *ns* electrons is a common property, characteristic of the elements, starting with periods IV or III of the periodic table. A typical example illustrating this phenomenon is the decrease in the valence angles in hybrids of the elements of group IV and VI: 107° in NH_3 and 105° in H_2O; 94° and 92° in PH_3 and AsH_3; 92°, 91°, and 90° in H_2S, H_2Se, and H_2Te. Until 1958 this was considered as a proof of the nearly pure *p* character of the A–H bonds in compounds of P, As, S, Se, and Te. Doubts later arose in the justification of this conclusion since it was shown that the NQR data for $H_2^{33}S$ and $HD^{33}S$ can be explained by assigning the S–H bonds 73.5% *p* character and 10.5% *s* character, but, in return, by allowing a 16% impurity of 3*d* orbitals of the S atom [189] (although the interpretation of the NQR data is ambiguous). Nevertheless, it must be pointed out that even if this interpretation is correct, the appearance of the 3*d* orbital impurity as one goes from H_2O to H_2S is accomplished at the expense of reducing the contribution from the *s* and not the *p* orbitals. Thus, in this case the NQR data also attest to the relative inertness of the 3*s* electrons of the sulfur atom. For group IV elements the hybrids AH_4 have a symmetrical tetrahedral configuration, so that the geometrical form of the molecules says nothing about the degree of participation of the *s* orbitals in the bond. When, however, radicals$\cdot AH_3$ of the hybrids of C, Si, Ge, and Sn were investigated by the EPR method [190], it was shown the hybrids of Si, Ge, and Sn have a pyramidal configuration. At the same time the radical$\cdot CH_3$ has a plane trigonal form, which also attests to an increase in the *s* character of the bonds as one goes from heavy group IV elements to light.

From an *a priori* point of view two independent factors can apparently be identified that determine the stated inertness of the *s* orbitals. In part the increase in the inertness of the *ns* electrons with an increase in *n* can be explained by a relative decrease in the resonance integrals in which the valence *ns* orbital enters, compared with the resonance integrals in which the *np* orbitals enter. Such a viewpoint is confirmed, in particular, by a calculation of the corresponding overlap integrals $\langle ns|1s\rangle$, $\langle np|1s\rangle$ (with the Hartree–Fock functions), as done by Merrill and Randic for hybrids of elements of groups V and VI of the periodic table (see [16]). According to the data of Merrill and Randic, the ratio $\langle ns|1s\rangle : \langle np|1s\rangle$ as one goes from NH_3 to PH_3 decreases by 40%; as one goes from H_2O to H_2S, by 65%.

Another obvious factor leading to the "inert pair" phenomenon is the general decrease of all the resonance integrals compared with the energy for promoting an *ns* electron into the valence *np* state. As is clearly seen from Table 3, as one goes from the C atom to the Sn atom the promotion energy $\mathscr{E}_{\text{promot}} = \alpha_A^{(p)} - \alpha_A^{(s)}$ decreases only slightly, whereas the resonance integrals decrease much more;

see Table 4. (The transition from Si to Ge is even accompanied by an increase in $\mathscr{E}_{\text{promot}}$ since the promotion energy is lowest for silicon of all the group IV elements.)

It is hard to say which of these two factors–which supplement each other, of course–plays the major role. However, with regard to the band structure, as we will see, both lead to one result. In this sense one can say that the characteristic change in the type of band structure as one goes from diamond to silicon to germanium and gray tin is the consequence of the "inert pair" phenomenon itself.

4.4.2 Inert Pair and the Change in the Band Structure in the Series of Group IV Elements

Let us first consider the question of the position of the $\Gamma_{2'}^c$ and Γ_{15}^c levels in the conduction band, after writing Eq. (4.2) for this in the form of the sum of two terms:

$$\varepsilon(\Gamma_{15}^c) - \varepsilon(\Gamma_{2'}^c) = \left\{\mathscr{E}_{\text{promot}} + \frac{8}{3}\beta_\pi - \frac{16}{3}\Delta\beta\right\} + \frac{8}{3}\beta_0 \tag{4.51}$$

The first term is always positive, while the second is always negative. Here, as always, β_0 denotes the arithmetic average of the resonance parameters β_{ss} and β_{pp} and $\Delta\beta$ is a quantity that characterizes the deviation of β_{ss} from β_{pp}; $\Delta\beta = (1/2)\{\beta_{pp} - \beta_{ss}\}$ (it is obvious that $\Delta\beta < 0$). It is seen from Eq. (4.51) that an increase in $|\Delta\beta|$ (i.e., a relative decrease in $|\beta_{ss}|$ compared with $|\beta_{pp}|$) or a relative increase in $\mathscr{E}_{\text{promot}}$ compared with β_0 leads to an increase of the entire right side of Eq. (4.51); thus for a sufficiently inert *ns* orbital the level difference $\varepsilon(\Gamma_{15}^c) - \varepsilon(\Gamma_{2'}^c)$ becomes positive. Just such a situation occurs as one goes from silicon to germanium and gray tin, when the value of $|\beta_0|$ decreases approximately the same as $(-\Delta H_{\text{at}})$, whereas the value of $\mathscr{E}_{\text{promot}}$ even increases from 7.2 eV for Si to ≈8 eV for Ge and to 7.5-8 eV for Sn.

The movement of the edge of the conduction band from the point $\Delta_{\min}$, close to X_1^c, to the center of the band $\Gamma_{2'}^c$ is explained in a similar manner. To do this we write the difference $\varepsilon(X_1^c) - \varepsilon(\Gamma_{2'}^c)$ in a form similar to Eq. (4.51):

$$\varepsilon(X_1^c) - \varepsilon(\Gamma_{2'}^c) = \frac{1}{2}\left\{\mathscr{E}_{\text{promot}} - \frac{4}{\sqrt{3}}\beta_{sp} - 8\Delta\beta\right\} + 4\beta_0 \tag{4.52}$$

It is seen from this that with an increase in $|\Delta\beta|$ or with a relative increase in $\mathscr{E}_{\text{promot}}$ the entire sum on the right side of Eq. (4.52) increases. With an increase in the inertness of the *ns* electrons the minimum of the conduction band ultimately is shifted from the point X to the center Γ of the first zone.

4.4.3 Narrowing of Forbidden Band and the Transition from Semiconductors to Semimetals in Group IV

Crystals of group IV elements are an especially convenient object for studying the question, often discussed in the literature, of the factors determining the narrowing of the forbidden band and the appearance of "metallic" properties as the atomic number increases. We shall consider this question here, assuming for simplicity that the absolute minimum of the conduction band lies at the point $\Gamma_{2'}^{c}$. This is true for Ge* and Sn, as well as the hypothetical tetrahedral form of lead [103]. Thus the entire discussion will refer to the "lower" part of group IV of the periodic table, in which the transition from semiconductors to semimetals occurs.

If it is assumed that $E_g = \varepsilon(\Gamma_{2'}^{c}) - \varepsilon(\Gamma_{25'}^{v})$, then we have the formula

$$E_g = \left\{ -\mathscr{E}_{\text{promot}} + \frac{8}{3}\beta_\pi + \frac{8}{3}\Delta\beta \right\} - \frac{16}{3}\beta_0 \tag{4.53}$$

which is completely analogous to Eqs. (4.51) and (4.52) except that the first term in Eq. (4.53) is negative, while in Eqs. (4.51) and (4.52) it is positive. Accordingly, it is not hard to see that with an increase in the inertness of the *ns* orbitals the value of E_g should decrease. As such within the framework of the EO LCAO or the tight binding approximation (they give identical results at the center Γ of the first zone), the cause of the narrowing of the forbidden band is the relative decrease in $|\beta_{ss}|$ and relative increase in $\mathscr{E}_{\text{promot}} = \alpha_A^{(p)} - \alpha_A^{(s)}$.

It is convenient to investigate the effect of the latter of these two factors in the transition process from semiconductors to semimetals separately in a much simplified model† that, nevertheless, correctly provides the basic features of the phenomenon.

Let us examine the simplest of the possible versions of the EO method in which only the interaction of the EOs for the first-neighbor bonds is considered, so that the parameters $\beta_t^{(+)}$, $\beta_t^{(-)}$, $\beta_g^{(+)}$, $\beta_g^{(-)}$, describing the interaction of the second-neighbor bonds, are assumed to be zero. In addition, for the parameters $\alpha_A^{(+)}$ and $\alpha_A^{(-)}$ in Eq. (3.44) we shall take into account only their principal Coulomb components:

$$\beta_A^{(\pm)} = \frac{1}{8}\left\{\alpha_A^{(s)} - \alpha_A^{(p)}\right\} = -\frac{1}{8}\mathscr{E}_{\text{promot}} \tag{4.54}$$

Then Eqs. (3.23)-(3.26) for the dispersion laws assume the form:

*Strictly speaking, for germanium E_g is determined by the indirect transition $\Gamma_{15'}^{v} \to L_1^{c}$. However, the absolute minimum at L for Ge is only 0.18 eV below the minimum at Γ.

†A similar model was examined by Leman and Friedel [191] within the framework of the tight binding method.

$\Delta = [100]$ direction

$$\varepsilon_{1,2}^{(\pm)}(q) = \alpha^{(\pm)} + 2\beta_A^{(\pm)} \pm 4\beta_A^{(\pm)} \cos\frac{q}{2} \tag{4.55}$$

$$\varepsilon_{3,4}^{(\pm)}(q) = \alpha^{(\pm)} - 2\beta_A^{(\pm)} \tag{4.56}$$

$\Lambda = [111]$ direction

$$\varepsilon_{1,2}^{(\pm)}(q) = \alpha^{(\pm)} + 2\beta_A^{(\pm)} \pm 2\beta_A^{(\pm)} \sqrt{1+3\cos q} \tag{4.57}$$

$$\varepsilon_{3,4}^{(\pm)}(q) = \alpha^{(\pm)} - 2\beta_A^{(\pm)} \tag{4.58}$$

On the basis of Eqs. (4.55)-(4.58) it is easy to prove that the following inequalities always are true regardless of the numerical values of the parameters for the conduction band in this model

$$\varepsilon(\Gamma_{2'}^c) = \alpha^{(-)} + 6\beta_A^{(-)} < \varepsilon(\Gamma_{15}^c) = \alpha^{(-)} - 2\beta_A^{(-)} \tag{4.59}$$

$$\varepsilon(\Gamma_{2'}^c) = \alpha^{(-)} + 6\beta_A^{(-)} < \varepsilon(X_1^c) = \alpha^{(-)} + 2\beta_A^{(-)} \tag{4.60}$$

$$\varepsilon(\Gamma_{2'}^c) = \alpha^{(-)} + 6\beta_A^{(-)} < \varepsilon(\bar{\bar{L}}^c) = \alpha^{(-)} \tag{4.61}$$

$$\varepsilon(\Gamma_{2'}^c) = \alpha^{(-)} + 6\beta_A^{(-)} < \varepsilon(\bar{L}) = \alpha^{(-)} + 4\beta_A^{(-)} \tag{4.62}$$

so that such a model can certainly not be applicable for describing diamond and silicon, although it is suitable, in principle, for describing crystals for which the absolute minimum of the conduction band is at the center Γ of the first zone. [It was already mentioned that for describing C and Si it is necessary that the inequality $\beta_g^{(-)} + \beta_t^{(-)} > 0$ be satisfied, which is incompatible with this model.]

In the model being considered we have

$$E_g = \{\alpha^{(-)} - \alpha^{(+)}\} + 6\beta_A^{(-)} + 2\beta_A^{(+)} \tag{4.63}$$

for the forbidden band gap or, taking Eqs. (3.48) and (4.54) into consideration

$$E_g = -\mathcal{E}_{\text{promot}} - 2\beta \tag{4.64}$$

where β is the interaction integral* of the two sp^3 orbitals within one A–A bond.

*By reintroducing the parameter β, we eliminate the necessity of considering the integrals β_{ss}, etc., separately. Let us also note once more that Eqs. (4.63) and (4.64) are valid not only in the EO LCAO method, but also in the tight binding method.

As now seen from Eq. (4.64) the forbidden band gap E_g goes to zero if the following inequality is satisfied:

$$\beta = -\frac{1}{2}\mathscr{E}_{\text{promot}} \tag{4.65}$$

It is interesting that despite the extreme simplicity of the model, the relation (4.65) quite accurately describes the conditions necessary for the transition from a semiconductor to a semimetal such as α-Sn. For the Sn atom the value of $\mathscr{E}_{\text{promot}}$ is equal to ≈8 eV (or, possibly more precisely, ≈7.5 eV). On the other hand, assuming a value of −8.3 eV for the resonance integral β in diamond (see Sec. 3.4.2), we have $\beta(\alpha\text{-Sn}) = \beta(\text{diamond})\,\Delta H_{\text{at}}(\alpha\text{-Sn})/\Delta H_{\text{at}}(\text{diamond}) = -3.5$ eV, which is, in fact, close to the value $(-1/2)\mathscr{E}_{\text{promot}}(\text{Sn}) = -3.75$ eV.

Finally, in connection with Eqs. (4.53), (4.63), and (4.64) let us discuss two points of view with regard to the cause of the decrease of E_g with an increase in atomic number that are quite prevalent in the literature on solid state chemistry.

It is often assumed that the decrease in E_g for elements with large atomic numbers is due to the decrease in the atomic ionization potentials since a decrease in the energy to remove an electron from atoms reduces the work necessary to remove the electron from the entire system. In fact, the ionization potentials of the atoms decrease in the series C-Si-Ge-Sn. However, as seen from Eqs. (4.53) and (4.64), the atomic ionization potentials $I_1 \approx -\alpha_A^{(p)}$ enter into the formula for E_g only in the combination $\alpha_A^{(p)} - \alpha_A^{(s)}$. Therefore a case is possible when $I_1(\text{A})$ increases, whereas E_g decreases because of a corresponding lowering of the s level $\alpha_A^{(s)}$ (or because of an increase in $|\Delta\beta|$). We encounter such a case as we go from α-Sn to Pb. The ionization potential of Pb [$I_1(\text{Pb}) = 7.4$ eV] is higher than the ionization potential of Sn [$I_1(\text{Sn}) = 7.3$ eV]. At the same time (according to the OPW calculations of Herman [103]) in the hypothetical diamond-like modification of Pb the $\Gamma_{2'}^{c}$ level is located about 4 eV lower than in Sn (it is "submerged" by 4 eV in the valence band).

According to another point of view, the value of E_g is related to the "degree of localization of electrons at the bonds," and it is assumed that an increase in the degree of localization leads to an increase in E_g. Such an interpretation, of course, is insufficiently clear since nothing is said about how to measure this "degree of localization." For example, it is clear that it makes no sense to measure the magnitude of the overlap integrals between different equivalent orbitals since (in the one-electron model) all the "true" EOs $\varphi^{(\pm)}$, localized at different A—B bonds, are orthogonal, and in this sense the degree of localization of the electrons at the bonds is equal to zero for all diamond-like crystals. Nevertheless, one can still attempt to impart some physical meaning to such an interpretation by assuming that the degree of localization of the bonds is measured by the magnitude of the interaction integrals of the different EOs and will be smaller, the larger the values of the parameters β_A, β_t, β_g, etc. In such a formulation the latter interpretation is actually confirmed by Eq. (4.63), although the integrals $\beta_A^{(+)}$, $\beta_A^{(-)}$ in this formula have a purely "atomic" origin and, of course, do not characterize the behavior of the bonds in the crystal.

However, the situation becomes more complicated if we switch from the simplified model considered above to a more complete theory that takes account of the interaction of the EOs for second-neighbor bonds. In this case we obtain

$$E_g = \alpha^{(-)} - \alpha^{(+)} + 6\beta_A^{(-)} + 12\beta_g^{(-)} + 6\beta_t^{(-)} + 2\beta_A^{(+)} + 4\beta_g^{(+)} - 6\beta_t^{(+)} \tag{4.66}$$

or, taking Eqs. (3.50) and (3.51) into consideration

$$\begin{aligned} E_g &= \alpha^{(-)} - \alpha^{(+)} + 6\beta_A^{(-)} + 2\beta_A^{(+)} - 8\beta_g^{(+)} - 12\beta_t^{(+)} \\ &= \alpha^{(-)} - \alpha^{(+)} + 6\beta_A^{(-)} + 2\beta_A^{(+)} - 20\beta_g^{(+)} + \frac{8}{3}\beta_\pi \end{aligned} \tag{4.67}$$

It is seen from this that when the bonds of remote neighbors are taken into account, such a decrease in the "localization of the bonds" can lead, on the other hand, to an increase in E_g and not a decrease for appropriate values of the parameters.

4.5 FEATURES OF BAND STRUCTURE OF GERMANIUM AND SOME OTHER QUESTIONS

4.5.1 Features of Band Structure of Germanium

The investigation of the relationship between the band structure and the phenomenon of the "inert pair of *s* electrons," conducted in the previous section, was insufficiently complete since in speaking of the edge of the conduction band we compared only the levels at the center Γ and at the edge X of the first zone. Here we shall also include in the discussion the behavior of the levels at the edge L [187, 188]. Then, as shown below, the quite high inertness of the valence *ns* electrons actually results in the fact that the absolute minimum of the conduction band is localized at Γ, whereas for a sufficiently "active" *ns* orbital the edge of the conduction band lies at X. It is significant, however, that in the intermediate case the edge of the conduction band should be localized not at the edge X and not at the center Γ, but at the edge L of the first zone. This is a unique, distinguishing feature of the band structure of germanium. It then follows that from the viewpoint of chemical bond theory the type of band structure inherent in germanium appears in the group IV series of elements not by accident, but because it is "halfway" between the band structure of carbon and silicon and the band structure of gray tin [187, 188].

Let us consider again the two independent factors that determine the degree of inertness of the *s* orbital (see Sec. 4.4.1), and we shall characterize the effect of the first factor by the magnitude of the ratio $\Delta\beta/\beta_0$ and the effect of the second factor by the magnitude of the ratio $\mathscr{E}_{promot}/\beta_0$. In investigating the dependence of the band structure of a crystal with a diamond lattice on a continuous change in the parameters $\Delta\beta/\beta_0$ and $\mathscr{E}_{promot}/|\beta_0|$, we shall first study the effect of the first factor. As seen from Fig. 3.2 the distinguishing properties of the band structure of germanium consist of the fact that of the levels Γ^c_{15}, $\Gamma^c_{2'}$, X^c_1, $\overline{L}^c$, and $\overline{\overline{L}}^c$ the lowest is the $\overline{L}^c$ level; the $\Gamma^c_{2'}$ level lies below the Γ^c_{15} level. Thus for germanium the following four inequalities exist:

$$\varepsilon(X^c_1) - \varepsilon(\overline{L}^c) = \frac{1}{4}\mathscr{E}_{promot} + \frac{4}{3}\beta_0 - \frac{5}{3}\Delta\beta - \frac{1}{\sqrt{3}}\beta_{sp} - \frac{2}{3}\beta_\pi > 0 \quad (4.68)$$

$$\varepsilon(\Gamma^c_{2'}) - \varepsilon(\overline{L}^c) = -\frac{1}{4}\mathscr{E}_{promot} - \frac{8}{3}\beta_0 + \frac{7}{3}\Delta\beta + \sqrt{3}\,\beta_{sp} - \frac{2}{3}\beta_\pi > 0 \quad (4.69)$$

$$\varepsilon(\Gamma^c_{15}) - \varepsilon(\Gamma^c_{2'}) = \mathscr{E}_{promot} + \frac{8}{3}\beta_0 - \frac{16}{3}\Delta\beta + \frac{8}{3}\beta_\pi > 0 \quad (4.70)$$

$$\varepsilon(\overline{\overline{L}}^c) - \varepsilon(\overline{L}^c) = \frac{1}{2}\mathscr{E}_{promot} + \frac{4}{3}\beta_0 - \frac{8}{3}\Delta\beta - \frac{8}{3}\beta_\pi > 0 \quad (4.71)$$

which, obviously, are equivalent to the following four inequalities for the ratio $\Delta\beta/\beta_0$ (as before, κ denotes the ratio β_{sp}/β_0):

$$\left(\frac{\Delta\beta}{\beta_0}\right) > \frac{3}{20}\left(\frac{\mathscr{E}_{promot}}{\beta_0}\right) - \frac{\sqrt{3}}{5}\varkappa - \frac{2}{5}\left(\frac{\beta_\pi}{\beta_0}\right) + \frac{4}{5} \quad (4.72)$$

$$\left(\frac{\Delta\beta}{\beta_0}\right) < \frac{3}{28}\left(\frac{\mathscr{E}_{promot}}{\beta_0}\right) - \frac{3\sqrt{3}}{7}\varkappa + \frac{2}{7}\left(\frac{\beta_\pi}{\beta_0}\right) + \frac{8}{7} \quad (4.73)$$

$$\left(\frac{\Delta\beta}{\beta_0}\right) > \frac{3}{16}\left(\frac{\mathscr{E}_{promot}}{\beta_0}\right) + \frac{1}{2}\left(\frac{\beta_\pi}{\beta_0}\right) + \frac{1}{2} \quad (4.74)$$

$$\left(\frac{\Delta\beta}{\beta_0}\right) > \frac{3}{16}\left(\frac{\mathscr{E}_{promot}}{\beta_0}\right) - \left(\frac{\beta_\pi}{\beta_0}\right) + \frac{1}{2} \quad (4.75)$$

It is not hard to see that for reasonable values of the quantities κ, β_π/β_0 and $\mathscr{E}_{promot}/\beta_0$ the inequalities (4.72)-(4.75) will be consistent. Consequently, an interval $\{(\Delta\beta/\beta_0)_{min}, (\Delta\beta/\beta_0)_{max}\}$ exists such that for $(\Delta\beta/\beta_0)_{min} < (\Delta\beta/\beta_0) < (\Delta\beta/\beta_0)_{max}$ all four inequalities are satisfied simultaneously.*

*Let us assume, for example, that $(\Delta\beta/\beta_0) \approx 1/3$; this relationship is satisfied for all diamond-type crystals. Moreover, let us assume the quantity $|\mathscr{E}_{promot}/\beta_0|$ takes on any of the values that it can generally assume for a covalent crystal–from 1.7 (diamond) to 3.4 (germanium). Finally, let us assume the quantity $\kappa = \beta_{sp}/\beta_0$ also takes on any values in the interval from 0 to 1. Then the allowable intervals for the quantity $(\Delta\beta/\beta_0)$ will be the following: for $|\mathscr{E}_{promot}/\beta_0| = 1.7$ and $\kappa = 0$, it is $0.4 < (\Delta\beta/\beta_0) < 1.1$; for $|\mathscr{E}_{promot}/\beta_0| = 1.7$ and $\kappa = 0.5$, it is $0.3 < (\Delta\beta/\beta_0) < 0.9$; and for $|\mathscr{E}_{promot}/\beta_0| = 1.7$ and $\kappa = 1$, it is $0.3 < (\Delta\beta/\beta_0) < 0.7$. Similarly, for $|\mathscr{E}_{promot}/\beta_0| = 3.4$ for these same κ values we obtain, respectively, $0.2 < (\Delta\beta/\beta_0) < 0.8$; $0 < (\Delta\beta/\beta_0) < 0.7$ and $0 < (\Delta\beta/\beta_0) < 0.5$.

To investigate the dependence of the band structure on the ratio $(\mathscr{E}_{\text{promot}}/\beta_0)$ we rewrite the inequalities (4.72)–(4.75) in the equivalent form:

$$\left(\frac{\mathscr{E}_{\text{promot}}}{\beta_0}\right) < \frac{4}{\sqrt{3}}\varkappa + \frac{20}{3}\left(\frac{\Delta\beta}{\beta_0}\right) + \frac{8}{3}\left(\frac{\beta_\pi}{\beta_0}\right) - \frac{4}{3} \tag{4.76}$$

$$\left(\frac{\mathscr{E}_{\text{promot}}}{\beta_0}\right) > 4\sqrt{3}\,\varkappa + \frac{28}{3}\left(\frac{\Delta\beta}{\beta_0}\right) - \frac{2}{3}\left(\frac{\beta_\pi}{\beta_0}\right) - \frac{8}{3} \tag{4.77}$$

$$\left(\frac{\mathscr{E}_{\text{promot}}}{\beta_0}\right) < \frac{16}{3}\left(\frac{\Delta\beta}{\beta_0}\right) - \frac{8}{3}\left(\frac{\beta_\pi}{\beta_0}\right) - \frac{8}{3} \tag{4.78}$$

$$\left(\frac{\mathscr{E}_{\text{promot}}}{\beta_0}\right) < \frac{16}{3}\left(\frac{\Delta\beta}{\beta_0}\right) + \frac{16}{3}\left(\frac{\beta_\pi}{\beta_0}\right) - \frac{8}{3} \tag{4.79}$$

It is easy to see that in this case too for all the acceptable values of κ, $\Delta\beta/\beta_0$, and β_π/β_0 there exists an interval of those values of the quantity $(\mathscr{E}_{\text{promot}}/\beta_0)$ that satisfy the inequalities (4.76)–(4.79). Thus for $(\Delta\beta/\beta_0) \approx 0$ (diamond) and for $(\beta_\pi/\beta_0) \approx 1/3$ we have $-11.6 < \mathscr{E}_{\text{promot}}/\beta_0 < -4.4$ for $\kappa = 0$; $-8.1 < \mathscr{E}_{\text{promot}}/\beta_0 < -3.6$ for $\kappa = 0.5$; and $-4.7 < \mathscr{E}_{\text{promot}}/\beta_0 < -3.6$ for $\kappa = 1$.

Let us now change the value of $\Delta\beta/\beta_0$ or the value of $\mathscr{E}_{\text{promot}}/\beta_0$ in such a manner that either the value of $\Delta\beta/\beta_0$ exceeded the upper limit $(\Delta\beta/\beta_0)_{\max}$, or the value of $(\mathscr{E}_{\text{promot}}/\beta_0)$ was less than the lower limit $(\mathscr{E}_{\text{promot}}/\beta_0)_{\min}$. [Both limits are obviously determined by the second inequalities in the systems (4.72)–(4.75) and (4.76)–(4.79).] Then the sign of the rest of the inequalities in (4.72)–(4.75) and (4.76)–(4.79) remains unchanged, whereas the sign of the inequality (4.73) or the inequality (4.77) is reversed. Such a system of inequalities characterizes the band structure with an absolute minimum in the conduction band at the point $\Gamma_{2'}^{c}$, so that for a sufficiently large value of any of the quantities $(\Delta\beta/\beta_0)$ or $|(\mathscr{E}_{\text{promot}}/\beta_0)|$ the band structure that is typical of germanium does indeed become the band structure of gray tin. On the other hand, as already noted, for $(\Delta\beta/\beta_0) \approx 0$, for $\kappa = 0.5$ and for values of $|(\mathscr{E}_{\text{promot}}/\beta_0)| = 1.7$–$2.2$ the inequalities (4.40) exist. In short, for small values of the parameters $(\Delta\beta/\beta_0)$ and $|(\mathscr{E}_{\text{promot}}/\beta_0)|$ the germanium-type band structure becomes the structure characteristic of diamond and silicon.

4.5.2 Pressure Dependence of Band Structure and Band-to-Band Transitions

The relationship between the inertness of the valence *ns* electrons and the band structure considered above makes it possible to understand, at least in a qualitative form, the unique behavior of the band structure of covalent crystals in the presence of a hydrostatic pressure [192] (let us point out that the error permitted in this paper has been corrected here).

Let us assume that the potential V of the crystal is the sum of the atomic potentials. We shall assume that an increase in pressure leads only to a decrease in the interatomic distances R, but has no effect on the potential within an individual atom.* In the language of the tight binding method or the EO LCAO method this means that the hydrostatic pressure has no effect on the Coulomb integrals $\alpha_A^{(p)}$ and $\alpha_A^{(s)}$; it only affects the resonance integrals $\beta_{ss}, \ldots$, which change because of the decrease in R. The form of the dependence of the resonance integrals on R can actually be complicated. It is interesting, however, that many of the characteristic features in the behavior of the band structure can be understood even on the basis of the simplest assumption. Let us assume that the form of the dependence of β_{ij} on R for comparatively small changes in R is approximately the same as within the entire interval from the equilibrium value of R to zero; in other words, the average value of the derivative $\partial\beta_{ij}/\partial R$ for small ΔR is close to the average value within this entire interval. It is not hard to perceive,† then, that with a decrease in R the integrals β_{ss} and β_π will increase in absolute magnitude and will approach the Coulomb integrals $\alpha_A^{(s)}$ and $\alpha_A^{(p)}$. Similarly, the integral β_{sp} will first decrease in absolute magnitude and then, passing through zero, change sign and approach $-\alpha_A^{(p)}$:

$$\lim_{R\to 0} \beta_{ss} = \alpha_A^{(s)}, \quad \lim_{R\to 0} \beta_{pp} = -\alpha_A^{(p)}, \quad \lim_{R\to 0} \beta_\pi = \alpha_A^{(p)} \tag{4.80}$$

At the same time the integral β_{sp} will approach zero in view of the orthogonality of the atomic s and p functions:

$$\lim_{R\to 0} \beta_{sp} = 0 \tag{4.81}$$

Let us now examine how the position of the levels is changed with such a change in the resonance integrals. This approach resembles to some extent the use of correlation diagrams, relating the molecular levels to the levels of a "collective atom," for diatomic molecules (see, for example, [178]).

It follows from the relations (4.80) and (4.81) that as R decreases, the integral β_{ss} increases in absolute magnitude, whereas the integral β_{pp} for a change (small) in R decreases in absolute magnitude. In other words, the inertness of the ns orbital becomes less with an increase in pressure, which, in particular, should lead to a deepening of the minimum in the conduction band at X and for germanium should stimulate the transition to the diamond- and silicon-type band structure, while for α-Sn the transition is stimulated to the germanium-type band structure.

*Such a model of "rigid" atoms is usually used within the framework of the pseudopotential method for examining a similar problem; see, for example, the paper by Brust and Liu [193] as well as the review [105].

†This is clearly evident, for example, from the Wolfsberg–Helmholz approximation (see Sec. 3.3.2). Since $\beta_{ij} \approx S_{ij}[\alpha^{(i)} + \alpha^{(j)}]$, then the behavior of β_{ij} is determined by the behavior of S_{ij}, where as $R \to 0$, $S_{ij} \to \delta_{ij}$, except for the value of S_{pp}, which goes to -1. Let us assume here that the constant K in the Wolfsberg–Helmholz method is equal to one for infinite systems (see footnote on page 100); however, the arguments that follow remain valid even if it is assumed K has another value.

Actually, let us first consider the position of the X_1^c level and the $\bar{L}^c$ level (the lower levels in L) with respect to the $\Gamma_{2'}^c$ level. As seen from Eq. (4.52) the value of the difference $\varepsilon(X_1^c) - \varepsilon(\Gamma_{2'}^c)$ is determined by the behavior of the resonance term $4\{\beta_{ss} - (1/\sqrt{3})\beta_{sp}\}$. According to Eqs. (4.80) and (4.81) the value of $|\beta_{ss}|$ increases and $|\beta_{sp}|$ decreases with a decrease in R. Therefore one can expect that with an increase in pressure the difference $\varepsilon(X_1^c) - \varepsilon(\Gamma_{2'}^c)$ decreases and the X_1^c level drops with respect to the $\Gamma_{2'}^c$ level. The level $\bar{L}^c$ behaves in the same manner. In fact, the difference $\varepsilon(\bar{L}^c) - \varepsilon(\Gamma_{2'}^c)$ is

$$\varepsilon(\bar{L}^c) - \varepsilon(\Gamma_{2'}^c) = \frac{1}{4}\mathscr{E}_{\text{promot}} + \frac{5}{2}\beta_{ss} + \frac{1}{6}\beta_{pp} - \sqrt{3}\,\beta_{sp} - \frac{2}{3}\beta_{\pi} \qquad (4.82)$$

Here the positive term $-\sqrt{3}\beta_{sp}$ decreases with a decrease in R, while the negative term $(5/2)\beta_{ss} + (1/6)\beta_{pp} + (2/3)\beta_\pi$ increases the absolute magnitude since the value of β_{ss} and β_π increase in absolute magnitude. (The coefficient for β_{pp} is 15 times smaller than the coefficient for β_{ss} and a decrease in $|\beta_{pp}|$ has no effect on the nature of the change in the sum.)

In a similar manner one can attempt to predict the form of the absolute shift for typical conduction band levels with respect to the top of the valence band $\Gamma_{25'}^v$. According to Eq. (4.42) the expression for $\varepsilon(\Gamma_{2'}^c) - \varepsilon(\Gamma_{25'}^v)$ contains the resonance term $\{-4(\beta_{ss} + 1/3\beta_{pp}) + 8/3\beta_\pi\}$, where the first term $\{-4(\beta_{ss} + 1/3\beta_{pp})\}$ increases while the seond $(8/3\beta_\pi)$ decreases with pressure. It is easy to see, however, from numerical estimates and Eqs. (4.80) and (4.81) that the absolute increment of the first term is much greater than the increment of the second. Therefore on the whole the level $\Gamma_{2'}^c$ should rise with respect to the $\Gamma_{25'}^v$ level with an increase in pressure.

Conversely, the difference in the levels X_1^c and $\Gamma_{25'}^v$ is determined by the quantity $\{-(4/3)\beta_{pp} - (4\sqrt{3}/3)\beta_{sp} + (8/3)\beta_\pi\}$ [see Eq. (4.43)]. Here the first two resonance integrals decrease in absolute magnitude with a decrease in R, while the third integral increases in absolute magnitude. In summary, all three terms inside the braces change in the same direction, so that the expression on the whole increases in absolute magnitude while remaining negative. It is seen from Eq. (4.43) that in this case the X_1^c level approaches the top of the valence band $\Gamma_{25'}^v$.

These conclusions agree amazingly well with experiment.

The effect of pressure on the optical properties of the group IV elements (as well as the similar $A^{III}B^{V}$ compounds) has been investigated in detail [194–200]. As was shown, the minimums of the conduction band at X and L actually drop with respect to the minimum at Γ when a hydrostatic pressure is applied, and the $\Gamma_{2'}^c$ level (or Γ_1^c) rises, while the X_1^c level drops with respect to the maximum of the valence band $\Gamma_{25'}^v$ (or Γ_{15}^v).*

*Of the group IV elements the behavior of the $\Gamma_{2'}^c$ level was investigated for Ge [195, 196], and the behavior of the X_1^c (more precisely $\Delta_{\min}^c$) level was investigated for Ge and Si [197, 198]. With regard to the group III–V compounds see [199] as well as [194].

It is not hard to see that for crystals in which the absolute minimum of the conduction band lies at the center Γ of the zone such behavior of the levels with an increase in pressure should lead to a nonmonotonic change in the forbidden band gap E_g. In this case E_g should first increase because of the raising of the level at Γ and then decrease because of the lowering of the level at X with a simultaneous shifting of the absolute minimum away from the center of the first band toward its edge X. Such a phenomenon was actually observed for Ge crystals [197] (as well as for InP, AlSb, and GaSb [200]) at pressures of the order of 50 kbars. Finally, let us note that the integrals $\beta_{ss}, \ldots$ and the Coulomb integrals $\alpha_A^{(p)}$, $\alpha_A^{(s)}$ are quite close for the elements of groups III, IV, and V and periods III, IV, and V of the periodic table. The same is true for the interatomic distances in A^{IV} and $A^{III}B^{V}$ crystals and the corresponding compressibilities. Therefore the coefficient of the change of the energy with an increase in pressure for identical transitions should be approximately the same for different substances. This agrees with test data.

4.5.3 More about Factors Determining the Inertness of the *ns* Electrons in Heavy Atoms

To conclude the chapter we shall give one more example illustrating the application of band structure data in general and inorganic chemistry.

The "inert Sedgwick pair"—a phenomenon that detemines to a considerable extent the chemistry of elements with a large atomic number—has been mentioned a number of times above. Two possible causes of this phenomenon were indicated: the large value of the ratio $\mathscr{E}_{\text{promot}}$ (exchange integrals) and the relative smallness of the resonance integrals, in which the *ns* orbital takes part, compared with the resonance integrals in which the *ns* orbitals take part. Both of these viewpoints are discussed in the literature; however, on the basis of traditional "molecular" data from inorganic chemistry it is quite hard to establish the relative role of these two factors and to ascertain whether they exist in reality.*

Conversely, the structure of the bands in homopolar crystals provides a wealth of information for finding all the necessary parameters, especially since the spectroscopy of crystals has been developed to a much higher degree than uv spectroscopy for most molecular systems.

To find the resonance parameters we use the data for homopolar crystals.† This can be done, for example, by the same method as for diamond (see Sec.

*In particular, this is difficult to do because of the heteroatomicity of molecular systems containing more than two atoms. The partial ionicity of such molecules introduces an addition indeterminancy, so that even when appropriate test data and well-standardized Coulomb integrals are available, it is still quite an ambiguous procedure to find the resonance integrals.

†The widespread interpretation of the optical spectra for Si, Ge, and α-Sn [146, 147] was recently critiqued [83-85, 103], and the assignment of the $\Gamma_{25'}^{v} \rightarrow \Gamma_{15}^{c}$ transitions (necessary for finding the integrals β_{pp}) is now ambiguous. Here, to be specific, we shall adhere to the interpretation of Herman, whose EC OPW data are used in Table 10.

TABLE 10. Values of Resonance Integrals (in eV) for Group IV Elements Obtained from Band Structure of Crystals[a]

Crystal	Parameter			
	$-\beta_{ss}$	$-\beta_{pp}$	$-\beta_{sp}$	$-\beta_{\pi}$
Diamond	4.3	5.9	2.5	1.6
Si	2.3	2.3	1.5	0.6
Ge	1.9	2.5	1.8	0.7
α-Sn	1.6	2.1	1.6	0.5

[a] According to the EC OPW data of Herman.

4.3.3). In this way we obtain the data shown in Table 10 and Fig. 4.2, which clearly illustrate the role of the different factors in the process of increasing the inertness of the *ns* orbital for group IV elements. As seen from Table 10 and Fig. 4.2, as one goes from Si to Ge and Sn, both a relative decrease in β_{ss} as well as a relative increase in $\mathcal{E}_{\text{promot}}$ occur. At the same time the ratio $\Delta\beta/\beta_0$ is smaller for Si than for C.

This fact can explain why Si, like C, has a valence of four in stable compounds with hydrogen and the halogens, even though the value of the ratio $\mathcal{E}_{\text{promot}}/\beta_0$ for Si is closer to Ge than to C.

Let us also note that the value of $\mathcal{E}_{\text{promot}}$, found from spectroscopic data for the atoms [13] and also from the band structure using the formula

$$\mathcal{E}_{\text{promot}}=\frac{1}{2}\{\varepsilon(\Gamma^c_{15})-\varepsilon(\Gamma^c_{2'})+E_v\} \tag{4.83}$$

were used in Fig. 4.2.

It is seen from Fig. 4.3 that these $\mathcal{E}_{\text{promot}}$ values correlate with the values obtained on the basis of atomic spectroscopy data. It is significant, however,

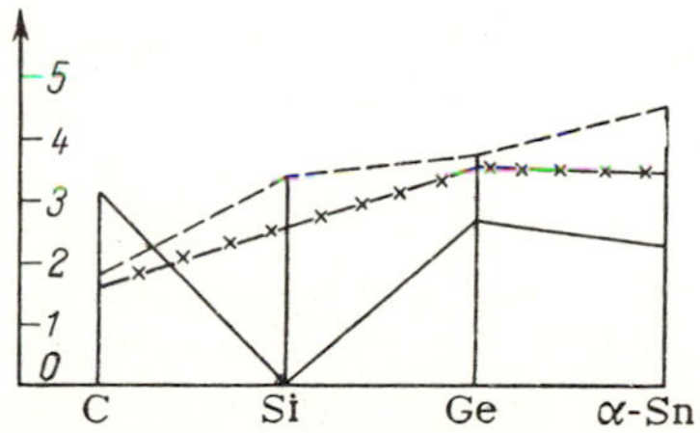

FIG. 4.2. Variation of the values of $\Delta\beta/\beta_0$ and $\mathcal{E}_{\text{promot}}/|\beta_0|$ in the group IV series of elements. Solid lines, values of $10\Delta\beta/\beta_0$; dashed lines, values of $\mathcal{E}_{\text{promot}}/|\beta_0|$, where $\mathcal{E}_{\text{promot}}$ is found from the atomic spectra; dashed line with cross, values of $\mathcal{E}_{\text{promot}}/|\beta_0|$, where $\mathcal{E}_{\text{promot}}$ is found from the band structure.

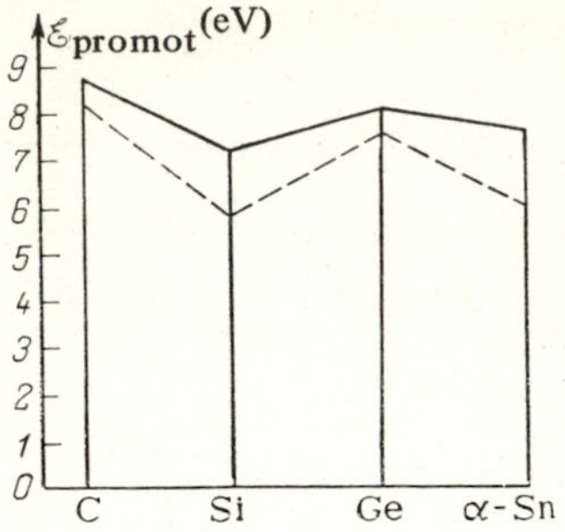

FIG. 4.3. Comparison of the promotion energies from atomic spectra and band structure. Solid lines, values found from atomic spectra; dashed lines, values found from the band structure of diamond according to EC OPW data [85] and also from the band structure of Si, Ge, and α-Sn according to relativistic OPW calculations [103].

that these values give a difference in the *s* and *p* levels for an atom in the sp^3 hybridization state, so that the data from the electron spectra of crystals represent the only possibility of its kind for finding the empirical values of the atomic levels for atoms in the corresponding valence state. It can be assumed that such a possibility, together with the possibility of an empirical estimate of the resonance parameters for the C–C, Si–Si, etc., bonds, will make it possible to advance the poorly developed theory of the uv spectra of the alkanes and their analogs A_nH_{2n+2} to the rest of the group IV elements.

5

Chemical Bond and Structure of Energy Bands in Partially Covalent Crystals with Sphalerite Lattice. Equivalent Orbital Methods and Valence Band

In this and the next chapters we shall turn from purely covalent crystals to the very similar class of covalent crystals—"tetrahedral" $A^N B^{8-N}$ crystals with a zinc sulfide (sphalerite) lattice. One of our main goals will be to investigate those qualitative characteristics that the partial ionicity of the crystal imparts to the energy band structure. Here, as before (and for the same reasons), we shall first consider the valence band, the structure of which reflects the general properties of all tetrahedral partially ionic crystals.

5.1 GENERAL DESCRIPTION OF $A^N B^{8-N}$ CRYSTALS

The tetrahedral $A^N B^{8-N}$ crystals with the sphalerite structure (see Fig. 3.1) include compounds of the various group IV elements (such as one of the modifications of SiC or the alloy SiGe) and compounds of the elements of groups III and V of the periodic table—the cubic modification of BN (borazone, an extremely hard substitute for diamond), BP, BAs, AlP, AlAs, AlSb, GaP, GaAs, GaSb, InP, InAs, and InSb. In addition, this group of crystals also includes compounds of elements of groups II and VI—BeS, BeSe, BeTe, ZnS, ZnSe, ZnTe, CdS, CdSe, CdTe, HgS, HgSe, HgTe—as well as compounds of groups I and VII: CuF, CuCl, CuBr, CuI, and AgI. Of all the compounds named, the "classical"

$A^{III}B^{V}$ semiconductors* and the $A^{II}B^{VI}$ compounds, as well as the compounds of zinc and cadmium, are obviously of special importance from a practical point of view.

Similar to diamond-type crystals, sphalerite-type crystals can be considered as a combination of two face-centered cubic sublattices, shifted with respect to each other. One of these sublattices contains the A^N atoms, while the other contains the B^{8-N} atoms; in short, two atoms are found in the unit cell of the crystal—one A^N atom and one B^{8-N} atom.

In view of the heteroatomicity the $A^N B^{8-N}$ crystals are partially ionic. In this situation, as experimental data show, the charges Z on the atoms in the $A^{III}B^{V}$ compounds are close to 0.5, while in the $A^{II}B^{VI}$ and $A^{I}B^{VII}$ compounds it is close to 0.7-1.0. The A^N atoms are positively charged and the B^{8-N} atoms negatively charged (see, for example, the reviews [205] and the bibliographies in them).

During crystal formation the A^N atom contributes N valence electrons to the common use and the B^{8-N} atom contributes $(8 - N)$ electrons. Therefore, starting from the "origin" of the electrons taking part in the chemical bond, the bond in $A^N B^{8-N}$ crystals for $N < 4$ is a donor-acceptor bond, where the atom A^N is the acceptor and the B^{8-N} atom is the donor.† Let us point out that in view of the donor-acceptor nature of the bond in $A^N B^{8-N}$ crystals for $N < 4$ the covalency of the crystal is not equivalent to the effective charges becoming zero. In fact, as follows from a comparison of Eqs. (2.57) and (2.58), for homoatomic purely covalent A^{IV} crystals the covalence parameter is $\lambda = 1$. Therefore it is also appropriate to consider tetrahedral $A^N B^{8-N}$ crystals as covalent in the limiting case when $\lambda = 1$, that is when there are four valence electrons on each atom. In this case, however, each A^N atom must accept $(4 - N)$ electrons, and each B^{8-N} atom must provide them. Then the charge distribution will be given by the formula $A^{(4-N)-}B^{(4-N)+}$. This, of course, can also be seen from the relation (2.63) if we set $\lambda = 1$ in it.

For a more detailed description of the electron structure of an $A^N B^{8-N}$ crystal let us note that the configuration of valence electrons for the ground state of the A^{III} atom has the form ns^2np, while the ground state of the B^V atom it is $n's^2n'p^3$; the valence AOs then are both s and p orbitals of both kinds of atoms. In a similar manner the valence configurations of B^{VI} and B^{VII} atoms have, respectively, the form $n's^2np^4$ and $n's^2n'p^5$; here again the valence orbitals will be the s and p orbitals of the B atoms. For the A^{II} and A^{I} atoms, however, the valence configurations in the ground state will be ns^2 and ns. Nevertheless,

*See [201] for a detailed review of the properties of these semiconductors.

†This, of course, does not mean that the B^{8-N} atoms must be charged positively, and the A^N negatively. Such a charge distribution must follow from the donor-acceptor nature of the bond only within the framework of the valence bond method. In the one-electron model the sign of the charge on the atom is not determined by its acceptor or donor nature, except in the special case when this atom forms only one bond.

the tetrahedral coordination for these atoms definitely indicates that their excited *np* AOs also take in the bond. Thus, in all $A^N B^{8-N}$ crystals the bond is formed by both the *s* and *p* orbitals of the B^{8-N} atoms and by the analogous orbitals* of the A atoms.

On the whole crystals of the compounds $A^N B^{8-N}$ are very similar to crystals of the group IV elements, and this similarity is especially marked for isoelectron crystals,† i.e., those in which an identical number of electrons is found in the formular unit AB or A_2^{IV}.

The band structure of the sphalerite $A^N B^{8-N}$ crystals is shown in Figs. 5.1 and 5.2. It also resembles the band structure of the group IV elements (see Fig. 3.2), so that all that has been said in Sec. 3.1 applies, almost without changes, to the $A^N B^{8-N}$ crystals too. The only difference here is caused by the

*In all probability the *d* orbitals of the metal also take part in the bond for the Cu and Ag halides. The band structure of these crystals is somewhat more complicated in nature and we shall not discuss it in detail here.

†For example, the interatomic distances in diamond (1.542 Å), Si (2.35 Å), Ge (2.45 Å), and α-Sn (2.80 Å) are nearly exactly identical to the interatomic distances in their isoelectron analogs: BN, 1.56 Å [202]; AlP, 2.35 Å; GaAs, 2.43 Å; and InSb, 2.80 Å [120].

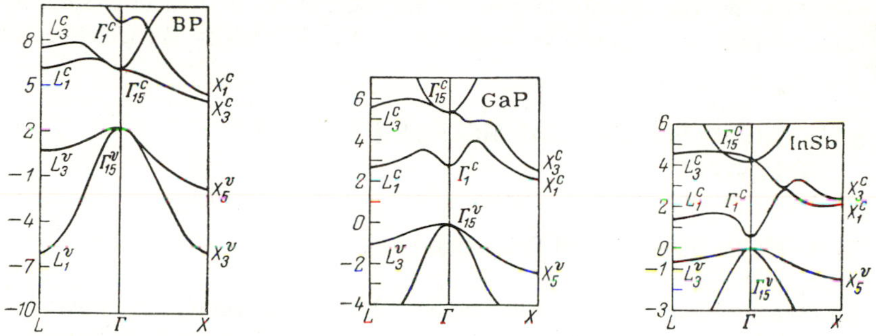

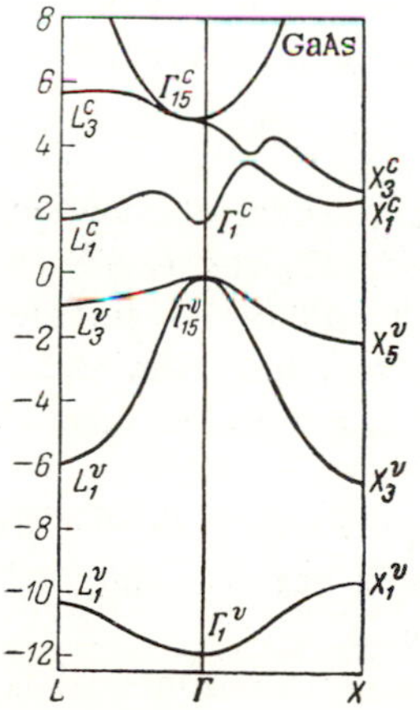

FIG. 5.1. Typical examples of band structure of tetrahedral $A^N B^{8-N}$ crystals. Vertical series of $A^{III}B^{V}$ compounds: BP, GaP, GaAs, InSb.

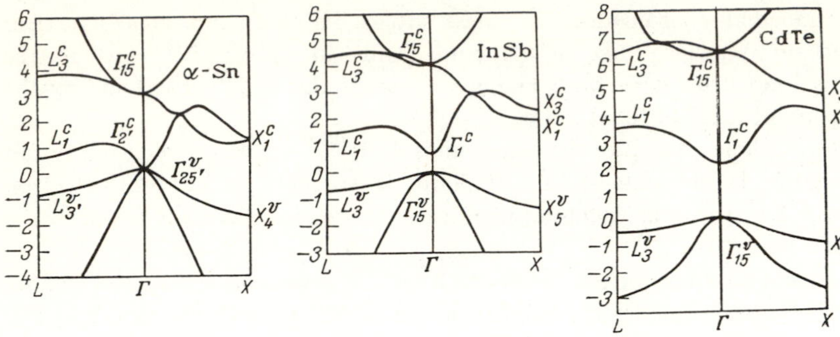

FIG. 5.2. Typical examples of band structure of tetrahedral $A^N B^{8-N}$ crystals. Isoelectron series: α-Sn, InSb, CdTe.

heteroatomicity of these crystals, which leads to the disappearance of the center of symmetry in the middle of each A–B bond. For this reason the sequence of levels has a different notation, as is evident from a comparison of Figs. 5.1 and 3.2, since these levels now correspond to other space group representations. However, the multiplicity of the levels is basically not changed with the exception of the double levels X_1^v and X_1^c, each of which is split in the heteroatomic crystal into two single levels–X_1^v, X_3^v and X_1^c, X_3^c (see Sec. 6.5.4).

5.2 EQUIVALENT ORBITAL METHOD FOR $A^N B^{8-N}$ CRYSTALS

5.2.1 Matrix Elements and Dispersion Laws

As already indicated previously (see Sec. 2.2.1), the unitary equivalence of the Bloch functions and the two-center EOs, localized along the A–B bonds, is true for any partially ionic crystals $A^N B^{8-N}$ with a tetrahedral arrangement of the bonds $>\!\mathrm{A}\!<$, $>\!\mathrm{B}\!<$. The equivalent orbital method, which was discussed especially for covalent crystals in such a manner that the homoatomicity of the latter was used only to find the matrix elements (3.22) and (3.48)–(3.51), (3.62)–(3.65), can also be applied to these crystals. To do this it is sufficient to repeat, nearly word for word, all the arguments presented in Secs. 3.2.1 and 3.2.2. In the final analysis this leads to the secular equations (3.9) and (3.10) that define the dispersion laws.

The only difference from the homopolar case here will be in the fact that in heteroatomic crystals the sublattice of B atoms is actually and not just arbitrarily differentiated from the sublattice of A atoms.

Therefore, for crystals with a sphalerite lattice instead of one integral β_A it is necessary to introduce two resonance integrals, β_A and β_B, that correspond to the interaction of adjacent EOs having a common A or B atom (the number of resonance integrals for the second-neighbor EOs, as seen from Fig. 3.3, does not change here).

Thus instead of the matrix elements (3.22) we now obtain the more general expressions [164, 165]

$$
\begin{aligned}
H_{11}^{(\pm)} &= \alpha^{(\pm)} + 2\beta_t^{(\pm)} \{\cos (q_x + q_y) + \cos (q_x + q_z) + \cos (q_y + q_z)\} \\
H_{22}^{(\pm)} &= \alpha^{(\pm)} + 2\beta_t^{(\pm)} \{\cos (q_x + q_y) + \cos (q_x - q_z) + \cos (q_y - q_z)\} \\
H_{33}^{(\pm)} &= \alpha^{(\pm)} + 2\beta_t^{(\pm)} \{\cos (q_x + q_z) + \cos (q_x - q_y) + \cos (q_z - q_y)\} \\
H_{44}^{(\pm)} &= \alpha^{(\pm)} + 2\beta_t^{(\pm)} \{\cos (q_y + q_z) + \cos (q_y - q_z) + \cos (q_x - q_z)\} \\
H_{12}^{(\pm)} &= \beta_{\mathrm{B}}^{(\pm)} + \beta_{\mathrm{A}}^{(\pm)} e^{i(q_x+q_y)} + 2\beta_g^{(\pm)} \left(e^{iq_x} + e^{iq_y}\right) \cos q_z \\
H_{13}^{(\pm)} &= \beta_{\mathrm{B}}^{(\pm)} + \beta_{\mathrm{A}}^{(\pm)} e^{i(q_x+q_z)} + 2\beta_g^{(\pm)} \left(e^{iq_x} + e^{iq_z}\right) \cos q_y \\
H_{14}^{(\pm)} &= \beta_{\mathrm{B}}^{(\pm)} + \beta_{\mathrm{A}}^{(\pm)} e^{i(q_y+q_z)} + 2\beta_g^{(\pm)} \left(e^{iq_y} + e^{iq_z}\right) \cos q_x \\
H_{23}^{(\pm)} &= \beta_{\mathrm{B}}^{(\pm)} + \beta_{\mathrm{A}}^{(\pm)} e^{i(q_z-q_y)} + 2\beta_g^{(\pm)} \left(e^{iq_z} + e^{-iq_y}\right) \cos q_x \\
H_{24}^{(\pm)} &= \beta_{\mathrm{B}}^{(\pm)} + \beta_{\mathrm{A}}^{(\pm)} e^{i(q_z-q_x)} + 2\beta_g^{(\pm)} \left(e^{-iq_x} + e^{iq_z}\right) \cos q_y \\
H_{34}^{(\pm)} &= \beta_{\mathrm{B}}^{(\pm)} + \beta_{\mathrm{A}}^{(\pm)} e^{i(q_y-q_x)} + 2\beta_g^{(\pm)} \left(e^{iq_y} + e^{-iq_x}\right) \cos q_z
\end{aligned}
\tag{5.1}
$$

Correspondingly, the dispersion laws (3.23)–(3.26) are replaced by the following [142, 164, 165]:

$\Delta = [100]$ direction

$$
\begin{aligned}
\varepsilon_{1,2}^{(\pm)}(q) &= \alpha^{(\pm)} + \{\beta_{\mathrm{A}}^{(\pm)} + \beta_{\mathrm{B}}^{(\pm)} + 2\beta_t^{(\pm)}\} + 4\{\beta_g^{(\pm)} + \beta_t^{(\pm)}\} \cos q \\
&\pm 2\sqrt{\{\beta_{\mathrm{A}}^{(\pm)} - \beta_{\mathrm{B}}^{(\pm)}\}^2 + 16\{\beta_{\mathrm{A}}^{(\pm)} + 2\beta_g^{(\pm)}\}\{\beta_{\mathrm{B}}^{(\pm)} + 2\beta_g^{(\pm)}\} \cos^2 \frac{q}{2}}
\end{aligned}
\tag{5.2}
$$

$$
\varepsilon_{3,4}^{(\pm)}(q) = \alpha^{(\pm)} - \{\beta_{\mathrm{A}}^{(\pm)} + \beta_{\mathrm{B}}^{(\pm)} - 2\beta_t^{(\pm)}\} - 4\{\beta_g^{(\pm)} - \beta_t^{(\pm)}\} \cos q
\tag{5.3}
$$

$\Lambda = [111]$ direction

$$
\begin{aligned}
\varepsilon_{1,2}^{(\pm)}(q) &= \alpha^{(\pm)} + \{\beta_{\mathrm{A}}^{(\pm)} + \beta_{\mathrm{B}}^{(\pm)} + 2\beta_g^{(\pm)} + 2\beta_t^{(\pm)}\} + 2\{\beta_g^{(\pm)} + 2\beta_t^{(\pm)}\} \cos 2q \\
&\pm \sqrt{\{[\beta_{\mathrm{A}}^{(\pm)} + \beta_{\mathrm{B}}^{(\pm)} + 2\beta_g^{(\pm)} + 2\beta_t^{(\pm)}] + 2[\beta_g^{(\pm)} - \beta_t^{(\pm)}] \cos 2q\}^2 + 3\{\beta_{\mathrm{A}}^{(\pm)} - \beta_{\mathrm{B}}^{(\pm)}\}^2 + 12\{\beta_{\mathrm{A}}^{(\pm)} + 2\beta_g^{(\pm)}\}\{\beta_{\mathrm{B}}^{(\pm)} + 2\beta_g^{(\pm)}\} \cos^2 q}
\end{aligned}
\tag{5.4}
$$

$$
\varepsilon_{3,4}^{(\pm)}(q) = \alpha^{(\pm)} - \{\beta_{\mathrm{A}}^{(\pm)} + \beta_{\mathrm{B}}^{(\pm)} + 2\beta_g^{(\pm)} - 4\beta_t^{(\pm)}\} - 2\{\beta_g^{(\pm)} - \beta_t^{(\pm)}\} \cos 2q
\tag{5.5}
$$

5.2.2 Expression of the Matrix Elements in the Equivalent Orbital Basis in Terms of the Elements in the Basis of Atomic Functions

Further differences from homopolar crystals arise as one goes from the matrix elements in the basis of EOs to the matrix elements in the basis of atomic orbitals since in the case of heteroatomic systems the partial ionicity leads to a more general form of (2.58) for expressing the EOs in terms of the atomic functions.

In addition, because of the heteroatomicity of the $A^N B^{8-N}$ compounds one must here distinguish between the integrals $\beta_{s_A p_B}$ and $\beta_{s_B p_A}$, corresponding to the interaction of the s orbital of one atom with the p orbital of the other:

$$\beta_{s_A p_B} = \langle s_A | \hat{H} | p_B \rangle, \quad \beta_{s_B p_A} = \langle s_B | \hat{H} | p_A \rangle \tag{5.6}$$

With these two comments taken into consideration we can repeat the derivation of Eqs. (3.48)-(3.51) almost word for word, again restricting our discussion to the interaction of the valence-bound atoms only. As a result the matrix elements α, β_A, β_B, β_t, β_g in the basis of atomic orbitals will have the following form [203]:

$$\alpha^{(+)} = \frac{1}{1+\lambda^2}\{\bar{\alpha}_B + \lambda^2\bar{\alpha}_A\} + \frac{2\lambda}{4(1+\lambda)^2}\{\beta_{ss} + \sqrt{3}\,(\beta_{s_A p_B} + \beta_{s_B p_A}) + 3\beta_{pp}\} \tag{5.7}$$

$$\beta_A^{(+)} = \frac{\lambda^2}{4(1+\lambda^2)}\{\bar{\alpha}_A^{(s)} - \bar{\alpha}_A^{(p)}\} + \frac{2\lambda}{4(1+\lambda^2)}\left\{\beta_{ss} + \frac{1}{\sqrt{3}}\,(3\beta_{s_A p_B} - \beta_{s_B p_A}) - \beta_{pp}\right\} \tag{5.8}$$

$$\beta_B^{(+)} = \frac{1}{4(1+\lambda^2)}\{\bar{\alpha}_B^{(s)} - \bar{\alpha}_B^{(p)}\} + \frac{2\lambda}{4(1+\lambda^2)}\left\{\beta_{ss} + \frac{1}{\sqrt{3}}\,(3\beta_{s_B p_A} - \beta_{s_A p_B}) - \beta_{pp}\right\} \tag{5.9}$$

$$\beta_g^{(+)} = \frac{2\lambda}{8(1+\lambda^2)}\left\{\beta_{ss} - \frac{1}{\sqrt{3}}\,(\beta_{s_A p_B} + \beta_{s_B p_A}) + \frac{1}{3}\beta_{pp}\right\} + \frac{2\lambda}{6(1+\lambda^2)}\beta_\pi \tag{5.10}$$

$$\beta_t^{(+)} = \frac{2\lambda}{8(1+\lambda^2)}\left\{\beta_{ss} - \frac{1}{\sqrt{3}}\,(\beta_{s_A p_B} + \beta_{s_B p_A}) + \frac{1}{3}\beta_{pp}\right\} - \frac{2\lambda}{3(1+\lambda^2)}\beta_\pi \tag{5.11}$$

$$\alpha^{(-)} = \frac{1}{1+\lambda^2}\{\lambda^2\bar{\alpha}_B + \bar{\alpha}_A\} - \frac{2\lambda}{4(1+\lambda^2)}\{\beta_{ss} + \sqrt{3}\,(\beta_{s_A p_B} + \beta_{s_B p_A}) + 3\beta_{pp}\} \tag{5.12}$$

$$\beta_{\mathrm{A}}^{(-)}=\frac{1}{4(1+\lambda^2)}\{\tilde{\alpha}_{\mathrm{A}}^{(s)}-\tilde{\alpha}_{\mathrm{A}}^{(p)}\}-\frac{2\lambda}{4(1+\lambda^2)}\left\{\beta_{ss}+\frac{1}{\sqrt{3}}(3\beta_{s_{\mathrm{A}}p_{\mathrm{B}}}-\beta_{s_{\mathrm{B}}p_{\mathrm{B}}})-\beta_{pp}\right\} \tag{5.13}$$

$$\beta_{\mathrm{B}}^{(-)}=\frac{\lambda^2}{4(1+\lambda^2)}\{\tilde{\alpha}_{\mathrm{B}}^{(s)}-\tilde{\alpha}_{\mathrm{B}}^{(p)}\}-\frac{2\lambda}{4(1+\lambda^2)}\left\{\beta_{ss}+\frac{1}{\sqrt{3}}(3\beta_{s_{\mathrm{B}}p_{\mathrm{A}}}-\beta_{s_{\mathrm{A}}p_{\mathrm{B}}})-\beta_{pp}\right\} \tag{5.14}$$

$$\beta_{g}^{(-)}=-\frac{2\lambda}{8(1+\lambda^2)}\left\{\beta_{ss}-\frac{1}{\sqrt{3}}(\beta_{s_{\mathrm{A}}p_{\mathrm{B}}}+\beta_{s_{\mathrm{B}}p_{\mathrm{A}}})+\frac{1}{3}\beta_{pp}\right\}-\frac{2\lambda}{6(1+\lambda^2)}\beta_{\pi} \tag{5.15}$$

$$\beta_{t}^{(-)}-\frac{2\lambda}{8(1+\lambda^2)}\left\{\beta_{ss}-\frac{1}{\sqrt{3}}(\beta_{s_{\mathrm{A}}p_{\mathrm{B}}}+\beta_{s_{\mathrm{B}}p_{\mathrm{A}}})+\frac{1}{3}\beta_{pp}\right\}+\frac{2\lambda}{3(1+\lambda^2)}\beta_{\pi} \tag{5.16}$$

Here, $\tilde{\alpha}_{\mathrm{A}}$, $\tilde{\alpha}_{\mathrm{B}}$ denote the average values of the Hamiltonian of the crystal in the hybrid sp^3 orbitals of the ions A, B:

$$\tilde{\alpha}_{\mathrm{A}}=\frac{1}{4}\tilde{\alpha}_{\mathrm{A}}^{(s)}+\frac{3}{4}\tilde{\alpha}_{\mathrm{A}}^{(p)} \tag{5.17}$$

$$\tilde{\alpha}_{\mathrm{B}}=\frac{1}{4}\tilde{\alpha}_{\mathrm{B}}^{(s)}+\frac{3}{4}\tilde{\alpha}_{\mathrm{B}}^{(p)} \tag{5.18}$$

$\tilde{\alpha}_{\mathrm{A}}^{(s)}$, $\tilde{\alpha}_{\mathrm{A}}^{(p)}$, $\tilde{\alpha}_{\mathrm{B}}^{(s)}$, and $\tilde{\alpha}_{\mathrm{B}}^{(p)}$ are the values of the electron energy in the s and p levels of these ions (in the field of the lattice, see below), and β_{ss}, $\beta_{s_{\mathrm{A}}p_{\mathrm{B}}}$, $\beta_{s_{\mathrm{B}}p_{\mathrm{A}}}$, β_{pp}, and β_{π} are the interaction integrals between the s and p orbitals of the valence-bound atoms.

Bearing in mind that the valence band of $\mathrm{A}^N\mathrm{B}^{8-N}$ crystals is rather insensitive to the nature of the approximation used for choosing the matrix elements (just as in the case of crystals of group IV elements), we shall also consider together with Eqs. (5.7)-(5.16) the simplified expressions for the matrix elements, analogous to Eqs. (3.62)-(3.65). Then, using the Wolfsberg-Helmholz approximation again (see Sec. 3.3.2) in combination with "Pauling's rule of angular parts," we obtain

$$\alpha=\frac{1}{1+\lambda^2}\{\tilde{\alpha}_{\mathrm{B}}+\lambda^2\tilde{\alpha}_{\mathrm{A}}+2\lambda\beta\} \tag{5.19}$$

$$\beta_{\mathrm{A}}=\frac{\lambda^2}{4(1+\lambda^2)}\{\tilde{\alpha}_{\mathrm{A}}^{(s)}-\tilde{\alpha}_{\mathrm{A}}^{(p)}\}+\frac{2\lambda}{4(1+\lambda^2)}\cdot\frac{\tilde{\alpha}_{\mathrm{A}}^{(s)}-\tilde{\alpha}_{\mathrm{A}}^{(p)}}{\tilde{\alpha}_{\mathrm{A}}+\tilde{\alpha}_{\mathrm{B}}}\beta \tag{5.20}$$

$$\beta_{\mathrm{B}}=\frac{1}{4(1+\lambda^2)}\{\tilde{\alpha}_{\mathrm{B}}^{(s)}-\tilde{\alpha}_{\mathrm{B}}^{(p)}\}+\frac{2\lambda}{4(1+\lambda^2)}\cdot\frac{\tilde{\alpha}_{\mathrm{B}}^{(s)}-\tilde{\alpha}_{\mathrm{B}}^{(p)}}{\tilde{\alpha}_{\mathrm{A}}+\tilde{\alpha}_{\mathrm{B}}}\beta \tag{5.21}$$

$$\beta_g = \frac{2\lambda}{6(1+\lambda^2)}\,\beta_\pi \tag{5.22}$$

$$\beta_t = -\frac{2\lambda}{3(1+\lambda^2)}\,\beta_\pi \tag{5.23}$$

Just as in Eqs. (3.62)-(3.65) the signs (+) and (−) are omitted here since Eqs. (5.19)-(5.23) will be used only for the valence band.

An important feature of Eqs. (5.7)-(5.16) is that for the purely covalent case (i.e., for $\lambda = 1$) the Coulomb integrals for both kinds of atoms enter into the parameters $\alpha^{(+)}$, $\beta_A^{(+)}$, . . . and $\alpha^{(-)}$, $\beta_A^{(-)}$, . . . , and consequently also into the corresponding dispersion laws with the same "weight." Conversely, in the case of a partially ionic structure ($\lambda < 1$) the parameters with the superscript "plus" contain the Coulomb integrals for the B atoms with a greater weight, while the parameters with a superscript "minus" contain these same integrals for the A atoms. Correspondingly, Eqs. (5.12)-(5.14) describe a partial transfer of charge from the anions B^{Z-} to the cations A^{Z+} as one goes from the valence band to the conduction band. As seen from Eqs. (5.7)-(5.16) and from the expressions for the dispersion laws, in the limiting case of a purely ionic structure $A^{N+}B^{N-}$* (i.e., for $\lambda = 0$), the valence band coincides with the s and p levels of the anion B^{N-}; the conduction band coincides with the s and p levels of the cation A^{N+} as it should on the basis of physical considerations. Thus Eqs. (5.7)-(5.16) in conjunction with the expressions for the dispersion law make it possible to trace the continuous transition from a purely covalent to a purely ionic crystal structure as the effective charge on the atoms changes.

5.3 EMPIRICAL DETERMINATION OF COULOMB AND RESONANCE INTEGRALS

5.3.1 Coulomb Integrals

As in purely covalent crystals Coulomb integrals $\tilde{\alpha}_A^{(s)}$, $\tilde{\alpha}_A^{(p)}$, $\tilde{\alpha}_B^{(s)}$, $\tilde{\alpha}_B^{(p)}$ in partially covalent crystals are also defined as the average values of the Hamiltonian of the crystal in the corresponding atomic functions. Here, however, each electron in the "vicinity" of each given atom (unlike the homopolar case) is acted upon not only by the intrinsic potential of this atom, but also by the Madelung potential of all the other ions of the lattice. Considering these ions to be points (since in our approximation we ignore the overlapping of the atomic functions) and assuming that their charge is equated to the effective charge Z on the atoms, one can write [203]

*Of course, this is never realized in actuality, although the $A^{I}B^{VII}$ compounds are quite close to purely ionic.

$$\begin{aligned}\tilde{\alpha}_{\mathrm{A}}^{(s)}(Z)&=\alpha_{\mathrm{A}}^{(s)}(Z)+\frac{MZ}{R}\\ \tilde{\alpha}_{\mathrm{A}}^{(p)}(Z)&=\alpha_{\mathrm{A}}^{(p)}(Z)+\frac{MZ}{R}\\ \tilde{\alpha}_{\mathrm{B}}^{(s)}(Z)&=\alpha_{\mathrm{B}}^{(s)}(Z)-\frac{MZ}{R}\\ \tilde{\alpha}_{\mathrm{B}}^{(p)}(Z)&=\alpha_{\mathrm{B}}^{(p)}(Z)-\frac{MZ}{R}\end{aligned} \tag{5.24}$$

where M is the Madelung constant and R is the smallest interatomic separation A–B (for the crystals with the sphalerite lattice being considered $M = 1.64$).

Here, $\alpha_{\mathrm{A}}^{(s)}$, etc., denote the "atomic" portions of the Coulomb integrals $\alpha_{\mathrm{A}}^{(s)}, \ldots$, determined by the intrinsic potentials of the corresponding atoms that, of course, also depend on Z.

To find the dependence on Z for the intrinsic Coulomb integrals $\alpha_{\mathrm{A}}^{(s)}, \ldots$, we expand them in power series in terms of Z:

$$\begin{aligned}\alpha_{\mathrm{A}}^{(s)}(Z)&=a_0+a_1Z+\ \ldots\\ \alpha_{\mathrm{A}}^{(p)}(Z)&=a_0'+a_1'Z+\ \ldots\\ \alpha_{\mathrm{B}}^{(s)}(Z)&=b_0+b_1Z+\ \ldots\\ \alpha_{\mathrm{B}}^{(p)}(Z)&=b_0'+b_1'Z+\ \ldots\end{aligned} \tag{5.25}$$

and retain a finite number of terms of the expansion. Then, substituting the values $Z = 0$, $Z = 1, \ldots$, simultaneously into the left and right sides of (5.25), one can express the coefficients $a_0, a_1, a_0', \ldots$ in terms of the Coulomb integrals $\alpha_{\mathrm{A}}^{(s)}(0)$, $\alpha_{\mathrm{A}}^{(s)}(1), \ldots, \alpha_{\mathrm{B}}^{(p)}(1), \ldots$, that correspond to the neutral atoms A^0, B^0 and the ions A^{1+}, $\mathrm{B}^{1-}, \ldots$ with an integer effective charge. Since the Z values for the crystals being considered are small and lie in the interval $0 < Z < 1$ [204, 205], it is sufficient to restrict the values to $Z = 0$ and $Z = 1$. Then the following expressions are obtained for the "atomic" portions of the Coulomb integrals*:

$$\begin{aligned}\alpha_{\mathrm{A}}^{(s)}(Z)&=\alpha_{\mathrm{A}}^{(s)}(0)+\{\alpha_{\mathrm{A}}^{(s)}(1)-\alpha_{\mathrm{A}}^{(s)}(0)\}\,Z\\ \alpha_{\mathrm{A}}^{(p)}(Z)&=\alpha_{\mathrm{A}}^{(p)}(0)+\{\alpha_{\mathrm{A}}^{(p)}(1)-\alpha_{\mathrm{A}}^{(p)}(0)\}\,Z\\ \alpha_{\mathrm{B}}^{(s)}(Z)&=\alpha_{\mathrm{B}}^{(s)}(0)+\{\alpha_{\mathrm{B}}^{(s)}(1)-\alpha_{\mathrm{B}}^{(s)}(0)\}\,Z\\ \alpha_{\mathrm{B}}^{(p)}(Z)&=\alpha_{\mathrm{B}}^{(p)}(0)+\{\alpha_{\mathrm{B}}^{(p)}(1)-\alpha_{\mathrm{B}}^{(p)}(0)\}\,Z\end{aligned} \tag{5.26}$$

*Let us note that a similar formula for the atomic ionization potentials was first stated by Moffitt [206].

in which the Coulomb integrals for the neutral atoms and the singly charged ions A^{1+}, B^{1-} can be determined from spectroscopic data similar to the way it was done in Sec. 3.4.1.*

Let us add only two comments to what has been said in this section, the first of which applies to finding the orbital ionization potentials for negative ions. According to the Slater method [14] that we have adopted, the orbital potential for the valence p orbital of the B^{1-} ion with an s^2p^{n+1} electron configuration is defined as the difference in the energies of the ion and the corresponding neutral atom:

$$\alpha_B^{(p)}(1) = \langle E(^{2S+1}L) \rangle_{s^2p^{n+1}} - \langle E(^{2S+1}L) \rangle_{s^2p^n} \tag{5.27}$$

($n = 3$ for group V and 4 for group VI).

Just as in Sec. 3.3.1, the last formula can obviously be rewritten in the form

$$\alpha_B^{(p)}(1) = E_0(s^2p^{n+1}) - E_0(s^2p^n) + \langle \mathcal{E}(^{2S+1}L) \rangle_{s^2p^{n+1}} - \langle \mathcal{E}(^{2S+1}L) \rangle_{s^2p^n} \tag{5.28}$$

where $E_0(s^2p^{n+1})$ and $E_0(s^2p^n)$ are the energies of the ground state of the ion B^{1-} and the neutral atom B^0, and $\mathcal{E}(^{2S+1}L)$ are the spectroscopic energies of the terms. Nevertheless, Eq. (5.28) cannot be used directly for an empirical estimate of $\alpha_B^{(p)}(1)$ since there are no experimental data on the spectroscopic energies of the terms for negative ions. It is known from spectroscopy, however, that the separaticn between the average terms of different configurations is usually close to the separation between the initial terms. Therefore we assume that

$$\alpha_B^{(p)}(1) = E_0(s^2p^{n+1}) - E_0(s^2p^n) = -F^B \tag{5.29}$$

Thus the orbital ionization potential for the p orbital of the singly charged negative ion B^{1-} is estimated approximately as the affinity to (one) electron taken with a minus sign. The corresponding orbital potential for the s orbital of the B^{1-} ion is written in the form

$$\alpha_B^{(s)}(1) = -F^B - \mathcal{E}(B^0 s^2p^n \rightarrow B^0 sp^{n+1}) \tag{5.30}$$

The second comment refers to the consideration of the valence state (see Sec. 3.4.1). Although in most cases it is sufficient to restrict ourselves to the $\alpha_A^{(s)}, \ldots$ values from Table 3 for the atoms in their ground state, for a more precise estimate of the sensitive levels of the conduction band (the Γ_1^c level in particular) it is necessary to take into consideration that the A^N atoms in the A^NB^{8-N} crystal are in the sp^3 hybridization state.† Actually, for elements of groups II and III (unlike elements of group IV, V, and VI) consideration of this as the valence state leads to an increase in the orbital ionization potentials $\alpha_A^{(s)}$. This is quite a natural circumstance; it is explained by the fact that in the ground state of the A^{III} and A^{II} atoms there is one electron in the p orbitals of the A^{III} atom and generally no electrons in the p orbitals of the A^{II} atom, so that a transition to the sp^3 hybridization in these cases

*In this situation, of course, $\alpha_A^{(p)}(0)$ for A^{III} atoms is simply equal to $(-I_1^A)$, and $\alpha_A^{(s)}(1) = -I_2^A$, where I_1, I_2 are the usual values of the first and second ionization potentials; exactly the same is true for the A^{II} atoms: $\alpha_A^{(s)}(0) = -I_1^A, \alpha_A^{(s)}(1) = -I_2^A$.

†Let us recall that according to Eqs. (5.12) and (5.13) it is exactly the A^{III} and A^{II} cations that make the major contribution to the state of the conduction band.

leads to a marked reduction of the screening of the charge of the atomic core for the s electrons. To introduce a correction for the valence state we proceed in the following manner.

Let

$$\nu = \frac{\text{population of } p \text{ orbitals of atom A}}{\text{population of } s \text{ orbitals of atom A}}$$

and $\alpha_{A,\nu}^{(s)}$ and $\alpha_{A,\nu}^{(p)}$ are the Coulomb integrals, calculated with the valence state of the atom taken into consideration. Then for an A^{III} atom and an arbitrary value of ν (i.e., an arbitrary population of p AOs), one can write the following expansion:

$$\alpha_{III,\,\nu}^{(s)} = \alpha_{III,\,1/2}^{(s)} + c\left(\nu - \frac{1}{2}\right) + \cdots \tag{5.31}$$

where $\alpha_{III}^{(s)}$ is the Coulomb integral for an atom in the ground state $A^{III}(s^2p)$. To find c one can now assume $\nu = 2$, which corresponds to the state $A^{III}(sp^2)$. Hence we have

$$\frac{3}{2}c = \alpha_{III}^{(s)}(\mathrm{A}\ sp^2) - \alpha_{III}^{(s)}(\mathrm{A}\ s^2p) \tag{5.32}$$

Moreover, considering the process

$$\begin{aligned} &\mathrm{A}\ (s^2p) \to A\ (sp^2) - \mathcal{E}\ (s^2p \to sp^2) \\ &\mathrm{A}\ (sp^2) \to \mathrm{A}^+\ (p^2) + \alpha_{III}^{(s)}\ (sp^2) + e \\ &\mathrm{A}^+\ (p^2) \to A^+\ (sp) + \mathcal{E}\ (sp \to p^2) \end{aligned} \tag{5.33}$$

and taking into consideration that $\alpha_{III}^{(s)} = E(\mathrm{A}s^2p) - E(\mathrm{A}^+sp)$, we have $c = \mathcal{E}(s^2p \to sp^2) \to \mathcal{E}(\mathrm{A}^+sp \to \mathrm{A}^+p^2)$. From this, using the appropriate spectroscopic data [13], we obtain the following values of the coefficient c (in eV): $c(\mathrm{B}) \approx -1$; $c(\mathrm{Al}) \approx c(\mathrm{Ga}) \approx c(\mathrm{In}) \approx -2/3$. Substituting these values into the right side of Eq. (5.31) for $\nu = 3$, we arrive at the refined values of the Coulomb integrals for an electron in the s orbital of the A^{III} atom when this atom is in the sp^3 hybridization state. It is not hard to see that these values of $\alpha_A^{*(s)}$ will be larger (in absolute magnitude) than the "old" values by ~2.5 eV for the boron atom and by ~1.5 eV for the Al, Ga, and In atoms. The values of the Coulomb integrals for an electron in the p orbital can be refined in this same way; however, these integrals are only a few tenths of an eV different from the previous values for all the atoms considered. The integrals $\alpha_A^{(s)}$ and $\alpha_A^{(p)}$ for the $(A^{III})^+$ ions change by about the same amount.

5.3.2 Principle of Isoelectronicity and the Resonance Integrals

For an empirical estimate of the resonance integrals we shall assume, following the papers [117, 142] in part, that $\beta(A^NB^{8-N}) = \beta(\mathrm{IV})$ or $\beta(A^{IV}B^{IV})$, so that

$$\begin{aligned} &\beta\ (\mathrm{BN}) \approx \beta(\text{diamond}),\ \beta\ (\mathrm{AlP}) \approx \beta\ (\mathrm{Si}), \ldots \\ &\beta\ (\mathrm{BP}) \approx \beta\ (\mathrm{SiC}) \approx \text{average}\ \{\beta\ (\mathrm{Si}),\ \beta\ (\mathrm{C})\}, \ldots \end{aligned} \tag{5.34}$$

Let us also assume that similar equations are also satisfied for the other resonance integrals [203]*:

$$
\begin{aligned}
&\beta_{ss}\,(\mathrm{BN}) \approx \beta_{ss}\,(\text{diamond}),\ \beta_{ss}\,(\mathrm{AlP}) \approx \beta_{ss}\,(\mathrm{Si}),\ \ldots\\
&\beta_{pp}\,(\mathrm{BN}) \approx \beta_{pp}\,(\text{diamond}),\ \beta_{pp}\,(\mathrm{AlP}) \approx \beta_{pp}\,(\mathrm{Si}),\ \ldots\\
&\beta_{\pi}\,(\mathrm{BN}) \approx \beta_{\pi}\,(\text{diamond}),\ \beta_{\pi}\,(\mathrm{AlP}) \approx \beta_{\pi}\,(\mathrm{Si}),\ \ldots \qquad (5.35)\\
&\beta_{s_A p_B}\,(\mathrm{BN}) \approx \beta_{s_B p_A}\,(\mathrm{BN}) \approx \beta_{sp}\,(\text{diamond}),\ \beta_{s_A p_B}\,(\mathrm{AlP}) \approx \beta_{s_B p_A}\,(\mathrm{AlP})\\
&\approx \beta_{sp}\,(\mathrm{Si}),\ \ldots
\end{aligned}
$$

Such a choice of parameters can be justified in several different ways.

First, there is a very close similarity (with regard to a large number of properties) between crystals of group IV elements and their isoelectron analogs among the $A^N B^{8-N}$ compounds, especially the $A^{III}B^{IV}$ compounds. This fact and, in particular, the nearly exactly identical interatomic spacings and the similarity of the heats of atomization [207] definitely indicate a similarity in the nature of the chemical bond in these isoelectron systems (the iso-electronicity principle). This makes the estimate (5.34) quite plausible. In fact, we customarily call the heat of atomization of the crystal minus the Madelung energy MZ^2/R of the attraction of the A^{Z+} and B^{Z-} ions plus the energy $E(A^{Z+}, B^{Z-})$ expended to form the ions the "covalent component" of the binding energy†:

$$E_{\mathrm{cov}} = -\Delta H_{\mathrm{at}} - MZ^2/R + E\,(A^{Z+},\ B^{Z-})$$

Then it is easy to see from data on the heats of atomization of $A^{III}B^{V}$ compounds [207] that even if the third term is ignored in the expression for E_{cov}, the value of $E_{\mathrm{cov}}(A^{III}B^{V})$ for $Z = 0.3$–0.5 [205] is about 70% of the value of $-\Delta H_{\mathrm{at}}$ for the isoelectron elements of group IV. Taking the third term, having a sign that is opposite to that of the second term, into account can only reduce the difference between E_{cov}(III–V) and $-\Delta H_{\mathrm{at}}$(IV).

In this regard let us note that the extremely close similarity of the nature of the chemical bond and the electron structure in isoelectron systems does not exist only for the tetrahedral $A^N B^{8-N}$ crystals. How well the pair of atoms $A^{III}B^{V}$ can simulate the isoelectron pair A^{IV}, B^{IV} is evident in the case of numerous data on the chemistry of boron and nitrogen compounds since the borazide analogs of many organic compounds have been produced recently.

*Within reasonable limits of accuracy it makes no difference how the averaging is done. Thus for $\beta(\mathrm{Si}) = -4$ eV and $\beta(\mathrm{Ge}) = -3$ eV we have $(1/2)\{\beta(\mathrm{Si}) + \beta(\mathrm{Ge})\} = -3.5$ eV, whereas $-\sqrt{\beta(\mathrm{Si})\cdot\beta(\mathrm{Ge})} = -3.46$ eV.

†In the more rigorous many-electron version of the theory (for example, in the Hartree–Fock version) the division of the total energy into "covalent" and "ionic" components is, of course, ambiguous.

(See, for example, [208] concerning the close analogies between organic and borazide compounds.)

The Wolfsberg-Helmholz approximation $\beta_{ij}(A^N B^{8-N}) = K/2 \times S_{ij}\{\alpha_A^{(i)} + \alpha_B^{(j)}\}$ provides more serious arguments in favor of the estimates (5.34) and (5.35). In fact, as seen from Table 4, the energy of electrons in the s and p levels of the A^N atoms is always higher, and in the s and p levels of the B^{8-N} atoms it is always lower than for the corresponding group IV elements, so that

$$\alpha_A^{(i)} + \alpha_B^{(j)} \approx \alpha_{IV}^{(i)} + \alpha_{IV}^{(j)} \tag{5.36}$$

In addition there is an approximate equality for the overlap integrals too:

$$S_{ij}(A^N B^{8-N}) \approx S_{ij}(A^{IV} B^{IV}) \tag{5.37}$$

It is explained by the fact that for the B^{8-N} atoms the atomic functions are more "condensed," while for the A^N atoms, conversely, they are more diffuse than for the group IV atoms, so that as one goes from the group IV elements to the isoelectron crystals $A^N B^{8-N}$ the amount of overlap remains the same. In fact, let us examine, for example, the overlap integrals between the AOs of the A^N and B^{8-N} atoms in crystals of $A^N B^{8-N}$ compounds, using the Slater atomic functions $\chi(\vartheta, \varphi, r) = AY(\vartheta, \varphi) \exp(-\xi\, r/n)$ for these integrals. It is known [209] that the value of the overlap integrals for the Slater AOs is determined by two parameters $p = (1/2)R(\mu_A + \mu_B)$ and $t = (\mu_A - \mu_B)/(\mu_A + \mu_B)$. Here R is the interatomic spacing in units of atomic length, and the quantity μ is equal to the quotient of the Slater charge ζ divided by the Slater effective quantum number n^* for the valence electrons: $\mu = \zeta/n^*$. Let us now compare the values of p and t for the isoelectron crystals of the group IV elements and the $A^N B^{8-N}$ compounds.

As it is easy to see, $p(A^{III}B^V) = p(A^{IV})$, and $t(A^{III}B^V) = 0.2$ for BN and 0.1 for the rest of the compounds, so that the values of the parameter p for the group IV elements and the $A^N B^{8-N}$ compounds are identical, and the t values are close ($t = 0$ for the group IV elements). Then, using the tables of [209], it is easy to see that in this case the values of all the overlap integrals will also be close.

As a third argument in favor of the estimates (5.34) and (5.35) we can make use of considerations based on perturbation theory (which will also be used below for a more precise estimate of the parameters β_{sApB} and β_{sBpA}). As first mentioned by Herman [210], the properties of tetrahedral $A^N B^{8-N}$ crystals with the sphalerite lattice agree well with the assumption that the potential $V(\mathbf{r})$ in a partially ionic crystal can be represented in the form of the sum of the potential $V_{IV}(\mathbf{r})$ for the isoelectron crystal of a group IV element (zero approximation) and the "heteropolar perturbation" $\Delta V(\mathbf{r})$

$$V(\mathbf{r}) = V_{IV}(\mathbf{r}) + \Delta V(\mathbf{r}) \tag{5.38}$$

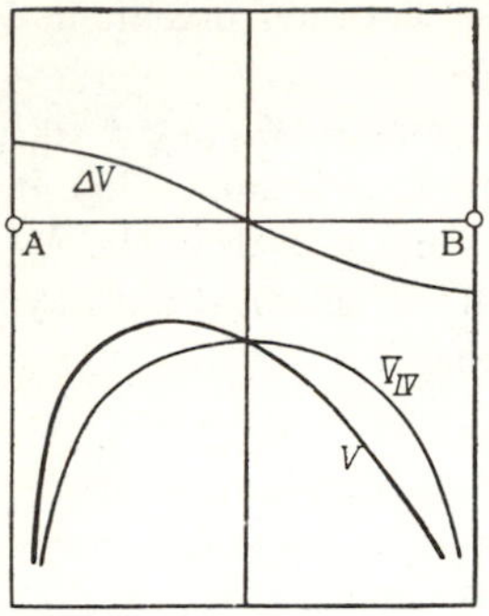

FIG. 5.3. Decomposition of the potential in a tetrahedral partially ionic crystal into symmetrical and antisymmetrical components.

where the potential $V_{\mathrm{IV}}(\mathbf{r})$ is symmetrical and the potential $\Delta V(\mathbf{r})$ is antisymmetrical with respect to the middle of any of the **A–B** bonds (Fig. 5.3):

$$V_{\mathrm{IV}}(-\mathbf{r}) = V_{\mathrm{IV}}(\mathbf{r}), \quad \Delta V(-\mathbf{r}) = -\Delta V(\mathbf{r}) \tag{5.39}$$

Let us now use the zero-approximation atomic orbitals (i.e., the s and p functions of the isoelectron crystal of a group IV element) as the basis functions for an $\mathrm{A}^N\mathrm{B}^{8-N}$ crystal. It is then obvious that in the zero approximation all the matrix elements of the Hamiltonian for the $\mathrm{A}^N\mathrm{B}^{8-N}$ crystal will be identical to the corresponding matrix elements for a group IV crystal. It is not hard to see, however, that because of the antisymmetry of the potential ΔV the equalities

$$\int s_{\mathrm{A}}^{\mathrm{IV}}\, \Delta V s_{\mathrm{B}}^{\mathrm{IV}}\, dv = 0, \quad \int p_{\mathrm{A}}^{\mathrm{IV}}\, \Delta V p_{\mathrm{B}}^{\mathrm{IV}}\, dv = 0 \tag{5.40}$$

are true. Therefore in the first approximation too (the first order of perturbation theory) only the Coulomb integrals $\alpha_{\mathrm{A}}^{(s)}$, $\alpha_{\mathrm{A}}^{(p)}$ change, while the resonance integrals remain unchanged, except for the integral β_{sp}. In the first approximation the latter corresponds to the two integrals $\beta_{s\mathrm{A}p\mathrm{B}}$ and $\beta_{s\mathrm{B}p\mathrm{A}}$, the difference between which is

$$\beta_{s_{\mathrm{A}}p_{\mathrm{B}}} - \beta_{s_{\mathrm{B}}p_{\mathrm{A}}} = \int \{ s_{\mathrm{A}}^{\mathrm{IV}} \hat{H} p_{\mathrm{B}}^{\mathrm{IV}} - s_{\mathrm{B}}^{\mathrm{IV}} \hat{H} p_{\mathrm{A}}^{\mathrm{IV}} \}\, dv$$

$$= \int \{ s_{\mathrm{A}}^{\mathrm{IV}}\, \Delta V p_{\mathrm{B}}^{\mathrm{IV}} - s_{\mathrm{B}}^{(\mathrm{IV})}\, \Delta V p_{\mathrm{A}}^{\mathrm{IV}} \}\, dv, \text{ where } \hat{H} = -\frac{1}{2}\Delta + V_{\mathrm{IV}} + \Delta V \tag{5.41}$$

It is not hard to show (see below), however, that the difference (5.41) is much less than both the corrections to the Coulomb integrals as well as the average value of the resonance integrals $(1/2)\{\beta_{s\mathrm{A}p\mathrm{B}} + \beta_{s\mathrm{B}p\mathrm{A}}\}$. Therefore in most cases the difference between the integrals $\beta_{s\mathrm{A}p\mathrm{B}}$ and $\beta_{s\mathrm{B}p\mathrm{A}}$ can be ignored.

Finally, let us note that the "isoelectronicity principle" in the specific form of relations (5.34) and (5.35) can be directly proved experimentally, at least for one type of resonance integrals [203].

Let us consider the level differences $\varepsilon(\Gamma^v_{15}) - \varepsilon(X^v_5)$, $\varepsilon(\Gamma^v_{15}) - \varepsilon(L^v_3)$ at the center and at the edges X and L of the first Brillouin zone; they are given by Eqs. (5.50) and (5.51) (see below). It is not hard to see that for the $A^{III}B^V$ compounds for any effective charges $0 < Z < 1$ the coefficient $2\lambda/(1 + \lambda^2)$ in Eqs. (5.50)-(5.54) changes only very slightly (from 0.87 to 0.97), so that one can assume $2\lambda/(1 + \lambda^2) \approx 0.9$ regardless of the value of Z. Similarly, the value of $2\lambda/(1 + \lambda^2) = 0.7$-$0.75$ for $A^{II}B^{VI}$ compounds for $0.5 < Z < 1$, so that within the limits of each class of substances A^{IV}, $A^{III}B^V$, and $A^{II}B^{VI}$ one can assume that the level difference $\varepsilon(\Gamma^v_{15}) - \varepsilon(X^v_5)$ or $\varepsilon(\Gamma^v_{15}) - \varepsilon(L^v_3)$ depends on the resonance integral β_π only, and not on the covalence parameter λ. Making use of this, we can compare the "experimental" values of the resonance integral β_π obtained from the band structure of the $A^N B^{8-N}$ compounds with similar values found from the band structure of group IV elements. This comparison (Table 11) shows that the relation $\beta_\pi(A^N B^{8-N}) = \beta_\pi(\text{IV})$ is fulfilled quite well in experiments.

In conclusion let us examine in more detail the question of the estimate of the resonance integrals β_{sApB}, β_{sBpA}, which we shall need below [212]. Although in most cases it is adequate to use the estimate $\beta_{sApB} \approx \beta_{sBpA} \approx \beta_{sp}(\text{IV})$ for these integrals, there are properties specific to heteroatomic crystals (for example, the characteristic splitting of the X_1 doublet into two levels X_1 and X_3) that cannot be studied without estimating the difference (5.41). It is easy to obtain such an estimate from Eq. (5.41) by using the Wolfsberg-Helmholz approximation. Hence

TABLE 11. Experimental Values of the Integral β_π for Isoelectron Analogs

Crystal	$-\beta_\pi$, (eV)	Crystal	$-\beta_\pi$, (eV)	Crystal	$-\beta_\pi$, (eV)	Crystal	$-\beta_\pi$, (eV)
Ge	0.6; 0.5 0.8 *; 0.6 *	GaAs	0.6; 0.55 0.9 *; 0.55 *	—	—	ZnSe	0.4; 0.4 — —
α-Sn	0.45; 0.45 —; 0.6 *	InSb	0.4; 0.35 — —	--	—	CdTe	0.35; 0.2 — ; —
SiGe	0.7; 0.6 0.8 *; 0.6 *	GaP	0.65; 0.55 0.8 *; —	AlAs	0.8 *; —	ZnS	0.7; 0.35 — ; —
SiSn	0.6; 0.55 —; 0.6 *	InP	0.55; 0.5 — ; —	AlSb	0.5; 0.45 0.8 *; 0.45 *	—	— —
GeSn	0.5; 0.5 —; 0.6 *	GaSb	0.5; 0.45 0.85 *; 0.6 *	InAs	0.45; 0.4 — ; —	ZnTe	0.4; 0.35 — ; —

*The first number in each line is found from the value of $\varepsilon(\Gamma^v_{15}) - \varepsilon(X^v_4)$; the second, from the value of $\varepsilon(\Gamma^v_{15}) - \varepsilon(L^v_3)$. The values of $-\beta_\pi$ without an asterisk are found from EP data [107]; the values with an asterisk are from the data of [211]. The data for SiGe and the hypothetical compounds SiSn and GeSn are obtained by averaging the data for the corresponding elements.

$$\beta_{s_A p_B} - \beta_{s_B p_A} = \frac{K}{2} \langle s_A^{IV} | p_B^{IV} \rangle \cdot \{\alpha_A^{(s)} + \alpha_B^{(p)} - \alpha_B^{(s)} - \alpha_A^{(p)}\} \tag{5.42}$$

whereas for an isoelectron crystal of a group IV element one can write

$$\beta_{sp} = \frac{K}{2} \langle s_A^{IV} | p_B^{IV} \rangle \{\alpha_{IV}^{(s)} + \alpha_{IV}^{(p)}\} \tag{5.43}$$

Eliminating $\langle s_A^{IV} | p_B^{IV} \rangle$ from (5.42) and (5.43), we obtain the relation

$$\beta_{s_A p_B} - \beta_{s_B p_A} = \beta_{sp}(IV) \cdot \frac{\{\alpha_A^{(s)} - \alpha_B^{(s)}\} - \{\alpha_A^{(p)} - \alpha_B^{(p)}\}}{\alpha_{IV}^{(s)} + \alpha_{IV}^{(p)}} \tag{5.44}$$

which also gives the desired estimate* for the difference of the resonance integrals β_{sApB} and β_{sBpA}.

This estimate can also be represented in a somewhat different, more convenient form by noting that according to the same rule of Herman [210] the potential V_{IV} is identical to the symmetrical portion of the potential in a partially ionic crystal:

$$2V_{IV}(\mathbf{r}) = V(\mathbf{r}) + V(-\mathbf{r}) \tag{5.45}$$

Then

$$\hat{H}_{IV} = \frac{1}{2}\left\{-\frac{1}{2}\Delta + V(\mathbf{r}) - \frac{1}{2}\Delta + V(-\mathbf{r})\right\} \tag{5.46}$$

from which

$$\begin{aligned} \alpha_{IV}^{(s)} &= \langle s_A^{(IV)} | \hat{H} | s_A^{IV} \rangle = \frac{1}{2}\{\alpha_A^{(s)} + \alpha_B^{(s)}\} \\ \alpha_{IV}^{(p)} &= \langle p_A^{(IV)} | \hat{H} | p_A^{IV} \rangle = \frac{1}{2}\{\alpha_A^{(p)} + \alpha_B^{(p)}\} \end{aligned} \tag{5.47}$$

so that finally

$$\beta_{s_A p_B} - \beta_{s_B p_A} = \beta_{sp}(IV) \cdot \frac{\{\alpha_A^{(s)} - \alpha_B^{(s)}\} - \{\alpha_A^{(p)} - \alpha_B^{(p)}\}}{(1/2)\{\alpha_A^{(s)} + \alpha_B^{(s)} + \alpha_A^{(p)} + \alpha_B^{(p)}\}} \tag{5.48}$$

As follows from Eqs. (5.44) and (5.48) the difference (5.41) will actually be small since it is proportional not to the corrections $\Delta\alpha_{IV}^{(s)} = (1/2)\{\alpha_A^{(s)} - \alpha_B^{(s)}\}$, $\Delta\alpha_{IV}^{(p)} = (1/2)\{\alpha_A^{(p)} - \alpha_B^{(p)}\}$ themselves to the Coulomb integrals $\alpha_{IV}^{(s)}$, $\alpha_{IV}^{(p)}$, but to the difference of these corrections. (Assuming, in particular, that the perturbing antisymmetrical potential ΔV is

*Let us again call attention to the specific nature of the use of the Wolfsberg-Helmholz method and, in particular, to the fact that the proportionality constant K does not enter into the final formulas.

constant within the atoms A and B, we would obtain $\Delta\alpha_{IV}^{(s)} = \Delta\alpha_{IV}^{(p)}$ and $\beta_{sApB} - \beta_{sBpA} = 0$.) A calculation from Eq. (5.48) shows that the value of the difference (5.41) is ≈ 0.2 eV, which is about an order of magnitude less than the correction $\Delta\alpha_{IV}^{(s)}$, $\Delta\alpha_{IV}^{(p)}$, as well as the average value β_{sp} of the integrals β_{sApB}, β_{sBpA}.

5.4 STRUCTURE OF VALENCE BAND IN PARTIALLY COVALENT CRYSTALS

5.4.1 Fundamental Parameters of Valence Band Structure

As a multitude of experimental and theoretical data shows, the band structure of the $A^N B^{8-N}$ crystals varies regularly both in the "vertical" series (for example, the $A^{III}B^V$ or $A^{II}B^{VI}$ series of compounds) and also in the "horizontal" series (for example, the Ge-GaAs-ZnSe-CuBr series or the α-Sn-InSb-CdTe-AgI series). In the vertical series (similarly to the group IV elements) both the total width of the valence band E_v and the width of its p subband ΔE_p generally decrease monotonically.

At the same time in the horizontal (isoelectron) series the width of all the filled subbands of the valence band decreases as the compound components go farther from group IV. The total valence band width E_v is also not noticeably altered because of an expansion of the gap ΔE_v in the valence band. In this section we shall examine (in accordance with the purposes of our study, see Introduction) how this behavior of the valence band structure is related to the regular variation of the energy levels of the atoms, the parameters of the interaction of the atomic orbitals with each other, and the variation of the ionicity of the crystals along the isoelectron series.

As seen from Fig. 5.1 the structure of the valence band of the tetrahedral $A^N B^{8-N}$ crystal is described by the relative location of its characteristic levels Γ_{15}^v, Γ_1^v, X_5^v, $X_3^v = \bar{\bar{X}}^v$, $X_1 = \bar{X}^v$, L_3^v, $L_1^v = \bar{\bar{L}}^v$ and $L_1^v = \bar{L}^v$. [Here, as in chapter 4, for convenience we sometimes use the symbols $\bar{X}$, $\bar{L}$, $\bar{\bar{X}}$, $\bar{\bar{L}}$ to denote the s–p levels at the edges X and L of the first Brillouin zone, corresponding to those branches of the dispersion law (5.2)–(5.5) that become the Γ_1 and Γ_{15} levels, respectively, at the center of the zone.]

Using the dispersion laws (5.2)–(5.5) as well as Eqs. (5.7)–(5.16) and (5.16)–(5.23), one can express the characteristic parameters of the band structure $\varepsilon(\Gamma_{15}^v) - \varepsilon(\Gamma_1^v)$, $\varepsilon(\Gamma_{15}^v) - \varepsilon(X_5^v)$, etc., in terms of the Coulomb integrals, the interaction integrals of the different atomic orbitals, and in terms of the covalence parameter λ. In the "complete" version of the theory the corresponding expressions have the following form:

the total width of the valence band $E_v = \varepsilon(\Gamma_{15}^v) - \varepsilon(\Gamma_1^v)$

$$E_v = \varepsilon(\Gamma_{15}^v) - \varepsilon(\Gamma_1^v) = -4\{\beta_A^{(+)} + \beta_B^{(+)} + 4\beta_g^{(+)}\}$$

$$= \frac{\alpha_B^{(p)} - \alpha_B^{(s)} + \lambda^2\{\alpha_A^{(p)} - \alpha_A^{(s)}\}}{1+\lambda^2} - \frac{8\lambda}{1+\lambda^2}\left\{\beta_{ss} - \frac{1}{3}\beta_{pp} + \frac{2}{3}\beta_\pi\right\} \quad (5.49)$$

the width of the upper p band ΔE_p

$$\Delta E_p = \varepsilon(\Gamma_{15}^v) - \varepsilon(X_5^v) = 8\{\beta_t^{(+)} - \beta_g^{(+)}\} = -\frac{8\lambda}{1+\lambda^2}\beta_\pi \tag{5.50}$$

the width of the upper s-p band ΔE_1

$$\Delta E_1 = \varepsilon(\Gamma_{15}^v) - \varepsilon(X_3^v) = -2\{\beta_A^{(+)} + \beta_B^{(+)}\} + 8\beta_t^{(+)} - 2|\beta_A^{(+)} - \beta_B^{(+)}|$$

$$= \frac{\lambda^2}{1+\lambda^2}\{\alpha_A^{(p)} - \alpha_A^{(s)}\} + \frac{8\lambda}{3(1+\lambda^2)}\left\{\beta_{pp} - \frac{\sqrt{3}}{2}(\beta_{s_A p_B} + \beta_{s_B p_A}) - 2\beta_\pi\right\}$$

$$+ \frac{4\lambda}{\sqrt{3}(1+\lambda^2)}\{\beta_{s_B p_A} - \beta_{s_A p_B}\} \tag{5.51}$$

the width of the lower s-p band ΔE_2

$$\Delta E_2 = \varepsilon(X_1^v) - \varepsilon(\Gamma_1^v) = -2\{\beta_A^{(+)} + \beta_B^{(+)} + 8\beta_g^{(+)} + 4\beta_t^{(+)}\} - 2|\beta_A^{(+)} - \beta_B^{(+)}|$$

$$= \frac{\lambda^2\{\alpha_A^{(p)} - \alpha_A^{(s)}\}}{1+\lambda^2} - \frac{8\lambda}{1+\lambda^2}\left\{\beta_{ss} - \frac{1}{2\sqrt{3}}(\beta_{s_A p_B} + \beta_{s_B p_A})\right\}$$

$$+ \frac{4\lambda}{\sqrt{3}(1+\lambda^2)}\{\beta_{s_B p_A} - \beta_{s_A p_B}\} \tag{5.52}$$

the width of the gap in the valence band ΔE_v

$$\Delta E_v = \varepsilon(X_3^v) - \varepsilon(X_1^v) = 4|\beta_A^{(+)} - \beta_B^{(+)}|$$

$$= \frac{\alpha_B^{(p)} - \alpha_B^{(s)} - \lambda^2\{\alpha_A^{(p)} - \alpha_A^{(s)}\}}{1+\lambda^2} - \frac{8\lambda}{\sqrt{3}(1+\lambda^2)}\{\beta_{s_B p_A} - \beta_{s_A p_B}\} \tag{5.53}$$

the level difference δE_p

$$\delta E_p = \varepsilon(\Gamma_{15}^v) - \varepsilon(L_3^v) = 4\{\beta_t^{(+)} - \beta_g^{(+)}\} = -\frac{4\lambda}{1+\lambda^2}\beta_\pi \tag{5.54}$$

[In view of their cumbersome nature we do not write the expressions for the differences $\varepsilon(\Gamma_{15}^v) - \varepsilon(\bar{\bar{L}}^v)$, $\varepsilon(\Gamma_{15}^v) - \varepsilon(\bar{L}^v)$ here.]

Corresponding, for the "simplified" version of the theory we obtain

$$E_v = \frac{\{\alpha_B^{(p)} - \alpha_B^{(s)}\} + \lambda^2\{\alpha_A^{(p)} - \alpha_A^{(s)}\}}{1+\lambda^2}$$

$$-\frac{2\lambda}{1+\lambda^2}\cdot\frac{\alpha_A^{(s)} - \alpha_A^{(p)} + \alpha_B^{(s)} - \alpha_B^{(p)}}{\alpha_A + \alpha_B}\beta - \frac{16\lambda}{3(1+\lambda^2)}\beta_\pi \tag{5.55}$$

$$\Delta E_p = -\frac{8\lambda}{1+\lambda^2}\beta_\pi \tag{5.56}$$

$$\Delta E_1 = \frac{\lambda^2\{\alpha_A^{(p)}-\alpha_A^{(s)}\}}{1+\lambda^2} + \frac{2\lambda}{1+\lambda^2}\cdot\frac{\alpha_A^{(p)}-\alpha_A^{(s)}}{\alpha_A+\alpha_B}\beta - \frac{16\lambda}{3(1+\lambda^2)}\beta_\pi \tag{5.57}$$

$$\Delta E_2 = \frac{\lambda^2\{\alpha_A^{(p)}-\alpha_A^{(s)}\}}{1+\lambda^2} + \frac{2\lambda}{1+\lambda^2}\cdot\frac{\alpha_A^{(p)}-\alpha_A^{(s)}}{\alpha_A+\alpha_B}\beta \tag{5.58}$$

$$\Delta E_v = \frac{\alpha_B^{(p)}-\alpha_B^{(s)}-\lambda^2\{\alpha_A^{(p)}-\alpha_A^{(s)}\}}{1+\lambda^2} + \frac{2\lambda}{1+\lambda^2}\cdot\frac{\{\alpha_B^{(p)}-\alpha_B^{(s)}\}-\{\alpha_A^{(p)}-\alpha_A^{(s)}\}}{\alpha_A+\alpha_B}\cdot\beta \tag{5.59}$$

$$\delta E_p = -\frac{4\lambda}{1+\lambda^2}\beta \tag{5.60}$$

5.4.2 Qualitative Investigation of Valence Band Structure

Before turning to detailed calculations, let us draw several conclusions of a qualitative nature from Eqs. (5.49)-(5.54) and (5.55)-(5.60).

As follows from Eqs. (5.49) and (5.55) the width of the valence band E_v is always the sum of two quantities with the same sign: some "average" promotion energy*

$$\langle \mathscr{E}^v_{\text{promot}}\rangle = \frac{\alpha_B^{(p)}-\alpha_B^{(s)}+\lambda^2\{\alpha_A^{(p)}-\alpha_A^{(s)}\}}{1+\lambda^2} = \frac{\mathscr{E}_{\text{promot}}(B)+\lambda^2\mathscr{E}_{\text{promot}}(A)}{1+\lambda^2} \tag{5.61}$$

and a term having a "resonance" origin [the second term in Eq. (5.49) and the second and third terms in Eq. (5.55)]. Accordingly, E_v will always be larger than the average separation between the s and p levels for isolated ions A and B (compare Table 3 and Tables 12 and 13).

To draw conclusions concerning the variation of E_v in the vertical series, for example, for the family of $A^{III}B^{V}$ compounds, let us note that in the series from BN (or BP) to InSb the value of the resonance term is reduced by ~60% (by ~50%). Depending on the estimates of the resonance parameters, this amounts to a quantity of the order of 5-10 or 4-8 eV. At the same time the maximum possible increase in $\langle \mathscr{E}^v_{\text{promot}}\rangle$ amounts to ~1 eV in this case. Therefore for the

*Let us note that a somewhat different quantity $\mathscr{E}^c_{\text{promot}} = \{\mathscr{E}_{\text{promot}}(A) + \lambda^2\mathscr{E}_{\text{promot}}(B)\}/(1+\lambda^2)$ will enter into the corresponding expressions for the conduction band.

TABLE 12. Fundamental Parameters (in eV) of the Valence Band Structure for BN, BP, and AlP

Crystal		Experiment	Semi-empirical EO LCAO	OPW			APW [220]	EO	
				[162]	[89, 217]	[218, 219]		[142]	[165]
BN	ΔE_p	—	6	5.3	—	5.5	3.6	3	—
	δE_p	—	3	4.8	—	2	1.3	—	—
	ΔE_1	—	7	12	—	11	6.8	5.4	—
	ΔE_2	—	4	5.5	—	4.2	4.5	7.3	—
	ΔE_v	—	7	5.5	—	9.8	5.5	0	—
	E_v	15—22	18	23	—	25	17.8	12.6	—
BP	ΔE_p	—	5	—	4	4	—	—	—
	δE_p	—	2.5	—	1.7	1.5	—	—	—
	ΔE_1	—	6	—	8.4	9.5	—	—	—
	ΔE_2	—	3.5	—	4.6	5.5	—	—	—
	ΔE_v	—	4	—	2.3	0.7	—	—	—
	E_v	16.9	13.5	—	15.3	15.7	—	—	—
AlP	ΔE_p	—	4	2.5	2	—	—	—	—
	δE_p	—	2	5	0.8	—	—	—	—
	ΔE_1	—	7.4	3.3	5.3	—	—	3.5	7
	ΔE_2	—	4.6	1.5	2.3	—	—	3.7	8.6
	ΔE_v	—	I	6	3.8	—	—	0	3.7
	E_v	13.7	13	10	11.5	—	—	7.2	19.4

TABLE 13. Fundamental Parameters (in eV) of the Valence Band Structure for AlAs, GaP, and GaAs

Crystal		Data				EO	
		Experiment	Semiempirical EO LCAO	OPW [162]	EC OPW [103]	[142]	[165]
AlAs	ΔE_p	—	2.5	—	—	—	—
	δE_p	—	1.2	—	—	—	—
	ΔE_1	—	4.5	—	—	—	8.5
	ΔE_2	—	3	—	—	—	10.5
	ΔE_v	—	3	—	—	—	3.2
	E_v	—	10.5	—	—	—	22.2
GaP	ΔE_p	2.7	3	—	2.3	—	—
	δE_p	1.4	1.5	—	0.9	—	—
	ΔE_1	6.9	5.5	—	6.1	—	6.4
	ΔE_2	3.6	3.5	—	2.6	—	7.9
	ΔE_v	2.7	3	—	3.1	—	4.1
	E_v	11.4—13.6	12	—	11.8	—	18.4
GaAs	ΔE_p	2.5	3	2.5	2.3	1	—
	δE_p	0.8—1.4	1.2	2	0.9	—	—
	ΔE_1	6.4—7.3	5	6.5	5.5	3.8	7.2
	ΔE_2	1.1—3.1	2.5	2.5	1.7	4	9.4
	ΔE_v	11.3—14.2	4.5	2	5.2	0	3.3
	E_v	—	12	11	12.4	6.6	20.6

vertical series of $A^{III}B^{V}$ compounds a gradual decrease in E_v should be expected, just as occurs in actuality (Tables 12 and 13).

Any deviation from the monotonic nature of the variation of E_v can obviously be observed, then, when the transition from compounds of the lighter elements to compounds of heavier elements is accompanied by an increase in one of the terms on the right side of Eqs. (5.49) and (5.55) while the rest of the terms change by a quite small amount. A comparison of the promotion energies (Table 3) shows that $\mathcal{E}_{\text{promot}}$ is sometimes actually larger for the heavier elements than for the lighter, and the largest increase in $\mathcal{E}_{\text{promot}}$ is observed as one goes from aluminum to gallium: $\mathcal{E}_{\text{promot}}$(Al) is less than $\mathcal{E}_{\text{promot}}$(Ga) by ~1.5 eV, so that from this point of view the loss of the monotonic variation of E_v could be observed for any pair of compounds with the composition AlB^{V}, GaB^{V}. It is interesting that according to X-ray spectral investigations [213] such a breakdown in the monotonic behavior is actually observed as one goes from AlP to GaP.

Let us now consider the question of the variation of the width of the individual subbands ΔE_p, ΔE_1, ΔE_2 and the width of the gap ΔE_v in the valence band as a function of the location of the $A^N B^{8-N}$ crystal in the corresponding vertical or horizontal series. Let us start with the p band X_5-Γ_{15}-L_3^v.

As the Eqs. (5.50) and (5.56) for the level difference $\varepsilon(\Gamma_{15}^v) - \varepsilon(X_5^v)$ and the Eqs. (5.54) and (5.60) for the level difference $\varepsilon(\Gamma_{15}^v) - \varepsilon(L_3^v)$ show, both differences are proportional to the magnitude of the resonance integral $|\beta_\pi|$, which decreases monotonically along the vertical series, varying by more than a factor of two in the series from BN to InSb. At the same time the coefficient $2\lambda/(1 + \lambda^2)$ present in Eqs. (5.50), (5.54), (5.56), and (5.60), although it depends on the degree of ionicity, can change by only a very small amount (see below). Therefore it must be expected that both the total width of the p band $\Delta E_p = \varepsilon(\Gamma_{15}^v) - \varepsilon(X_5^v)$ and the difference $\delta E_p = \varepsilon(\Gamma_{15}) - \varepsilon(L_3^v)$ decrease in the vertical series. This is confirmed by data obtained by the EP method [107].

Let us mention another qualitative conclusion here that follows from Eqs. (5.50) and (5.54) or (5.56) and (5.60) and refers to the ratio of the quantities ΔE_p and δE_p. As seen from these formulas, for partially ionic crystals, just as for covalent, the equality

$$\frac{\Delta E_p}{\delta E_p} = \frac{\varepsilon(\Gamma_{15}^v) - \varepsilon(X_5^v)}{\varepsilon(\Gamma_{15}^v) - \varepsilon(L_3^v)} = 2 \tag{5.62}$$

exists. Direct experimental data [211], as well as the data from calculations using the EP method [107] and the OPW method with empirical corrections [103] show that the ratio $\Delta E_p/\delta E_p$ is to close to this value.

To estimate the effect of the degree of ionicity on the width of the p band, let us note that the ratio of the width of the band X_5^v-Γ_{15}^v-L_3^v in $A^N B^{8-N}$ crystals to the analogous quantity for group IV elements is determined by the coefficient $2\lambda/(1 + \lambda^2)$, which approaches zero with an increase in the degree of ionicity (i.e., as $\lambda \to 0$). Therefore the transition from purely covalent crystals to their isoelectron partially covalent analogs, as Eqs. (5.50) and (5.54) or (5.56)-(5.60) show, should be accompanied by a narrowing of the p band. For a quantitative estimate of this narrowing it is convenient to express the coefficient $2\lambda/(1 + \lambda^2)$ by means of the relation (2.63) directly in terms of the effective charge on the atom and in terms of the numbers N, $8-N$, defining the location of the components of the compound $A^N B^{8-N}$ in the periodic table. This leads to the relation

$$\frac{(\text{Width of band } X_5 - \Gamma_{15}^v - L_3^v)\ A^N N^{8-N}}{(\text{Width of band } X_4^v - \Gamma_{25'}^v - L_3^v)_{\text{IV}}} = \frac{1}{4}\sqrt{(N-Z)(8-N+Z)} \tag{5.63}$$

It is not hard to conclude from Eq. (5.63) that for $N=3$ ($A^{III}B^V$ compounds) and an effective charge of $0<Z<1$, the value of the ratio (5.63) lies within the interval 0.87–0.97, while for $N = 2$ ($A^{II}B^{VI}$ compounds) and $0.5<Z<1$ it lies within the interval 0.67–0.78. Such narrow limits for these intervals mean that the narrowing of the p band essentially does not depend so much on the actual value of the effective charge on the atoms,* as on the location of the atoms A^N and B^{8-N} in the periodic table. Therefore, without resorting to any experimental estimates of the degree of ionicity, we can estimate the amount of the narrowing of the X_5^v-Γ_{15}^v-L_3^v band at about 8% for an $A^{III}B^V$ compound and at 25-30% for $A^{II}B^{VI}$ compounds; this agrees well with EP data [107]: 15% for $A^{III}B^V$ compounds and 25-30% for $A^{II}B^{VI}$ compounds.

In a similar manner it is easy to prove that in isoelectron series one must expect a considerable decrease in ΔE_1 and ΔE_2 since according to Eqs. (5.41) and (5.42) or (5.57) and (5.58) the values of ΔE_1 and ΔE_2 also go to zero in the limit as $\lambda \to 0$. The nature of the behavior of ΔE_1 and ΔE_2 in the isoelectron series is particularly evident in the "simplified" version of the theory from Eqs. (5.57) and (5.58), which, by using the relations

$$\frac{\lambda^2}{1+\lambda^2} = \frac{1}{8}(N-Z) \tag{5.64}$$

$$\frac{2\lambda}{1+\lambda^2} = \frac{1}{4}\sqrt{(N-Z)(8-N+Z)} \tag{5.65}$$

can be rewritten in the form

$$\Delta E_1 = \frac{1}{8}(N-Z)\,\mathscr{E}_{\text{promot}}(A) + \frac{1}{4}\sqrt{(N-Z)(8-N+Z)}\left\{\frac{\mathscr{E}_{\text{promot}}(A)}{\alpha_A+\alpha_B}\beta - \frac{8}{3}\beta_\pi\right\} \tag{5.66}$$

*Within those limits, of course, within which actual effective charges can vary [205].

$$\Delta E_2 = \frac{1}{8}(N - Z)\,\mathscr{E}_{\text{promot}}(A) + \frac{1}{4}\sqrt{(N-Z)(8-N+Z)}\,\frac{\mathscr{E}_{\text{promot}}(A)}{\alpha_A + \alpha_B}\,\beta \tag{5.67}$$

It is seen from Eqs. (5.66) and (5.67) that both parameters ΔE_1 and ΔE_2 include a "Coulomb" part, decreasing as $N - Z$, and more slowly decreasing (as $\sqrt{N-Z}$) resonance terms. The Coulomb terms in the expressions for ΔE_1 and ΔE_2 are identical, and the resonance term in the expression for ΔE_1 is larger than in the expression for ΔE_2. Therefore in the isoelectron series the value of ΔE_1 should decrease more slowly than ΔE_2, but faster than ΔE_p. This is also confirmed by EP data [107].

Finally, let us consider the question of the gap in the valence band of $A^N B^{8-N}$. As mentioned by Hund and Mrowka [214], this gap (arising because of the splitting of the doubly degenerate X_1^v level in the diamond lattice into the two levels X_1^v and X_3^v) is a distinguishing feature of heteroatomic diamondlike crystals. Therefore the width of this gap $\Delta E_v = \varepsilon(X_3^v) - \varepsilon(X_1^v)$ can serve as some measure of the heteroatomicity of the system. This is evident from Eqs. (5.53) and (5.59), according to which the appearance of the gap ΔE_v is caused by three factors:

(a) the inequality of the promotion energies $\mathscr{E}_{\text{promot}}(A)$ and $\mathscr{E}_{\text{promot}}(B)$ for both components A and B of the crystal $A^N B^{8-N}$;

(b) the inequality of the resonance integrals β_{sApB} and β_{sBpA};

(c) the deviation of the covalence parameter λ from its "purely covalent" value of $\lambda = 1$.

Ignoring the second term in Eqs. (5.53) and (5.59), which is small compared with the first, we can rewrite ΔE_v in the form

$$\begin{aligned}\Delta E_v &\approx \left\{1 - \frac{1}{8}(N - Z)\right\}\mathscr{E}_{\text{promot}}(B) - \frac{1}{8}(N - Z)\,\mathscr{E}_{\text{promot}}(A)\\ &= \mathscr{E}_{\text{promot}}(B) - \frac{1}{8}(N - Z)\{\mathscr{E}_{\text{promot}}(B) + \mathscr{E}_{\text{promot}}(A)\}\end{aligned} \tag{5.68}$$

from which it is seen that ΔE_v is always less than $\mathscr{E}_{\text{promot}}(B)$ by an amount whose value decreases as one goes from group IV elements to the $A^{III}B^{V}$ and $A^{II}B^{VI}$ compounds. Since the value of $\mathscr{E}_{\text{promot}}(B)$ increases during such a transition, the gap $\varepsilon(X_3^v)$-$\varepsilon(X_1^v)$ in the valence band should expand. Thus taking $\mathscr{E}_{\text{promot}}$ from Table 3, we find from Eq. (5.68) that for the isoelectron series Ge-GaAs-ZnSe for effective charges of $Z(\text{GaAs}) = 0.5$, $Z(\text{ZnSe}) = 0.7$ [205], the gap ΔE_v increases from 0 (Ge) to ≈ 3 eV (GaAs) and then to ≈ 7 eV (ZnSe), whereas EP [107] and EC OPW data [88, 103] give close values for these compounds: ~ 4 eV and ~ 9 eV, respectively.

5.4.3 Preliminary Numerical Estimates of the Parameters of the Valence Band Structure

To illustrate the calculational possibilities of the semiempirical EO LCAO method, we shall give a numerical estimation of the parameters of the valence band (5.49)-(5.54) for certain $A^{III}B^{V}$ crystals. For the sake of simplicity and for comparison with the results of chapter 3 we use here the simplified version of the theory (5.55)-(5.60), which forces us to restrict the discussion to the valence band only; see chapter 6 for estimates for the complete band structure. We also adopt the simplifying assumption that all the effective charges in the $A^{III}B^{V}$ compounds are equal and amount to about 0.5.

Then using relations (5.24) and (5.26) and Table 3 to find the Coulomb integrals and relations (5.34) and (5.35) and Table 4 to determine the resonance parameters,* we obtain the $\Delta E_p, \ldots,$ values listed in Tables 12 and 13. These values in Tables 12 and 13 are compared with currently available experimental data obtained from X-ray spectra (the E_v value for BN [215], BP, AlP, and GaP [213]), photoelectron and X-ray-electron spectra for GaP and GaAs [216, 280], and from the data of EC OPW calculations for GaP and GaAs [103] and nonempirical OPW calculations in the rest of the cases when EC OPW data are lacking (BN [218, 219, 162], BP [217, 218, 219], AlP [89, 162], GaAs [162]).

As seen from Tables 12 and 13, despite the simplified nature of the theory, as well as the use of other simplifying assumptions, the parameter values obtained agree well with experiment and the data of the EC OPW calculations. Also the degree of their agreement with the "nonempirical" calculations by the OPW and APW methods is approximately the same as the degree of agreement of these calculations with each other.

The last two columns of Tables 12 and 13 contain data from previously cited calculations ([142] and [165]; see Sec. 3.6.1). A comparison of these data with the data from the preceding columns shows that (just as in the case of covalent crystals) the nonempirical estimate of the matrix elements within the framework of the EO method gives incorrect results even for the valence band, preventing (as in evident from [142, 165]) even a qualitative description of the conduction band structure.

*The values of β(IV), determined from Eq. (3.86) and starting from the value β(diamond) = −6.5 eV, are used in Tables 12 and 13, see Sec. 3.4.2.

6

Chemical Bond and Structure of Energy Bands in Partially Covalent Crystals with Sphalerite Lattice. Conduction Band and Complete Band Structure

6.1 NATURE OF CHEMICAL BOND AND THE CHANGE IN THE BAND STRUCTURE IN VERTICAL AND HORIZONTAL SERIES

6.1.1 Individual Features of Conduction Band in Tetrahedral Partially Covalent Crystals

In partially covalent crystals with the sphalerite lattice, just as in purely covalent crystals, the valence band depends only to a slight degree on the individual nature of the lattice, and the relative arrangement scheme of the levels in this band is common for all $A^N B^{8-N}$ compounds. Conversely, the relative arrangement of levels in the conduction band is highly dependent on the kind of crystal.

Actually, as experimental data show [146, 147, 211], among the $A^{III}B^V$ compounds the compounds of elements with a large atomic number, such as

InSb, InAs, CaSb, or CaAs, have an absolute minimum in the conduction band in the center of the first Brillouin zone–at the point Γ. In this situation the antibonding s level Γ_1^c at this point lies below the antibonding p level Γ_{15}^c.

On the other hand, for compounds of elements with a relatively small atomic number–boron, aluminum, nitrogen, and phosphorus–the absolute minimum of the conduction band probably lies at the X edge of the first zone, and the Γ_1^c and Γ_{15}^c levels are reversed in their order (although completely clearcut experimental data for the cubic modification of BN as well as for the crystals BP, BAs, and AlP are not available at the present time).

Similarly to the valence band structure, it is convenient to investigate the regular variation of the total band structure in partially covalent crystals by arranging the crystals into vertical or horizontal series. Considering, for example, Fig. 5.1, it is easy to see that for the vertical series of $A^{III}B^{V}$ compounds the nature of the variation of the band structure resembles the nature of the variation of the band structure in the group IV series of elements. Namely, as one goes from compounds of the lighter elements to compounds of the heavier, the following behavior is observed (along with a narrowing of the valence band, etc.; see Sec. 5.4.2):

(a) a general converging of the conduction band toward the valence band;

(b) a shift of the absolute minimum of the conduction band from the point X to the point Γ;

(c) an increase in the separation between the antibonding s and p levels Γ_1^c and Γ_{15}^c of the conduction band;

(d) a decrease in the forbidden band gap E_g, which for most $A^N B^{8-N}$ compounds is identical to the energy of the $\Gamma_{15}^v \to \Gamma_1^c$ transition.

Conversely, in the isoelectron series (Ge-GaAs-ZnSe, α-Sn-InSb-CdTe, the alloy SiGe-GaP-ZnS, the hypothetical alloy GeSn-InAs-CdSe) one can observe, as the components of the compound are farther removed from group IV (see Fig. 5.2):

(e) a general increase in the distance between the valence band and the conduction band; and

(f) an increase in the forbidden band gap E_g.

Below we will investigate–first qualitatively and then quantitatively–how this behavior of the band structure is related to the features of the chemical bond in $A^N B^{8-N}$ crystals and, in particular, to their degree of ionicity [221, 118].

6.1.2 Average Separation between Valence Band and Conduction Band for Vertical and Horizontal Series

Let us begin by examining the first and fifth of the band structure features listed above. Similarly to what was done for covalent crystals, let us take as a measure of the average separation between the conduction band and the valence band the difference $\langle \Delta\varepsilon \rangle$ of the average levels $\langle \varepsilon^c \rangle$, $\langle \varepsilon^v \rangle$ of these bands (see Sec. 2.6.1). Since the equivalent orbitals for partially ionic tetrahedral crystals have the form (2.58), it is not hard to see that

$$\langle \varepsilon^v \rangle = \langle \varphi^{(+)} | \hat{H} | \varphi^{(+)} \rangle = \frac{1}{1+\lambda^2} \{\alpha_B + \lambda^2 \tilde{\alpha}_A + 2\lambda\beta\} \tag{6.1}$$

$$\langle \varepsilon^c \rangle = \langle \varphi^{(-)} | \hat{H} | \varphi^{(-)} \rangle = \frac{1}{1+\lambda^2} \{\lambda^2 \alpha_B + \tilde{\alpha}_A - 2\lambda\beta\} \tag{6.2}$$

Consequently, the average distance between these bands is written in the following form:

$$\langle \Delta\varepsilon \rangle = \langle \varepsilon^c \rangle - \langle \varepsilon^v \rangle = \left(\frac{1-\lambda^2}{1+\lambda^2}\right) \left\{\alpha_A - \alpha_B + \frac{2MZ}{R}\right\} - \frac{4\lambda}{1+\lambda^2}\beta \tag{6.3}$$

An expression of similar form is also obtained if the distance between the valence band and the conduction band is characterized by the distance between analogous levels of both bands. Thus the distance between the bonding p level Γ_{15}^v and the antibonding p level Γ_{15}^c is expressed by the formula

$$\varepsilon(\Gamma_{15}^c) - \varepsilon(\Gamma_{15}^v) = \left(\frac{1-\lambda^2}{1+\lambda^2}\right) \left\{\alpha_A^{(p)} - \alpha_B^{(p)} + \frac{2MZ}{R}\right\} - \frac{16\lambda}{3(1+\lambda^2)} \{\beta_{pp} - 2\beta_\pi\} \tag{6.4}$$

and the distance between the bonding and antibonding s level Γ_1^v and Γ_1^c is given by the formula

$$\varepsilon(\Gamma_1^c) - \varepsilon(\Gamma_1^v) = \left(\frac{1-\lambda^2}{1+\lambda^2}\right) \left\{\alpha_A^{(s)} - \alpha_B^{(s)} + \frac{2MZ}{R}\right\} - \frac{16\lambda}{1+\lambda^2}\beta_{ss} \tag{6.5}$$

Now we can investigate to see how Eqs. (6.3)-(6.5) explain the tendency of the valence band and conduction band to converge in the vertical series.

For the case of purely covalent crystals (i.e., for $\lambda = 1$) the right sides of Eqs. (6.3)-(6.5) are similar to purely "resonance" terms. These terms decrease monotonically in parallel with the heats of atomization as one goes from light elements to heavy, thereby explaining the convergence of both bands. In the case of partially ionic crystals two other terms are added to these terms in Eqs. (6.3)-(6.5): The first is determined by the difference of the Coulomb integrals for isolated A^{Z+}, B^{Z-} ions; the second is proportional to the Madelung potential of the lattice. However, the appearance of these terms does not alter the tendency of the valence band and conduction band to converge [i.e., the tendency for the differences $\langle \varepsilon^c \rangle - \langle \varepsilon^v \rangle$, $\varepsilon(\Gamma_{15}^c) - \varepsilon(\Gamma_{15}^v)$, or $\varepsilon(\Gamma_1^c) - \varepsilon(\Gamma_1^v)$ to decrease]. This is readily evident for the example of $A^{III}B^V$ compounds. Actually, as one goes from BN to InSb the resonance terms on the right side of

Eqs. (6.3)-(6.5) are reduced, respectively, by ~7 and ~4 eV (see Tables 4 and 10 and Sec. 5.3.2). At the same time, using the data from Table 3 and calculating the Madelung potential of the lattice, it is not hard to see that the sum of the first two terms in these formulas either also decreases along with the second (in the $A_0^{III}B^V$ series of compounds, where B = N, P, As, Sb, and A_0^{III} is any group III element), remains constant (in the $A^{III}B_0^V$ series, where A^{III} = Al, Ga, In), or, finally, increases (as one goes from B to Al, Ga, and In) but only by 0.5-1 eV.

In a similar manner it is easy to see from Eqs. (6.3)-(6.5) that along the isoelectron series, for example, along the series Ge-GaAs-ZnSe, the nature of the variation of the differences $\langle\varepsilon^c\rangle - \langle\varepsilon^v\rangle$, $\varepsilon(\Gamma_{15}^c) - \varepsilon(\Gamma_{15}^v)$, and $\varepsilon(\Gamma_1^c) - \varepsilon(\Gamma_1^v)$ will be the opposite; thus the distance between the valence band and the conduction band should increase (feature "e" of the band structure, see Sec. 6.1.1). In fact, as one goes from the group IV elements to the $A^{III}B^V$ compounds and then to the $A^{II}B^{VI}$, the resonance terms in Eqs. (6.3)-(6.5) do not decrease very markedly. For partially ionic crystals they are multiplied by the factor $2\lambda/(1 + \lambda^2)$, which amounts to ~0.9-0.7 for charges $Z(A^{III}B^V) = 0.5$ and $Z(A^{II}B^{VI}) =$ 0.7-1.0, which are typical for crystals of these classes. Thus as one goes from the A^{IV} crystals to the $A^{III}B^V$ and $A^{II}B^{VI}$ compounds, the resonance terms are reduced by 10-30% or, in absolute quantities, by 1.5-2 eV for compounds of the elements from period II and by ~1 eV for all the other compounds. However, as can be seen on the basis of the data of Table 3, this decrease in the resonance terms is compensated,* and then some, by the appearance in Eqs. (6.3)-(6.5) of terms of the type

$$\left(\frac{1-\lambda^2}{1+\lambda^2}\right)\left\{\alpha_A - \alpha_B + \frac{2MZ}{R}\right\}$$

the value of which increases drastically in the isoelectron series.

Such qualitative conclusions are clearly illustrated by graphs, showing the dependence of the average energy of the transition $\Gamma_{15}^v \rightarrow \Gamma_{15}^c$ on the average

*We will represent the sum of the Coulomb (and Madelung) terms in Eqs. (6.3)-(6.5) by a and the corresponding combination of resonance integrals by b. Then the right sides of the three equations (6.3)-(6.5) can be rewritten in the general form: $(1 - \lambda^2)(1 + \lambda^2)^{-1}a + 2\lambda(1 + \lambda^2)^{-1}b = b + \{(1 - \lambda^2)a - (1 - \lambda^2)b\}(1 + \lambda^2)^{-1}$. The latter will be larger than b when a satisfies the inequality $a > b(1 - \lambda)(1 + \lambda)^{-1}$, which, for example, assumes the form $a > 0.35b$ for $A^{III}B^V$ compounds with $Z \simeq 0.5$. In fact, as seen from Tables 3, 4, and 11, the inequality $a > b$ is true for all three relations, so that the inequality $a > 0.35b$ is necessarily satisfied and the right side of Eqs. (6.3)-(6.5) always increases as one goes from the group IV elements to the $A^{III}B^V$ compounds. Similarly, it can be shown that the same thing occurs as one goes from the $A^{III}B^V$ compounds to the $A^{II}B^{VI}$ compounds.

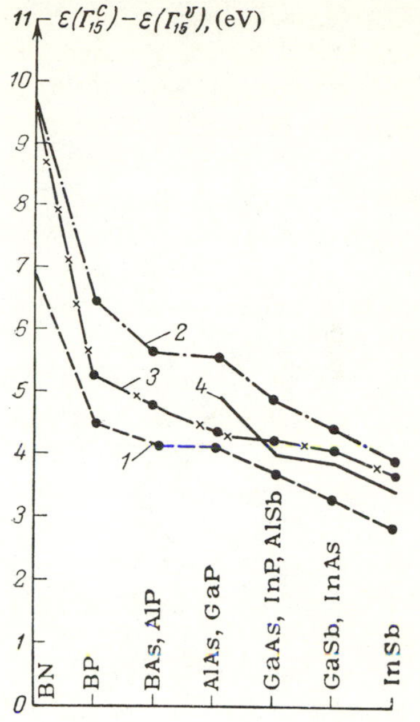

FIG. 6.1. The dependence of the energy of the transition $\Gamma_{15}^{v} \rightarrow \Gamma_{15}^{c}$ on the average atomic number for the vertical series of $A^{III}B^{V}$ compounds. Curve 1 corresponds to the estimate of the resonance parameters from the formula $\beta_{ij}(A^{III}B^{V}) = \beta_{ij}(IV) = \Delta H_{at}(IV) \times \beta_{ij}(\text{diamond}):\Delta H_{at}(\text{diamond})$, where the values of β_{ij}(diamond) are taken from data on hydrocarbons. Curve 2 corresponds to a similar estimate, where the β_{ij}(diamond) values are taken from the band structure of diamond [84, 85]. Curve 3 corresponds to an estimate of all the resonance integrals from the band structure of group IV elements [84]. Curve 4 is plotted from experimental data [146].

atomic number of the components for the vertical series of $A^{III}B^{V}$ compounds* (Fig. 6.1) and the dependence of the energy of the $\Gamma_{15}^{v} \rightarrow \Gamma_{15}^{c}$ transition on the location of the $A^{N}B^{8-N}$ crystal in the isoelectron series Ge-GaAs-ZnSe (Fig. 6.2). Here the different curves on the graph correspond to different methods of estimating the parameters for the purpose of demonstrating that the qualitative conclusions do not depend on the particular method used to make this estimate. [Let us now recall that the inadequacy of the spectroscopic data prevents an unambiguous determination of the resonance integrals $\beta_{ij}(IV)$ from the data for saturated $A_{2n}X_{2n+2}$ molecules.]

*Like any analogous class of two-component compounds, the $A^{III}B^{V}$ compounds form, generally speaking, some two-dimensional ensemble and are not confined to a one-dimensional sequence (except for the "lightest" compound BN and the "heaviest" InSb). Therefore, in order to trace the variation of the band structure as one goes from compounds of light elements to compounds of heavy elements, it is necessary to form from the two-dimensional ensemble of $A^{III}B^{V}$ compounds a one-dimensional ensemble by relating one point on the abscissa axis of each group of compounds to one average atomic number (the average of the atomic numbers of the components). In this case, obviously, the average values of the transition energy, which are obtained by averaging over all crystals with the same average atomic number, must also be plotted on the ordinate axis.

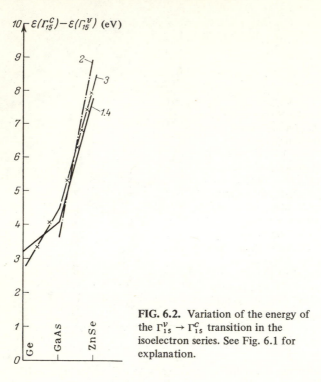

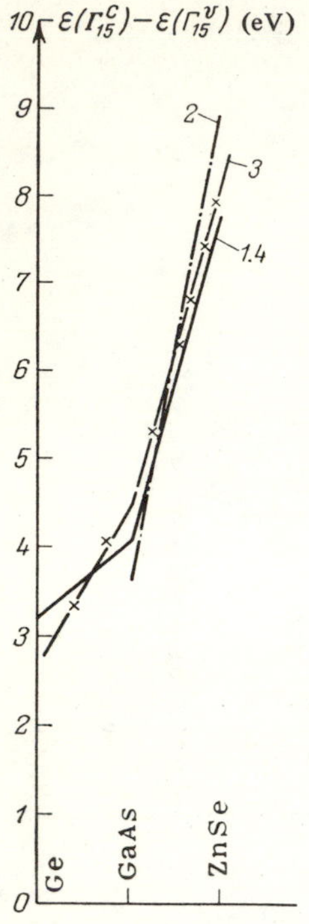

FIG. 6.2. Variation of the energy of the $\Gamma^{v}_{15} \rightarrow \Gamma^{c}_{15}$ transition in the isoelectron series. See Fig. 6.1 for explanation.

6.1.3 Edge of Conduction Band in $A^N B^{8-N}$ Compounds

Let us now turn to a discussion of the features "b" and "c" of the conduction band structure $A^N B^{8-N}$ compounds (see Sec. 6.1.1). As follows from the dispersion law (5.2)-(5.5) and Eqs. (5.12)-(5.16), the relative location of the levels Γ^c_1, Γ^c_{15}, $\bar{X}^c$, and $\bar{\bar{X}}^c$ is determined by the equations

$$\varepsilon(\Gamma^c_{15}) - \varepsilon(\Gamma^c_1) = \frac{\alpha^{(p)}_A - \alpha^{(s)}_A + \lambda^2\{\alpha^{(p)}_B - \alpha^{(s)}_B\}}{1+\lambda^2} + \frac{8\lambda}{1+\lambda^2}\left\{\beta_{ss} - \frac{1}{3}\beta_{pp} + \frac{2}{3}\beta_\pi\right\} \tag{6.6}$$

$$\varepsilon(\bar{X}^c) - \varepsilon(\Gamma^c_1) = \frac{\lambda^2}{1+\lambda^2}\{\alpha^{(p)}_B - \alpha^{(s)}_B\}$$
$$+ \frac{8\lambda}{1+\lambda^2}\left\{\beta_{ss} - \frac{1}{2\sqrt{3}}(\beta_{s_A p_B} + \beta_{s_B p_A})\right\} + \frac{4\lambda}{\sqrt{3}(1+\lambda^2)}\{\beta_{s_B p_A} - \beta_{s_A p_B}\} \tag{6.7}$$

$$\varepsilon(\bar{\bar{X}}^c) - \varepsilon(\Gamma^c_1) = \frac{1}{1+\lambda^2}\{\alpha^{(p)}_A - \alpha^{(s)}_A\}$$
$$+ \frac{8\lambda}{1+\lambda^2}\left\{\beta_{ss} - \frac{1}{2\sqrt{3}}(\beta_{s_A p_B} + \beta_{s_B p_A})\right\} - \frac{4\lambda}{\sqrt{3}(1+\lambda^2)}\{\beta_{s_B p_A} - \beta_{s_A p_B}\} \tag{6.8}$$

It is not hard to see that Eqs. (6.6)-(6.8) are completely analogous to the corresponding formulas (4.2), (4.4) for group IV elements, except that the role of the promotion energy of the s electron into the p level $\mathscr{E}_{\text{promot}}(\text{IV})$ is now played either by one of the two quantities $\mathscr{E}_{\text{promot}}(\text{A}) = \alpha_{\text{A}}^{(p)} - \alpha_{\text{A}}^{(s)}$, $\mathscr{E}_{\text{promot}}(\text{B}) = \alpha_{\text{B}}^{(p)} - \alpha_{\text{B}}^{(s)}$ or by the "average" promotion energy

$$\langle \mathscr{E}_{\text{promot}}^{c} \rangle = \frac{\mathscr{E}_{\text{promot}}(\text{A}) + \lambda^2 \mathscr{E}_{\text{promot}}(\text{B})}{1+\lambda^2} \tag{6.9}$$

(This average energy is different from the average promotion energy for the valence band; see footnote on page 181.)

Therefore the relative arrangement of the levels Γ_1^c, Γ_{15}^c, $\overline{X}^c$, and $\overline{\overline{X}}^c$ in tetrahedral $A^N B^{8-N}$ crystals depends on the values of $\mathscr{E}_{\text{promot}}(\text{A})$, $\mathscr{E}_{\text{promot}}(\text{B})$, and $\langle \varepsilon_{\text{promot}}^{c} \rangle$ and on the resonance integrals $\beta_{ss}(A^N B^{8-N})$, . . . in the same manner as the location of the levels Γ_1^c, Γ_{15}^c, and X_1^c in purely covalent crystals of group IV elements depends on the value of $\mathscr{E}_{\text{promot}}(\text{IV})$ and the resonance integrals $\beta_{ss}(\text{IV})$, ...

In chapter 4 it was stated that the variation of the band structure as one goes from carbon and silicon to germanium and gray tin is a consequence of the inertness of the valence *ns* electrons (Sedgwick inert pair). This inertness of the valence *ns* orbitals is also a characteristic for other elements in the lower part of the periodic table.* Just as for the group IV elements, it can be ascribed either to a relative increase in the promotion energies compared with the resonance integrals, or to a relative decrease in the resonance integrals, including the *ns* orbitals, compared with the integrals including the *np* orbitals (see Sec. 4.4.1). Accordingly, repeating the arguments presented above in chapter 4, it is easy to prove that for elements with a well-pronounced inertness of the *ns* orbital: For elements for which the ratios $|\mathscr{E}_{\text{promot}}(\text{A})/\beta_{ij}|$ and $|\mathscr{E}_{\text{promot}}(\text{B})/\beta_{ij}|$ are large or the ratios β_{ss}/β_{pp} are small, the right sides in Eqs. (6.6)-(6.8) will be positive. Thus, in this case the inequalities

$$\varepsilon(\Gamma_1^c) < \varepsilon(\Gamma_{15}^c); \quad \varepsilon(\Gamma_1^c) < \varepsilon(\overline{X}_v^c); \quad \varepsilon(\Gamma_1^c) < \varepsilon(\overline{\overline{X}}^c) \tag{6.10}$$

will be satisfied; this is actually observed for all compounds of heavy elements (GaAs, GaSb, InAs, InSb, ZnS, ZnSe, CdS, CdSe). In this situation, as is immediately evident from Eq. (6.6), an increase in the inertness of the *ns* electrons should lead to an increase in the difference $\varepsilon(\Gamma_{15}^c) - \varepsilon(\Gamma_1^c)$.

On the other hand, as also already stated, for elements with a small atomic number, especially for elements of the second period of the periodic table, the

*See Sec. 4.4.1 concerning the inertness of the *ns* electrons for elements of groups V and VI. See also [122] for a general discussion of the problem, especially for elements of groups III and II.

valence ns orbitals are extremely prone to participate in the bonds.* Accordingly, going to compounds of the light elements leads to a relative increase in the resonance integrals β_{ss} and a relative decrease in the promotion energies; this is readily evident from a comparison of Table 3 with Table 10. It follows from Eqs. (6.7) and (6.8) that this behavior of β_{ss} and $\mathcal{E}_{\text{promot}}$ should be accompanied by a tendency for an increase in the minimum at or near the edge X of the valence band, just as in the case of covalent crystals.

For a more definite conclusion let us assume, in accordance with what has been said in Sec. 5.3.2, that the resonance integrals in covalent crystals and in their isoelectron partially ionic analogs are similar in value, and let us consider those $A^N B^{8-N}$ compounds that contain light elements of periods II and III in their composition. Then it is not hard to see that for a moderate degree of ionicity the minimum at point X of the first Brillouin zone for these partially ionic crystals should be deeper than the minimum at the point Γ. In fact, for purely covalent crystals the presence of such an absolute minimum essentially means that the contribution of the resonance integrals on the right side of Eq. (4.4) is quite large compared with the value of $(1/2)\mathcal{E}_{\text{promot}}(\text{IV})$. As one goes from covalent crystals to their isoelectron partially ionic analogs, and therefore from Eq. (4.4) to Eqs. (6.7) and (6.8), this contribution decreases by the amount $[1 - 2\lambda(1+\lambda^2)^{-1}]$. For $A^{III}B^{V}$ compounds with $Z \sim 0.5$ this quantity amounts to about 0.1, so that in this case the contribution of the resonance parameters on the right side of Eqs. (6.7) and (6.8) remains nearly constant (we ignore the difference $\beta_{sBpA} - \beta_{sApB}$ in this situation; see Sec. 5.3.2). As seen from Table 3, the Coulomb terms $\mathcal{E}_{\text{promot}}(A)/(1+\lambda^2)$ and $\lambda\mathcal{E}_{\text{promot}}(B)/(1+\lambda^2)$ behave approximately the same in Eqs. (6.7) and (6.8). Thus the ratio between the Coulomb and resonance terms as one goes from the group IV elements to the $A^{III}B^{V}$ compounds changes hardly at all. It can be assumed that the relative arrangement of the levels should also remain unchanged.

Actually, as band structure data show, the bottom of the conduction band in $A^{III}B^{V}$ compounds, containing elements of periods III and II, lies at the point X, as occurs for GaP [211], AlAs [211, 222, 243], AlSb [211, 224], BP [225, 226], AlP [211, 227, 243], and probably for BAs [228] and the cubic modification of BN (although only calculated data are currently available for BN [162, 219, 220, 229-231].

Another consequence, which follows from the appreciable participation of the valence ns electrons in the bond, involves, as is seen from the relations (6.6), a possible inversion of the antibonding s and p levels Γ_1^c and Γ_{15}^c. Although there are no adequately reliable experimental data yet available for the band structure of BN, BP, and BAs, it is fully probable that the level arrangement is this:

*In particular, nitrogen and boron atoms are apparently prone to being used in the bonds of their 2s electrons to approximately the same extent as carbon atoms. In this regard see the footnote in Sec. 5.3.2 concerning the similarity between the chemistry of carbon and the chemistry of borazide compounds.

$\varepsilon(\Gamma_1^c) > \varepsilon(\Gamma_{15}^c)$, similar to the arrangement of these levels in diamond and silicon crystals; this arrangement is found in the cubic modification of BN* (see numerical estimates in Sec. 6.2) and also in two other boron compounds.† In any case the tendency for the Γ_1^c level to rise relative to the Γ_{15}^c level as the average atomic number decreases in $A^{III}B^V$ compounds is observed quite clearly (see Fig. 6.3). The curves given in Figs. 6.1-6.4 clearly illustrate how the relative location of the levels Γ_1^c, Γ_{15}^c, and X_{min} varies when we move up or down along the vertical series of $A^{III}B^V$ compounds.

With regard to this last statement, let us point out a more subtle fact. For group IV elements the inequalities

$$\varepsilon(\Gamma_{2'}^c) > \varepsilon(\Gamma_{15}^c), \quad \varepsilon(\Gamma_{2'}^c) > \varepsilon(X_1^c) \tag{6.11}$$

and

$$\varepsilon(\Gamma_{2'}^c) < \varepsilon(\Gamma_{15}^c), \quad \varepsilon(\Gamma_{2'}^c) < \varepsilon(X_1^c) \tag{6.12}$$

are always satisfied simultaneously: The first pair of inequalities is for diamond and silicon; the second pair is for germanium and gray tin. However, as seen from Eqs. (6.6)-(6.8), this situation can be altered for partially ionic crystals. Actually, with an increase in the degree of ionicity (i.e., as $\lambda \to 0$) the quantity $\lambda^2\mathscr{E}_{promot}(B)/(1 + \lambda^2)$, entering into the right side of Eq. (6.7), decreases as $\sim\lambda^2$ and, consequently, faster than the quantity $8\lambda|\beta_{ss}|/(1 + \lambda^2)$, whereas the value of $\langle\mathscr{E}_{promot}^c\rangle = \{\mathscr{E}_{promot}(A) + \lambda^2\mathscr{E}_{promot}(B)\}(1 + \lambda^2)^{-1}$ approaches a finite limit of $\mathscr{E}_{promot}(A)$. Therefore for partially ionic crystals with "not too inert" an *ns* orbital,‡ rather than for group IV elements, a situation is possible whereby when the inequality $\varepsilon(\Gamma_1^c) < \varepsilon(\Gamma_{15}^c)$ is fulfilled, the inequality $\varepsilon(\Gamma_1^c) > \varepsilon(\overline{X}^c)$ is true. We encounter exactly this situation in the case of GaP [211] and, possibly, in AlP as well [89, 162, 232].

In discussing the relative location of the minimums at the center and at the edges of the first zone, until now we have been speaking only of levels at points Γ and X. Therefore for a complete explanation of feature "b" of the band structure (see Sec. 6.1.1) we must still discuss the question of the relative location of the levels at points Γ and L. This question is somewhat more complicated to examine in general form because of the cumbersome formulas defining the energy of the levels at L. However, without resorting to a numerical estimate of the parameters one can show in general form that for sufficiently small resonance integrals (one of the two factors determining the inertness of the *ns* electrons) the inequalities $\varepsilon(\Gamma_1^c) < \varepsilon(\overline{L})$ and $\varepsilon(\Gamma_1^c) < \varepsilon(\overline{\overline{L}})$ will be satisfied.

Let us consider the difference

*According to calculations of [162, 220, 229, 230]. Let us note, however, that in the OPW calculation of [231] (see also [218, 219]) the opposite level arrangement is obtained: $\varepsilon(\Gamma_1^c) < \varepsilon(\Gamma_{15}^c)$, unlike the results of a previous calculation [230] of the same authors.

†According to SC OPW calculations [93, 217-219].

‡Otherwise, as we have seen, the inequalities (6.10) will always be satisfied.

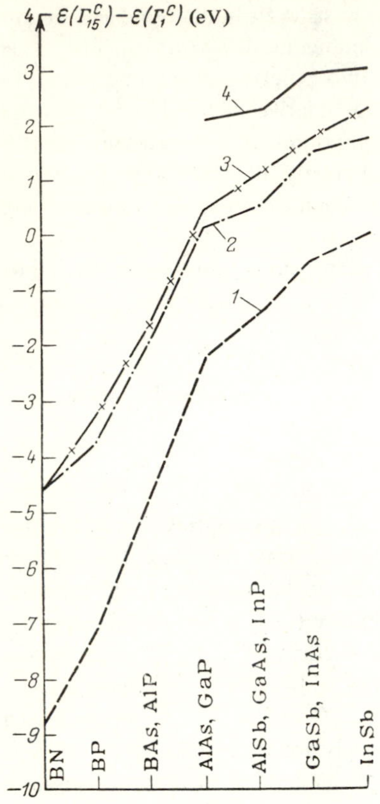

FIG. 6.3. Variation of the level difference $\varepsilon(\Gamma_{15}^c) - \varepsilon(\Gamma_1^c)$ in the vertical series of $A^{III}B^V$ compounds. See Fig. 6.1 for explanation.

$$\varepsilon(\bar{L}^c) - \varepsilon(\Gamma_1^c) = -2\beta_A^{(-)} - 2\beta_B^{(-)} - 12\beta_g^{(-)} - 8\beta_t^{(-)} - \sqrt{\{\beta_A^{(-)} + \beta_B^{(-)} + 4\beta_t^{(-)}\}^2 + 3\{\beta_A^{(-)} - \beta_B^{(-)}\}^2} \tag{6.13}$$

As follows from Eqs. (5.13)–(5.16), for small values of all the resonance parameters the sign of the difference (6.13) is determined by the sign of the expression

$$\delta = -2\beta_A^{(-)} - 2\beta_B^{(-)} - \sqrt{\{\beta_A^{(-)} + \beta_B^{(-)}\}^2 + 3\{\beta_A^{(-)} - \beta_B^{(-)}\}^2} \tag{6.14}$$

in which the values of the parameters $\beta_A^{(-)}$, $\beta_B^{(-)}$ (if the resonance integrals are small) are determined by their "Coulomb" components $\{\alpha_A^{(s)} - \alpha_A^{(p)}\}/4(1 + \lambda^2)$ and $\lambda^2\{\alpha_B^{(s)} - \alpha_B^{(p)}\}/4(1 + \lambda^2)$. Therefore for small resonance integrals we have

$$\delta > (-2\beta_A^{(-)} - 2\beta_B^{(-)} - 2\sqrt{\{\beta_A^{(-)}\}^2 + \{\beta_B^{(-)}\}^2} > (-2\beta_A^{(-)} - 2\beta_B^{(-)} - 2\{|\beta_A^{(-)}| + |\beta_B^{(-)}|\}) = 0 \tag{6.15}$$

This proves the inequality $\varepsilon(\Gamma_1^c) < \varepsilon(\bar{L}^c)$ and the inequality $\varepsilon(\Gamma_1^c) < \varepsilon(\bar{\bar{L}}^c)$ along with it

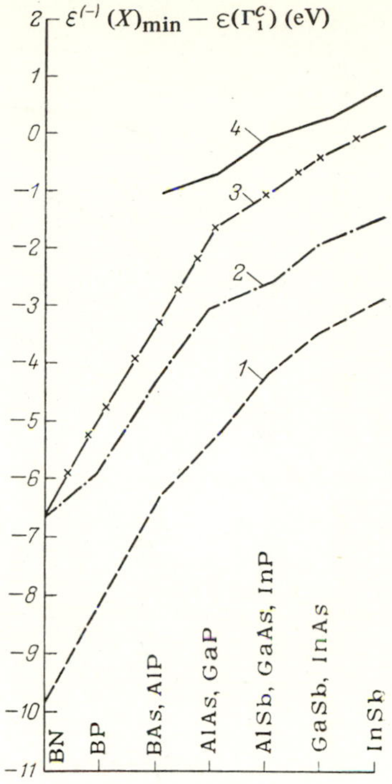

FIG. 6.4. Variation of the level difference $\varepsilon^{(-)}(X)_{min} - \varepsilon(\Gamma_1^c)$ in the vertical series of $A^{III}B^{V}$ compounds. See Fig. 6.1 for explanation.

since for small resonance parameters $\varepsilon(\overline{L}^c) < \varepsilon(\overline{\overline{L}}^c)$ always. Thus, we see that for a sufficiently inert ns orbital the antibonding s level Γ_1^c should actually be lower than all the levels in L; this actually occurs. This relative lowering of the Γ_1^c level as the average atomic number increases in the $A^{III}B^{V}$ series of compounds is clearly illustrated in Fig. 6.5. As seen from curves 1-3, a decrease in the resonance parameters (regardless of the method used to choose them) always leads to an increase in the difference $\varepsilon(L^c)_{min} - \varepsilon(\Gamma_1^c)$. The theoretical curves, 1, 2, 3 are approximately parallel to the experimental curve, so that the increase in the theoretical values of the level difference $\varepsilon(L^c)_{min} - \varepsilon(\Gamma_1^c)$ proceeds at about the same rate as in actuality (although for the specific parameters used here the s level Γ_1^c does not drop below the L_{min}^c level; see Sec. 6.2).

6.1.4 Forbidden Band in Vertical and Horizontal Series of $A^N B^{8-N}$ Compounds

To conclude this section let us turn to the question of the variation of the forbidden band gap in the horizontal and vertical series. To be specific we will assume that E_g is determined by the transition $\Gamma_{15}^v \to \Gamma_1^c$, which is actually the case for most of the diamond-like crystals that have been investigated experimentally.

To investigate the behavior of E_g in this case, let us write the difference in the levels Γ_1^c and Γ_{15}^c in the following form:

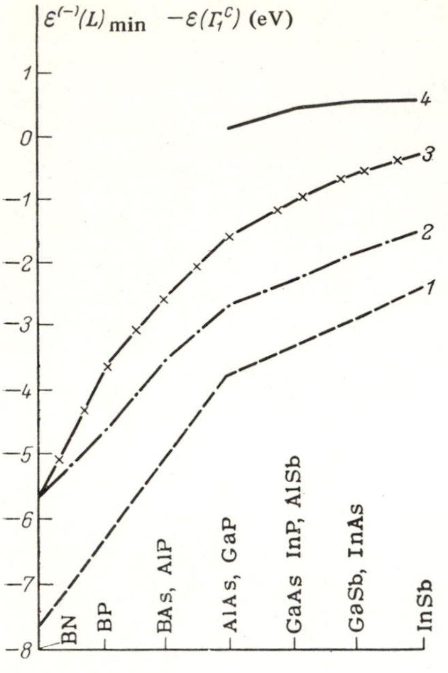

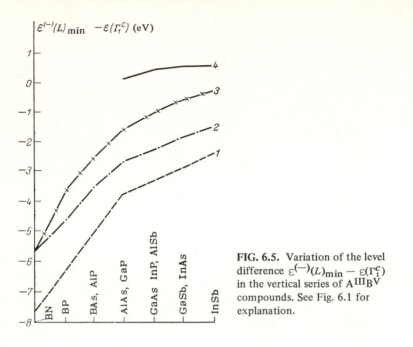

FIG. 6.5. Variation of the level difference $\varepsilon^{(-)}(L)_{\min} - \varepsilon(\Gamma_1^c)$ in the vertical series of $A^{III}B^V$ compounds. See Fig. 6.1 for explanation.

$$\varepsilon(\Gamma_1^c) - \varepsilon(\Gamma_{15}^v) = \{\varepsilon(\Gamma_{15}^c) - \varepsilon(\Gamma_{15}^v)\} - \{\varepsilon(\Gamma_{15}^c) - \varepsilon(\Gamma_1^c)\} \tag{6.16}$$

We have already seen that in the vertical series the energy difference $\varepsilon(\Gamma_{15}^c) - \varepsilon(\Gamma_{15}^v)$ should decrease and the difference $\varepsilon(\Gamma_{15}^c) - \varepsilon(\Gamma_1^c)$ should increase (the inert pair!). Such a variation of both terms on the right side of Eq. (6.16) obviously leads to a decrease in E_g. This is confirmed by test data.

In this situation the ratio

$$\frac{E_g}{\varepsilon(\Gamma_{15}^v \to \Gamma_{15}^c)} = 1 - \frac{\varepsilon(\Gamma_{15}^c) - \varepsilon(\Gamma_1^c)}{\varepsilon(\Gamma_{15}^c) - \varepsilon(\Gamma_{15}^v)} \tag{6.17}$$

should also obviously decrease since the numerator of the fraction on the right side of Eq. (6.17) increases while the denominator decreases. Therefore the forbidden band will decrease faster than the average separation between the conduction band and the valence band. The data given in Table 14 show that this effect is actually observed experimentally.

A somewhat more detailed analysis leads to the conclusion that in the horizontal (isoelectron) series, such as, for example, Ge-GaAs-ZnSe, the value of $E_g = \varepsilon(\Gamma_1^c) - \varepsilon(\Gamma_{15}^v)$ should increase. To understand this let us turn once more to the relation (6.16). As we have already seen in Sec. 6.1.2, the energy of the $\Gamma_{15}^v \to \Gamma_{15}^c$ transition, i.e., the first term on the right side of Eq. (6.16), increases along the isoelectron series. A tentative numerical estimate of the parameters shows that the transition from group IV elements to $A^{III}B^V$ and $A^{II}B^{VI}$

TABLE 14. Measurement of the Energy of the Transition $\Gamma_{15}^{v} \rightarrow \Gamma_{15}^{c}$ and the Forbidden Band Gap for the Vertical Series of $A^{III}B^{V}$ and $A^{II}B^{VI}$ Compounds

Crystal	Transition[a]				
	$\Gamma_{15}^{v} \rightarrow \Gamma_{15}^{c}$	$\Gamma_{15}^{v} \rightarrow \Gamma_{15}^{c}$ average	$\Gamma_{15}^{v} \rightarrow \Gamma_{1}^{c}$	$\Gamma_{15}^{v} \rightarrow \Gamma_{1}^{c}$ average	$\dfrac{E_g(\Gamma_{15}^{v} \rightarrow \Gamma_{1}^{c})_{cp}}{\varepsilon(\Gamma_{15}^{v} \rightarrow \Gamma_{15}^{c})_{cp}}$
GaP	4.9	4.9	2.7	2.7	0.55
GaAs, InP	4.2; 4.1	4.15	1.55; 1.3	1.4	0.35
GaSb, InAs	3.8; 4.3	4.05	0.7; 0.3	0.5	0.12
InSb	3.4	3.4	0.2	0.2	0.06
ZnTe	4.6	4.6	2.3	2.3	0.5
CdTe, HgSe	5.1	5.1	2.15; 0.6	1.4	0.25
HgTe	4.1	4.1	0.02	0.02	0.005

[a]Data on transition energies are taken from the review [146]. All energies are given in eV.

compounds is accompanied by an increase in the level difference $\varepsilon(\Gamma_{15}^{c}) - \varepsilon(\Gamma_{15}^{v})$ by ~1 and ~4 eV, respectively (see Fig. 6.2).

Let us now estimate the behavior of the second term in Eq. (6.16), after rewriting it in the following form:

$$\varepsilon(\Gamma_{15}^{c}) - \varepsilon(\Gamma_{1}^{c}) = \frac{1}{2}\{\mathscr{E}_{\text{promot}}(A) + \mathscr{E}_{\text{promot}}(B)\}$$
$$- \frac{4-N+Z}{8}\{\mathscr{E}_{\text{promot}}(B) - \mathscr{E}_{\text{promot}}(A)\}$$
$$+ \frac{1}{4}\sqrt{(N-Z)(8-N+Z)}\left\{4\beta_{ss} - \frac{4}{3}\beta_{pp} + \frac{8}{3}\beta_{\pi}\right\} \tag{6.18}$$

It is easy to see that the expression within the first set of braces on the right side of this relation changes hardly at all since the sum of the promotion energies $\mathscr{E}_{\text{promot}}(A) + \mathscr{E}_{\text{promot}}(B)$ is nearly constant in the isoelectron series. At the same time the other two terms in Eq. (6.18) vary along the horizontal series, but they compensate one another to a considerable extent. Based on the date of Table 3, it is not hard to see that going from the group IV elements (Ge, Sn) to the $A^{II}B^{VI}$ compounds (ZnS, ZnSe, CdS, CdSe) leads to an increase in the difference $\mathscr{E}_{\text{promot}}(B) - \mathscr{E}_{\text{promot}}(A)$ from zero to 4-5 eV. Therefore for the $A^{II}B^{VI}$ compounds the second term amounts to about 1.5 eV (absolute magnitude). The third term in Eq. (6.18) changes by this same amount as one goes from the A^{IV} to the $A^{II}B^{VI}$ crystals. As before, we will assume that the resonance integrals for the $A^{N}B^{8-N}$ compounds and for the group IV elements are quite similar in value. Then the value of the sum of the resonance integrals in Eq. (6.18)—without the coefficient $(1/4)\sqrt{(N-Z)(8-N+Z)}$—will amount to

~5 eV. As one goes from the group IV elements to the $A^{II}B^{VI}$ compounds, the coefficient $1/4\sqrt{(N-Z)(8-N+Z)}$ changes by 25-30%, so that the third term in the result actually changes by ~1.5 eV. Thus a simultaneous consideration of Eqs. (6.16) and (6.18) shows that E_g increases in the isoelectron series: This actually occurs in real crystals.

6.2 CALCULATION OF THE COMPLETE BAND STRUCTURE OF $A^N B^{8-N}$ COMPOUNDS

In the previous section the question of the dependence of the band structure in partially ionic crystals on the nature of the bond was considered primarily from a qualitative point of view. Here we will consider numerical estimates of the complete band structure of tetrahedral partially ionic crystals [233] for the example of the $A^{III}B^V$ compounds. The appropriate calculations are easily performed, using the same formulas that were used for the qualitative study of the band structure. For the sake of simplicity we will use the tentative effective charge values of $Z = 0.5$ for all the $A^{III}B^V$ compounds, we will ignore the valence state of the atoms in the Coulomb integrals (Table 3), and we assume the resonance integral β_{sBpA} is equal to the integral β_{sApB} (see Sec. 5.3.2). To find the rest of the resonance parameters as well as the average value β_{sp} of the integrals β_{sApB} and β_{sBpA}, we will, as usual, start from Eq. (5.35).

The second column of Table 15 lists the estimates of the complete band structure of the cubic modification of BN obtained in this manner. The resonance integrals found from data for hydrocarbons (see Sec. 4.2.2-4.2.3) were used to find these estimates. For comparison, the third column of this same table lists the analogous estimates using the resonance parameters found from the band structure of diamond (see Table 10), while the rest of the columns list the currently available experimental* and other calculated data.

It is seen from Table 15 that the use of atomic and molecular data in the band structure calculations leads to satisfactory results for the partially ionic crystals too. In every case the data for the cubic modification of BN obtained in this manner agree well with the results of OPW and APW calculations. At the same time, as shown by valence band calculations and the graphs in Figs. 6.1-6.5 for the rest of the $A^N B^{8-N}$ compounds, the estimates of the resonance parameters on the basis of an investigation of the bond in molecules are also able, in principle, to provide a correct description of the band structure, although precise calculations of the conduction band require the parameters found from the uv spectra of appropriate molecules or from the spectroscopic data for saturated $A_n^{IV}X_{2n+2}$ compounds. These data, however, are lacking at

*The estimate $E_v = \varepsilon(\Gamma_{15}^v) - \varepsilon(\Gamma_1^v) \approx$ 15-22 eV for BN was obtained in [215] on the basis of X-ray spectra, the same as the estimate of the forbidden band gap of ~6 eV [234]. The experimental estimate of $E_g \approx 8$ eV is given in [220], and the transition energy estimate of $X_5^v \rightarrow X_1^c \approx 14.5$ eV is given in [235, 236] on the basis of optical data.

TABLE 15. Band Structure of BN from Data of Semiempirical EO LCAO Method and from Other Data[a]

Levels	Semi-empirical EO LCAO	Semi-empirical EO LCAO	Experi-ment	OPW [229]	OPW [162]	OPW [219]	APW [173]
Γ^c_{15}	6.9	9.3		12.6	7.7	7.5	8.9
Γ^c_1	15.6{15.2}	13.9{12.4}		18.1	8.7	7	10.9
Γ^v_{15}	0.0	0.0	0.0	0.0	0.0	0.0	0.0
Γ^v_1	—27.4	—23.2	15—22	18.7	23.2	—26.0	—17.8
X^c_5	13.2	15.3		—	—	—	—
$\overline{\overline{X}}^c$	6.0	7.3	} 6 ÷ 8	9.2	3.4	7	7.2
$\overline{X}^c$	6.5 {5.0}	7.8 {6.3}		11.3	7.4	3.4	10.9
X^v_5	—6.3	—6.0		—2.8	—6.2	—6.1	—3.6
$\overline{\overline{X}}^v$	—5.3	—4.1		—1.7	—10.6	—11.1	—6.8
$\overline{X}^v$	—12	—10.8		—13.2	—18.0	—20.0	—13.3
$\overline{\overline{X}}^c - X^v_5$	12.3	13.3	14.5	10.9	9.6	13.1	10.8
$\overline{X}^c - X^v_5$	12.8{11.3}	13.8{12.3}		14.1	13.6	9.5	14.5
L^c_3	10	12.3		13.3	6.8	9.5	10.4
$\overline{\overline{L}}^c$	10	11.9		12.8	8.8	—	9.9
$\overline{L}^c$	8.1	8.2		—	—	6.8	—
L^v_3	—3.1	—3.0		—1.2	—5.5	—2.5	—1.3
$\overline{\overline{L}}^v$	—6.8	—6.0		—5.1	—12.3	—13.5	—8
$\overline{L}^v$	—16.2	—14.0		—14.6	—19.9	—18.5	—14.3
$\overline{\overline{L}}^c - L^v_3$	13.1	14.9		14	14.3	—	11.2
$\overline{L}^c - L^v_3$	11.2	11.2		—	—	9.2	—

[a] All energies are given in eV.

the present time. Therefore to illustrate the calculation capabilities of the semiempirical EO LCAO method we borrow the resonance parameters from the band structure of the group IV elements (see Table 10).

The corresponding calculation results are listed in Tables 16-19, where, as in Table 15, they are compared with experimental data (in parentheses)* and calculated data. For compounds of In and Ga the experimental level energies in the conduction band are taken from optical spectra [146] and in the valence band from X-ray-electron spectra [280]; the data for AlSb are taken from [39]

*Another satisfactory system of experimental estimates is given in [107] and, in general, it differs little from the estimates of [146]. Let us note, however, that the most recent experimental data—for example, the energy of the $\Gamma^v_{15} \to \Gamma^c_{15}$ transition (4.8–5.0 eV for GaAs [216] and 3.4 eV for GaSb [237]), obtained by a photoemission method, or the energy of the $\Gamma^v_{15} \to X^c_1$ transition (1.6 eV for AlSb [238])—differ markedly from the data given in [107] and in [146].

TABLE 16. Band Structure of BP from Data of Semiempirical EO LCAO Method and from Other Data[a]

Levels	Semiempirical EO LCAO	OPW [217]	OPW [217]	OPW [219]
Γ_{15}^{c}	5.2	3.6	3.1	3.4
Γ_{1}^{c}	8.4 {6.9}	6.8	7.3	8.2
Γ_{15}^{v}	0.0	0.0	0.0	0.0
Γ_{1}^{v}	−16.2	−15.3	−15.7	15.6
X_{5}^{c}	9.0	—	—	13.6
$\overline{\overline{X}}^{c}$	4.8	1.9	0.9	2.5
$\overline{X}^{c}$	3.6 {2.1}	2.2	1.2	3.5
X_{5}^{v}	−3.8	−3.9	−4.3	4.1
$\overline{\overline{X}}^{v}$	−4.1	−8.4	−8.7	9.5
$\overline{X}^{v}$	−7.5	−10.7	−10.8	10.2
$\overline{\overline{X}}^{c}-X_{5}^{v}$	8.6	5.9	5.2	6.6
$\overline{X}^{c}-X_{5}^{v}$	7.4	6.2	5.5	7.6
L_{3}^{c}	7.13	5.2	4.4	5.2
$\overline{\overline{L}}^{c}$	7.0	3.8	3.6	4.8
$\overline{L}^{c}$	4.7	—	—	7.5
L_{3}^{v}	−1.9	−1.7	−1.8	−1.5
$\overline{\overline{L}}^{v}$	−4.1	−8.7	−9.3	9.5
$\overline{L}^{v}$	−10.2	−12.2	−12.2	−12.2
$\overline{\overline{L}}^{c}-L_{3}^{v}$	8.9	6.9	6.2	6.3
$\overline{L}^{c}-L_{3}^{v}$	6.6	5.5	5.4	9.0

[a]All energies are given in eV. There are experimental data for only the Γ_{1}^{v} (−16.9 eV) and X^{c} (1.9–2.1 eV) levels.

and for AlAs from [245–247]. (See also [282].*) The calculated data are taken from EC OPW data [103] and EP data [107] and, when unavailable, from nonempirical OPW calculations. Only calculation results are given for BP, AlP, and BAs (with the Slater and Kohn–Sham exchange potentials for the first two).†

*The experimental data currently available for compounds in In and Ga are extremely fragmentary. The width of the valence band (16.9 eV [213]) and forbidden band gap (1.9–2.1 eV [234, 239]) have been determined for BP from X-ray spectra; the latter value agrees well with an estimate from optical properties (2.0–2.1 eV [240, 241]). The value of E_g (tentatively 0.8–2.0 eV [241]) has been determined for BAs; experimental values of the Γ_{1}^{c} (3.6 eV) and X_{1}^{c} (2.5–2.7 eV) level energies are known [242–244]. See also [281].

†The calculation results with the Slater potential are given for BAs and AlAs. Let us point out that when the Kohn–Sham potential is used, the energy values of the sensitive Γ_{1}^{c}, $\overline{X}^{c}$ levels in these crystals are reduced by ~1 eV.

TABLE 17. Band Structure of BAs and AlP from Data of Semiempirical EO LCAO Method and OPW Method[a]

Levels	BAs		AlP			
	Semiempirical EO LCAO	OPW [92]	Semiempirical EO LCAO	OPW [225]	OPW [89]	OPW [89]
Γ^c_{15}	5.1	3.6	4.3	7.2	4.8	4.2
Γ^c_1	7.4 {6}	4.5	5.3 {4.0}	6.8	3.3	3.3
Γ^v_{15}	0.0	0.0	0.0	0.0	0.0	0.0
Γ^v_1	−15.6	−15.1	−14.5	−10.0	−11.5	−11.7
X^c_5	9.1		6.7	13.6	—	—
$\bar{\bar{X}}^c$	5.1	1.9	3.7	5.5	2.9	2.0
$\bar{X}^c$	3.9 {2.4}	2.2	2.2 {1.2}	4.8	2.1	1.0
X^v_5	−4.0	−3.8	−2.4	−1.4	−2.1	−2.3
$\bar{\bar{X}}^v$	−4.5	−8.3	−4.0	−2.9	−5.5	−5.7
$\bar{X}^c$	−7.7	−11.1	−7.2	−8.2	−9.1	9
$\bar{\bar{X}}^v - X^v_5$	9.2	5.7	6.1	8.4	5.0	4.3
$\bar{X}^v - X^v_5$	7.9	7.2	4.6	7.7	4.2	3.3
L^c	7.1	5.2	5.5	6.8	5.3	4.4
$\bar{\bar{L}}^c$	7.4	2.9	5.2	10.7	8.7	7.6
$\bar{L}^c$	4.5	—	3.2	6.7	3.0	2.6
L^v_3	−2.0	−1.7	−1.2	−2.7	−0.8	−0.8
$\bar{\bar{L}}^v$	−5.2	−8.3	−3.8	−3.7	−5.5	−6.0
$\bar{L}^v$	−9.8	−12.3	−9.4	−8.2	−9.8	−9.7
$\bar{\bar{L}}^c - L^v_3$	9.4	4.6	6.4	13.4	9.5	8.4
$\bar{L}^c = L^v_3$	6.5	—	4.4	9.4	3.8	3.4

[a]Energies in eV.

As seen from Tables 15-19, even with the simplifying assumptions we have introduced the data of the semiempirical EO LCAO method agree well on the whole with experiment and with data of EC OPW and EP calculations, and their degree of agreement with the data of nonempirical methods is about the same as the agreement among the nonempirical methods.

It is not hard to show that a refinement in the approach leads to a further improvement in the results of the EO LCAO calculations. To accomplish this refinement, we take into consideration that

(a) the effective charge values are different for the different $A^{III}B^{V}$ compounds and for compounds of elements from periods III and V they may be closer to $Z = 1/3$ than to $Z = 1/2$*;

*In particular, a value of $Z = 1/3$ leads to very satisfactory agreement with experiment for the value of the spin orbital splitting in the valence band [248].

TABLE 18. Band Structure[a] of AlAs, AlSb, GaP, and GaAs from Data of Semiempirical EO LCAO Method and from Other Data

Levels	AlAs		AlSb		GaP		GaAs	
	Semiempirical EO LCAO	Experimental OPW [249]	Semiempirical EO LCAO	Experimental EP	Semiempirical EO LCAO	Experimental EC OPW	Semiempirical EO LCAO	Experimental EC OPW
Γ_{15}^c	4.2	4.6 (4.4)	3.6	4.1 (3.9)	4.5	4.7 (4.9)	4.5	4.6 (4.2)
Γ_1^c	4.05 {3.0}	2.5 (3.2)	2.5 {1.5}	1.9 (1.8)	3.8 {2.8}	2.8	3.3 {2.3}	1.54 (1.55)
Γ_{15}^v	0.0	0.0	0.0	0.0	0.0	0.0	0.0	0.0
Γ_1^v	—13.6	—11.5	—13.9	—	—13.90	—11.8	—13.2	—12.4
X_5^c	6.8	—	5.7	—	7.0	—	7.2	—
$\bar{\bar{X}}^c$	3.9	2.9	2.6	2.5	4.1	2.5	4.6	2.5
$\bar{X}^c$	2.4 {1.7}	2.4 (2.2)	1.5 {0.8}	2.0 (1.9)	2.24	2.3	2.6	1.9
X_5^v	—2.5	—2.0	—2.1	—1.9	—2.54	—2.3	—2.7	—2.3
$\bar{\bar{X}}^v$	—4.3	—5.2	—4.0	—	—4.5	—6.1	—4.9	—5.5
$\bar{X}^v$	—7.4	—9.6	—8.0	—	—7.5	—9.2	—7.8	—10.7
$\bar{\bar{X}}^c - X_5^v$	6.5	4.9 (5.0)	4.7	4.3 (4.6)	6.6	4.8 (5.5)	7.3	4.8 (5.1)
$\bar{X}^c - X_5^v$	5.0	4.4 (4.6)	3.6	3.9 (4.2)	4.7	4.6 (5.2)	5.3	4.2 (4.6)
L_3^c	5.5	—	4.7	—	5.7	5.5	5.8	5.3
$\bar{\bar{L}}^c$	5.5	5.2	4.2	4.6	5.7	—	6.2	—
$\bar{L}^c$	2.6	2.6	1.5	2	2.4	2.5	2.5	2.0
L_3^v	—1.3	—0.8	—1.1	—0.8	—1.27	—0.9	—1.4	—0.9
$\bar{\bar{L}}^v$	—4.0	—5.2	—3.8	—	—4.32	—6.0	—4.7	—5.6
$\bar{L}^v$	—9.2	—10.1	—9.7	—	—9.42	—10.0	—9.3	—11.1
$\bar{\bar{L}}^c - L_3^v$	6.8	6	5.3	5.3 (5.1)	6.97	6.4	7.5	6.4
$\bar{L}^c - L_3^v$	3.9	3.4 (3.9)	2.5	2.8 (2.9)	3.67	3.4	3.8	2.5

[a] All energies in eV.

TABLE 19. Band Structure[a] of GaSb, InP, InAs, and InSb from Data of Semiempirical EO LCAO Method and from Other Data

Levels	GaSb		InP		InAs		InSb	
	Semiempirical EO LCAO	Experimental EC OPW	Semiempirical EO LCAO	Experimental EP	Semiempirical EO LCAO	Experimental EP	Semiempirical EO LCAO	Experimental EP
Γ_{15}^{c}	3.7	3.4 (3.8)	4.45	4.6 (4 1)	4.4	4.6 (4.3)	3.7	3.7 (3.4)
Γ_{1}^{c}	1.6 {0.6}	1 (0.7)	3.5 {2.5}	1.6 (1.3)	2.9 {1.9}	0.5 (0.3)	1.4 {0.4}	0.6 (0.2)
Γ_{15}^{v}	0.0	0.0	0.0	0.0	0.0	0.0	0.0	0.0
Γ_{1}^{v}	—13.5	—11.1	—13.8	—12.8	—12.5	—	—12.9	—
X_{5}^{c}	6.0	—	6.6	—	6.6	—	5.7	—
$\bar{\bar{X}}^{c}$	3.0	1.5	3.8	2.5	4.1	2.5	2.7	2.5
$\bar{X}^{c}$	1.5	1.3	2.2 {1.5}	2.2	2.4 {1.8}	2.1	1.57 {0.6}	2.0
X_{5}^{v}	—2.3	—2.2	—2.1	—2.0	—2.3	—1.9	—1.9	—1.5
$\bar{\bar{X}}^{v}$	—4.5	—5.2	—4.1	—	—4.4	—	—4.1	—
$\bar{X}^{v}$	—8.2	—9.8	—7.2	—	—7 4	—	—8.0	—
$\bar{\bar{X}}^{c} - X_{5}^{v}$	5.3	3.7 (4.6)	5.9	4.5	6.4	4.3 (4.9)	4.7	3.8 (4.5)
$\bar{X}^{c} - X_{5}^{v}$	3.8	3.6 (4.2)	4.3	4.2 (4.8)	4.7	3.9 (4.5)	3.5	3.5 (4.0)
L_{3}^{c}	4.9	4.2	5.5	5.2	5.5	5.0	4.6	4.6
$\bar{\bar{L}}^{c}$	4.7	—	5.2	—	5.6	—	4.4	—
$\bar{L}^{c}$	1.2	1.1	2.3	2.0	2.3	1.6	1.1	1.6
L_{3}^{v}	—1.1	—0.9	—1.1	—0.9	—1.1	—0.8	—0.9	—0.6
$\bar{\bar{L}}^{v}$	—4.3	—5.5	—3.8	—	—4.2	—	—3.8	—
$\bar{L}^{v}$	—9.8	—9.9	—9.1	—	—8.9	—	—9.5	—
$\bar{\bar{L}}^{c} - L_{3}^{v}$	5.9	5.1 (5.7)	6.3	6.0 (6.9)	6.6	5.7 (6.9)	5.3	5.1 (5.8)
$\bar{L}^{c} - L_{3}^{v}$	2.3	2.0 (1.7)	3.4	2.8	3.5	2.3 (2.3)	2.1	2.1 (1.9)

[a] All energies in eV.

(b) taking the valence state into consideration leads to a lowering of the ns levels for A^{III} atoms by ~2.5 eV (boron atoms) and by ~1.5 eV (Al, Ga, and In atoms) (see Sec. 5.3.1);

(c) the resonance integral β_{sBpA} is not identical to β_{sApB} and the difference of these integrals, according to the estimates (5.44), (5.48), is equal to ~0.15–0.2 eV.

It is then easy to see from Eqs. (6.4) and (6.6) that a change in the charge from $Z = 1/2$ to $Z = 1/3$, like the other two factors, has practically no effect on the location of the Γ_{15}^c level, while consideration of the valence state lowers the Γ_1^c ns level with respect to the Γ_{15}^c level by the amount $\{\alpha_A^{(s)} - \alpha_A^{(s)*}\}/(1 + \lambda^2) \approx (2/3)\{\alpha_A^{(s)} - \alpha_A^{(s)*}\} \approx 1.5$ eV for boron compounds and ~1 eV for Al, Ga, and In compounds. At the same time, as follows from Eq. (6.7), the $\overline{X}^c$ level in the B, Al, Ga, and In compounds, when the refinements (a)-(c) are taken into consideration, is displaced upward with respect to the Γ_1^c level by ~0.3 eV—primarily because of the increase in the difference $\{\alpha_B^{(p)} - \alpha_B^{(s)}\}$ when Z is reduced (see Table 3).

The refined energies of the Γ_1^c and $\overline{X}^c$ levels are given in the braces in the columns of Tables 15–19. As is seen from these tables, they actually agree better with the experimental data (the level energies can be refined still further by a more careful estimation of the Coulomb integrals, by replacing the tentative Z values with more precise values of the effective charges, etc.).

6.3 RELATIONSHIP BETWEEN STRUCTURE OF ENERGY BANDS IN PARTIALLY COVALENT AND IN PURELY COVALENT CRYSTALS

6.3.1 Analogy between Band Structure of Group IV Elements and $A^N B^{8-N}$ Semiconductors

In chapter 5 and also in Sec. 6.1 we have already touched upon the question of the relationship between the band structure in covalent and in partially ionic crystals. Here we will consider this in more detail and from a different point of view [250].

As already mentioned many times there is a close similarity between the crystals of group IV elements and the crystals of the rest of the $A^N B^{8-N}$ semiconductors in many respects and, in particular, a similarity in the structure of the energy bands.

The comparison of Fig. 3.2 with Figs. 5.1 and 5.2 that has already been done shows that the band structure of partially ionic crystals can essentially be considered to be the band structure of the group IV elements, but deformed somewhat because of the heteroatomicity (and heteropolarity) of the $A^N B^{8-N}$ semiconductors. It is natural to assume that this fact not only can be expressed as some qualitative statement, but it can also be used to obtain specific formulas that would make it possible to relate the band structure parameters of partially ionic crystals with the corresponding parameters for purely covalent crystals.

6.3.2 Heteropolar Perturbation and the Herman Theory

This point of view was first set forth by Herman [210]. Considering the Hamiltonian (1.9) for a tetrahedral $A^N B^{8-N}$ semiconductor with an arbitrary potential V, Herman wrote it in the form

$$-\frac{1}{2}\Delta+V=-\frac{1}{2}\Delta+V_{IV}+\Delta V \tag{6.19}$$

where V_{IV} is the potential for the isoelectron semiconductor A^{IV}, and ΔV is an added potential, which Herman suggested be considered as a perturbation ("heteropolar perturbation"). Herman postulated that the perturbing potential ΔV satisfies two requirements:

(a) the potential ΔV is antisymmetrical with respect to the middle of any of the A–B bonds (see Sec. 5.3.2);

(b) in any isoelectron series of $A^N B^{8-N}$ compounds the potential ΔV increases proportionally to the distance the components of the compound are from group IV, so that, for example, for ZnSe the perturbing potential is twice as large and for CuBr three times as large as for GaAs:

$$\Delta V=\lambda_H \Delta V_0 \tag{6.20}$$

where ΔV_0 is the perturbation that appears as one goes from the crystal A^{IV} to the isoelectron crystal $A^{III}B^V$. Thus, in Eq. (6.20) λ_H–the "Herman ionicity parameters"*–is a measure of the heteropolarity of the $A^N B^{8-N}$ crystal and is assumed to be equal to zero for group IV elements, one for $A^{III}B^V$ compounds, two for $A^{II}B^{VI}$ compounds, and three for $A^I B^{VII}$ compounds.

Based on Eqs. (6.19) and (6.20), one can express the matrix elements of the potential V in any basis in terms of the matrix elements of the potentials V_{IV} and ΔV_0 and in terms of the Herman ionicity parameter. Correspondingly, on the basis of postulates (a) and (b) Herman developed a theory (the "heteropolar perturbation theory") that makes it possible to express transitions in a partially ionic $A^N B^{8-N}$ crystal in terms of the analogous transitions in an isoelectron covalent crystal. Thus in the simplest case of the transition $\Gamma_{15}^v \to \Gamma_{15}^c$ it is not hard to obtain the following expression using the Herman method [146]

$$\Delta\varepsilon=\Delta\varepsilon_{IV}+\frac{2\lambda_H^2 A^2}{\Delta\varepsilon_{IV}} \tag{6.21}$$

where $\Delta\varepsilon$ is the energy of the $\Gamma_{15}^v \to \Gamma_{15}^c$ transition in a partially ionic crystal, $\Delta\varepsilon_{IV}$ is the energy of the $\Gamma_{25'}^v \to \Gamma_{15}^c$ transition in the corresponding A^{IV} isoelectron crystal, and A is a matrix element of the transition between the $\Gamma_{25'}^v$ and Γ_{15}^c states for the "unit" perturbation ΔV_0:

$$A=\langle \psi(\Gamma_{25'}^v)\,|\,\Delta V_0\,|\,\psi(\Gamma_{15}^c)\rangle$$

Another, somewhat more rigorous approach to the problem of the relationship between the transitions in partially ionic and covalent crystals was suggested by Cardona [251]. For

*The parameter λ_H should not be confused with the parameter λ used by us!

the $\Gamma_{15}^{v} \rightarrow \Gamma_{15}^{c}$ transition cited above this approach gives the formula

$$\Delta\varepsilon = \Delta\varepsilon_{IV}\sqrt{1+\frac{4\lambda_H^2 A^2}{\Delta\varepsilon_{IV}^2}} \tag{6.22}$$

[Equation (6.22) is obtained in a two-band approximation, according to which the perturbation ΔV shifts the $\Gamma_{25'}^{v}$ and Γ_{15}^{c} states.]

An indisputable advantage of the Herman theory is the ability to interrelate the transition energies in an isoelectron series. Thus Eq. (6.21) does a good job of presenting the parabolic nature of the dependence of the energy of the $\Gamma_{15}^{v} \rightarrow \Gamma_{15}^{c}$ transition on the Herman ionicity parameter.

However, the Herman theory (like the Cardona theory too) is not closed in the sense that is does not make it possible to determine the proportionality constant A. This matrix element of the theory is simply taken from experimental data. Thus it allows one to find, for example, how much the value of $\varepsilon(\Gamma_{15}^{v} \rightarrow \Gamma_{15}^{c})$ increases for ZnSe and CuBr compared with Ge if it is known how much it increases for GaAs. However, the theory does not allow the difference $\varepsilon(\Gamma_{15}^{v} \rightarrow \Gamma_{15}^{c})_{GaAs} - \varepsilon(\Gamma_{25'}^{v} \rightarrow \Gamma_{15}^{c})_{Ge}$ to be determined. As we will see below, the semiempirical EO LCAO method makes it possible to develop an approach that removes this drawback.

6.3.3 Relationship between Band Structure Parameters of Partially Ionic and Covalent Crystals

To find the relationship between the band structure parameters in A^{IV} and $A^N B^{8-N}$ semiconductors, we will compare pairwise the formulas for the analogous transitions in covalent and in partially ionic crystals, for example, (4.14) and (6.4). It is easy to see that they contain the same characteristic combinations of the resonance integrals (if the difference between the parameters $\beta_{sA pB}$ and $\beta_{sB pA}$ is ignored).

Let us now assume, as always, that each of these integrals has approximately the same value for an $A^N B^{8-N}$ crystal as for an isoelectron A^{IV} crystal. Then we can eliminate these combinations from each pair of formulas and can obtain immediately the relationship between the transition energies in both types of crystals. For example, for the $\Gamma_{15}^{v} \rightarrow \Gamma_{15}^{c}$ transition considered above this relationship is expressed by the formula

$$\Delta\varepsilon = \left(\frac{1-\lambda^2}{1+\lambda^2}\right)\left\{\alpha_A^{(p)} - \alpha_B^{(p)} + \frac{2MZ}{R}\right\} + \frac{2\lambda}{1+\lambda^2}\Delta\varepsilon_{IV} \tag{6.23}$$

or the equivalent formula

$$\Delta\varepsilon = \frac{1}{4}(4-N+Z)\left\{\alpha_A^{(p)} - \alpha_B^{(p)} + \frac{2MZ}{R}\right\} + \frac{1}{4}\sqrt{(N-Z)(8-N+Z)}\,\Delta\varepsilon_{IV} \tag{6.24}$$

which are obtained by eliminating the combination $(8/3)\{\beta_{pp} - 2\beta_\pi\}$ from the pair of equations (4.14) and (6.4).

Equations (6.23) and (6.24) are typical for the approach formulated above. Just as in the Cardona and Herman relations [(6.21) and (6.22)], these formulas make it possible to relate transitions in partially ionic crystals with the corresponding transitions in purely covalent crystals. However, unlike Eqs. (6.21) and (6.22) they do not contain an undefined constant. Therefore they not only make it possible to correlate the transitions in the isoelectron series, but also allow one to find the absolute values of the energies of these transitions.

The data on the energies of the $\Gamma_{15}^{v} \to \Gamma_{15}^{c}$ transition, obtained by the EO LCAO method from Tables 15-19, as well as the graphs of Fig. 6.2 can be used as a numerical illustration for Eqs. (6.23) and (6.24). As seen from these data, the results of the transition energy calculation agree fairly well with experiment.

It is not hard to obtain similar relations for other transitions too. Thus, for the $\Gamma_{15}^{v} \to \Gamma_{1}^{c}$, $X_5^{v} \to \overline{X}^{c}$, $X_5^{v} \to \overline{\overline{X}}^{c}$ transitions we obtain

$\Gamma_{15}^{v} \to \Gamma_{1}^{c}$ transition

$$\Delta\varepsilon = \left(\frac{1-\lambda^2}{1+\lambda^2}\right)\left\{\alpha_A^{(p)} - \alpha_B^{(p)} + \frac{2MZ}{R}\right\} - \frac{\mathscr{E}_{\text{promot}}(A) + \lambda^2\mathscr{E}_{\text{promot}}(B)}{1+\lambda^2}$$
$$+ \frac{2\lambda}{1+\lambda^2}\{\mathscr{E}_{\text{promot}}(IV) + \Delta\varepsilon_{IV}\} \tag{6.25}$$

$X_5^{v} \to \overline{X}^{c}$ transition

$$\Delta\varepsilon = \left(\frac{1-\lambda^2}{1+\lambda^2}\right)\left\{\alpha_A^{(p)} - \alpha_B^{(p)} + \frac{2MZ}{R}\right\} - \frac{\mathscr{E}_{\text{promot}}(A)}{1+\lambda^2}$$
$$+ \frac{2\lambda}{1+\lambda^2}\left\{\frac{1}{2}\mathscr{E}_{\text{promot}}(IV) + \Delta\varepsilon_{IV}\right\} \tag{6.26}$$

$X_5^{v} \to \overline{\overline{X}}^{c}$ transition

$$\Delta\varepsilon = \left(\frac{1-\lambda^2}{1+\lambda^2}\right)\left\{\alpha_A^{(p)} - \alpha_B^{(p)} + \frac{2MZ}{R}\right\} - \frac{\lambda^2\mathscr{E}_{\text{promot}}(B)}{1+\lambda^2}$$
$$+ \frac{2\lambda}{1+\lambda^2}\left\{\frac{1}{2}\mathscr{E}_{\text{promot}}(IV) + \Delta\varepsilon_{IV}\right\} \tag{6.27}$$

Let us point out, finally, that relations such as (6.23)-(6.27) can be derived not only for band-to-band transitions, but also for the other band structure parameters. Let us eliminate, for example, the integrals β_π from Eqs. (3.96) and (3.99) for covalent crystals and from Eqs. (5.56) and (5.60) for partially ionic crystals; then we obtain the relation

$$\Delta\varepsilon = \frac{2\lambda}{1+\lambda^2}\Delta\varepsilon_{IV} \tag{6.28}$$

which makes it possible to relate (within the framework of the EO LCAO method) the level difference $\varepsilon(\Gamma_{15}^v) - \varepsilon(X_5^v)$ for a partially ionic crystal to the corresponding level difference $\varepsilon(\Gamma_{25'}^v) - \varepsilon(X_4^v)$ for a covalent crystal. This same relationship obviously exists between the level difference $\varepsilon(\Gamma_{15}^v) - \varepsilon(L_3^v)$ in partially ionic crystals and the level difference $\varepsilon(\Gamma_{25'}^v) - \varepsilon(L_{3'}^v)$ in purely covalent crystals.

Let us note that the last example illustrates one more feature of the approach considered in this section. In calculating the band structure in $A^N B^{8-N}$ crystals, we found the resonance integrals from the band structure of covalent crystals and then used them for the calculation of partially ionic crystals. Therefore at first glance the approach being described is equivalent to the method used previously in the calculation of the band structure. (Elimination of the resonance terms means in essence that they are found for the covalent crystals and substituted into the formulas for the partially ionic crystals.) However, the two approaches are not in fact equivalent since different values of the resonance integrals can correspond to different parameters of the band structure; this is ignored in the previous approach with its single system of resonance parameters. Actually, as is seen, for example, from the EC OPW and EP data [103, 107] (see also Table 11), a value of β_π about 10-15% smaller than for describing the $\varepsilon_{3,4}(\mathbf{k})$ p level at X must be taken to describe this level at point L. Almost this same degree of difference occurs both for covalent and for partially ionic crystals and is obviously taken into account automatically by Eq. (6.28).

6.3.4 Herman Postulates and Empirical Values of Coulomb Integrals

In concluding this section let us examine the question of Herman's two postulates (see Sec. 6.3.2), especially since the first of them was used in Sec. 5.3.2. We will see that the semiempirical approach to the band structure makes it possible to give these postulates some justification.

Let us consider the first postulate. Although it is hardly possible on the basis of semiempirical considerations to prove the antisymmetrical nature of the perturbing potential $\Delta V = V - V_{\mathrm{IV}}$ itself, it is easy to accomplish such a proof in a more indirect manner by using the Coulomb integrals $\alpha^{(p)}$ and $\alpha^{(s)}$ (the orbital ionization potentials).

Actually, let us assume the symmetrical component of the potential $V = V_{A^N B^{8-N}}$ in a partially ionic crystal is equal to the potential V_{IV} in a crystal of the corresponding group IV element. Then taking the atomic orbitals for the group IV elements as the basis s and p functions, we obtain

$$\tilde{\alpha}_A^{(s)} \approx \left\langle s_{\mathrm{IV}}^A \middle| -\frac{1}{2}\Delta + V \middle| s_{\mathrm{IV}}^A \right\rangle = \left\langle s_{\mathrm{IV}}^A \middle| -\frac{1}{2}\Delta + V_{\mathrm{IV}} + \Delta V \middle| s_{\mathrm{IV}}^A \right\rangle$$

$$= \alpha_{\mathrm{IV}}^{(s)} + \Delta\alpha_A^{(s)}$$

$$\tilde{\alpha}_A^{(p)} \approx \left\langle p_{\mathrm{IV}}^A \middle| -\frac{1}{2}\Delta + V \middle| p_{\mathrm{IV}}^A \right\rangle = \left\langle p_{\mathrm{IV}}^A \middle| -\frac{1}{2}\Delta + V_{\mathrm{IV}} + \Delta V \middle| p_{\mathrm{IV}}^A \right\rangle$$

$$= \alpha_{\mathrm{IV}}^{(p)} + \Delta\alpha_A^{(p)}$$

$$\bar{\alpha}_{\mathrm{B}}^{(s)} \approx \left\langle s_{\mathrm{IV}}^{\mathrm{B}} \left| -\frac{1}{2}\Delta + V \right| s_{\mathrm{IV}}^{\mathrm{B}} \right\rangle = \left\langle s_{\mathrm{IV}}^{\mathrm{B}} \left| -\frac{1}{2}\Delta + V_{\mathrm{IV}} + \Delta V \right| s_{\mathrm{IV}}^{\mathrm{B}} \right\rangle$$

$$= \alpha_{\mathrm{IV}}^{(s)} + \Delta\alpha_{\mathrm{B}}^{(s)}$$

$$\bar{\alpha}_{\mathrm{B}}^{(p)} \approx \left\langle p_{\mathrm{IV}}^{\mathrm{B}} \left| -\frac{1}{2}\Delta + V \right| p_{\mathrm{IV}}^{\mathrm{B}} \right\rangle = \left\langle p_{\mathrm{IV}}^{\mathrm{B}} \left| -\frac{1}{2}\Delta + V_{\mathrm{IV}} + \Delta V \right| p_{\mathrm{IV}}^{\mathrm{B}} \right\rangle$$

$$= \alpha_{\mathrm{IV}}^{(p)} + \Delta\alpha_{\mathrm{B}}^{(p)}$$

where, obviously,

$$\Delta\alpha_{\mathrm{A}}^{(s)} = -\Delta\alpha_{\mathrm{B}}^{(s)}, \quad \Delta\alpha_{\mathrm{A}}^{(p)} = -\Delta\alpha_{\mathrm{B}}^{(p)} \tag{6.29}$$

It follows from this that in such a case the condition

$$\frac{1}{2}\{\bar{\alpha}_{\mathrm{A}}^{(s)} + \bar{\alpha}_{\mathrm{B}}^{(s)}\} \approx \alpha_{\mathrm{IV}}^{(s)}$$

$$\frac{1}{2}\{\bar{\alpha}_{\mathrm{A}}^{(p)} + \bar{\alpha}_{\mathrm{B}}^{(p)}\} \approx \alpha_{\mathrm{IV}}^{(p)}$$

must be satisfied, which in the approximation being used here reduces to similar equations for the Coulomb integrals $\alpha_{\mathrm{A}}^{(s)}(Z), \ldots$

$$\begin{aligned} \frac{1}{2}\{\alpha_{\mathrm{A}}^{(s)} + \alpha_{\mathrm{B}}^{(s)}\} &\approx \alpha_{\mathrm{IV}}^{(s)} \\ \frac{1}{2}\{\alpha_{\mathrm{A}}^{(p)} + \alpha_{\mathrm{B}}^{(p)}\} &\approx \alpha_{\mathrm{IV}}^{(p)} \end{aligned} \tag{6.30}$$

TABLE 20. Comparison of Average Orbital Ionization Potentials[a] of Atoms for $A^N B^{8-N}$ Compounds and Orbital Ionization Potentials of Atoms for Isoelectron A^{IV} Crystals

Compounds	Z=0		Z=1		Element	Z=0	
	$\frac{1}{2}\{\alpha_A^{(s)}+\alpha_B^{(s)}\}$	$\frac{1}{2}\{\alpha_A^{(p)}+\alpha_B^{(p)}\}$	$\frac{1}{2}\{\alpha_A^{(s)}+\alpha_B^{(s)}\}$	$\frac{1}{2}\{\alpha_A^{(p)}+\alpha_B^{(p)}\}$		$\alpha_{IV}^{(s)}$	$\alpha_{IV}^{(p)}$
BN	—19.8	—10.7	—18.8	—9.7	C	—19.5	—10.7
BeO	—20.8	—10.9	—17.6	—7.8	C	—19.5	—10.7
AlP	—15.0	—8.1	—13.1	—7.1	Si	—15.0	—7.8
GaAs	—15.1	—7.5	—13.8	—7.1	Ge	—15.6	—7.5
ZnSe	—15.1	—7.8	—13.9	—6.8	Ge	—15.6	—7.5
InSb	—14.1	—6.6	—	—	Sn	—14.5	—7.0

[a]All values in eV.

The data of Table 20, composed on the basis of Table 3, show that the last equality is actually fulfilled for the isoelectron analogs, regardless of the numerical values of the effective charges. This supports the first of Herman's postulates.

In a similar manner it is easy to prove the second postulate too, although this proof requires some estimate of the effective charges. Tentatively, let us take a value of $Z = 0.5$ for the $A^{III}B^{V}$ compounds and $Z = 1.0$ for the $A^{II}B^{VI}$ compounds. It is not hard to see then that in most cases the difference in the Coulomb integrals $\tilde{\alpha}_A^{(p)} - \tilde{\alpha}_B^{(p)} = \alpha_A^{(p)} - \alpha_B^{(p)} + 2MZ/R$ will be approximately twice as large for the $A^{II}B^{VI}$ compounds as for the $A^{III}B^{V}$ compounds, just as Herman assumed (although this relationship is better satisfied not for the individual values of this difference, but for values averaged over compounds with the same average atomic number) (see Table 21).

6.4 BAND THEORY AND EMPIRICAL CORRELATIONS FOR DIAMOND-LIKE SEMICONDUCTOR CRYSTALS

6.4.1 Relationship between Forbidden Band Gap and Physical and Chemical Properties of $A^N B^{8-N}$ Semiconductors

In this section we return to a discussion of the qualitative aspect of the theory and examine in more detail the relationship between the electron structure and other physical and chemical characteristics of tetrahedral $A^N B^{8-N}$ crystals [252].

In conjunction with the problem of looking for semiconductor compounds the question of such correlation relationships was considered on an empirical basis by a large number of researchers; primary attention was paid to

TABLE 21. Comparison of Differences in Coulomb Integrals[a] $\tilde{\alpha}_A^{(p)} - \tilde{\alpha}_B^{(p)}$ for $A^{III}B^{V}$ and $A^{II}B^{VI}$ Compounds

$A^{III}B^{V}$ Compounds	Values of $\alpha_A^{(p)} - \alpha_B^{(p)} + 2MZ/R$	$A^{II}B^{VI}$ Compounds	Values of $\alpha_A^{(p)} - \alpha_B^{(p)} + 2MZ/R$
GaP	5.9	ZnS	10.4
AlAs	5.2	ZnS	10.4
GaAs	5.7	ZnSe	9.1
InP	5.4	CdS	9.6
AlSb	2.9	CdS	9.6
InAs	5.3	CdSe	8.4
GaSb	3.4	CdSe	8.4
(1/2) {GaP + AlAs}	5.5	ZnS	10.4
GaAs	5.7	ZnSe	9.1
(1/2) {InP + AlSb}	4.1	CdS	9.6
(1/2) {InAs + GaSb}	4.3	CdSe	8.4

[a]All values in eV.

FIG. 6.6. The dependence of E_g on the heat of atomization in bond calculation. Data on heats of atomization of $A^{III}B^{V}$ compounds are taken from [207]. The E_g values for InSb, InAs, InP, and AlSb are taken from [146]; for GaSb, GaAs, and GaP, from new photoemission data [103]; and for BP, from [234, 238–240]. The heat of atomization for BP is unknown; however, by analogy with other $A^{III}B^{V}$ compounds $Q_{at}(BP) \approx (1/2)\{Q_{at}(\text{diamond}) + Q_{at}(\text{Si})\} = 60$ kcal/mol.

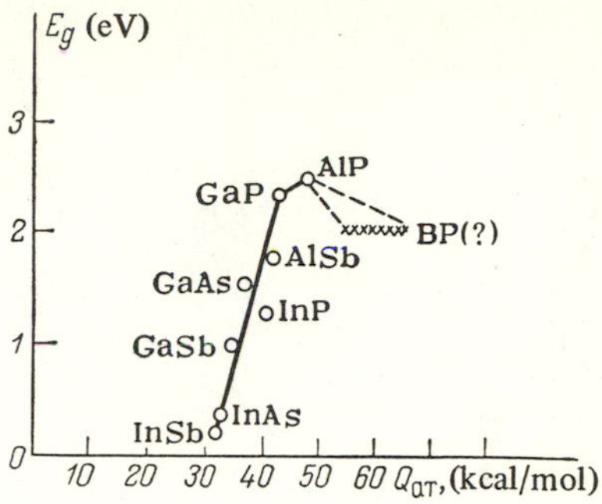

correlations between various properties and the forbidden band gap E_g. Of the correlations found in such a manner, the following are mentioned especially frequently:

(a) correlation between E_g and the heat of atomization (in an analysis of binding) [207, 253, 254];

(b) correlation between E_g and the Pauling "binding energy" [255];

(c) correlation between E_g and the difference ΔX in the electronegativities* $X(A)$ and $X(B)$ of the components A and B of a compound ([256–258]; see also Pearson's paper in [35] and the references in [36]);

(d) correlation between E_g and the reciprocal of the smallest interatomic distance $1/R$ [259].

These correlation relationships are presented in graphical form in Figs. 6.6–6.9 with modern test data taken into consideration.

It is seen from the graphs in Figs. 6.6–6.9 that the forbidden band gap E_g shows a tendency to increase with an increase in the heat of atomization Q_{at}, the Pauling binding energy D, or the electronegativity difference ΔX and to decrease with an increase in the interatomic distance R. Nevertheless, it must be noted that these tendencies in the variation of E_g are by no means universal even within a series of similar (in chemical and crystal-chemical respects) compounds. Thus the correlation between E_g and ΔX is rather indefinite in nature for the $A^{III}B^{V}$ compounds and is generally absent for the Cu and Ag halides.

Although these correlations undoubtedly have a practical use in a number of cases, none of them has been given a sufficiently satisfactory theoretical interpretation.

Thus the usual interpretation of the first two correlation relationships is based on the intuitive, fairly plausible idea of the proportionality between E_g and the strength of the

*The comparative merits of the different systems of electronegativities (Pauling [31], Gordy [260], Haissinsky [261], etc.) are sometimes discussed in the literature in considering the correlation between E_g and ΔX. This question has no significance for the problem to be discussed by us, and the Pauling scale of electronegativities according to Allred's data [262] is used in Fig. 6.8.

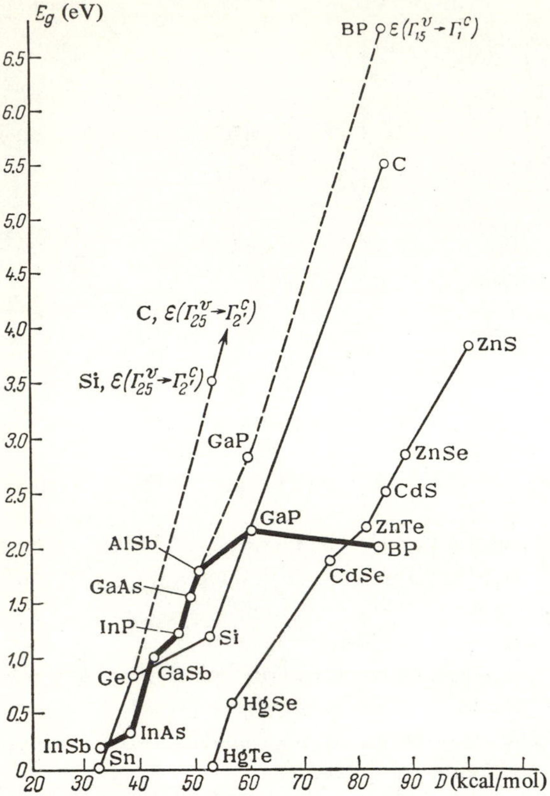

FIG. 6.7. Relationship between E_g and the Pauling binding energy. See explanation for Fig. 6.6. D values are taken from [255], corrected with modern thermochemical data [121] taken into consideration.

two-center A–B bonds in a crystal. In this case E_g is identified either with the "removal energy" for an electron from the bond, or with the "excitation energy" of the corresponding localized orbital. It is easy to see that such an interpretation is not convincing from the viewpoint of band theory. Actually, as we have already stated, the strength of the two-center bonds is determined by the average energy of the electrons in the crystal (i.e., by the average level of the valence band), whereas E_g corresponds to the transition between individual one-electron levels that, moreover, are different for different crystals. The widespread interpretation of the relationship between E_g and ΔX on the basis of a diatomic "hydrogen-like" model [263] and a one-dimensional [264–266] or three-dimensional [267, 268] model with a Kronig–Penney type of δ-shaped potential has a similar defect. It is not hard to see that such an interpretation is also unsatisfactory since the energy spectra of these simplified models is markedly different from the level scheme in a real crystal.

Here we will examine how the correlations (a)-(d) are explained from the viewpoint of band theory (within the framework of the semiempirical EO LCAO

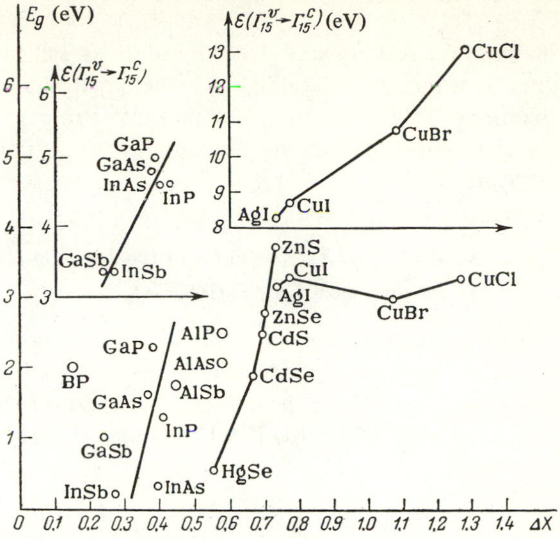

FIG. 6.8. Relationship between E_g and electronegativity difference. The explanation for the E_g values and the energy of the band-to-band transitions for $A^{III}B^{V}$ compounds is same as in Fig. 6.6. The experimental data for the $A^{II}B^{VI}$ compounds are taken from [88]; for the $A^{I}B^{VII}$ compounds, from [103]. The electronegativity values (Pauling scaling) are taken from [262].

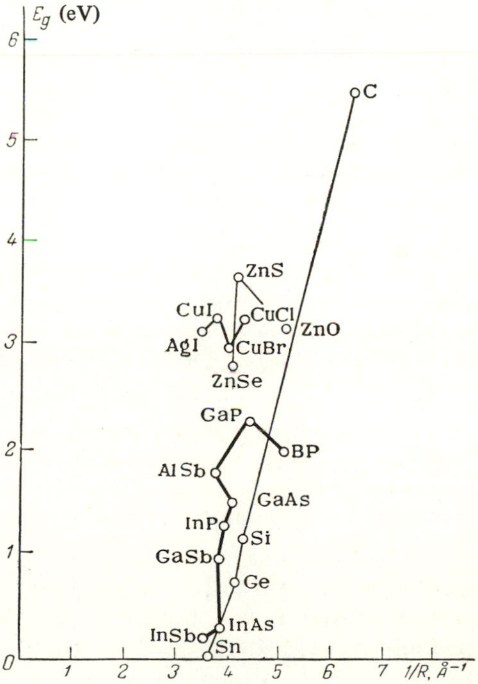

FIG. 6.9. Relationship between E_g and the reciprocal of the interatomic spacing $1/R$. See Figs. 6.6–6.8 for explanation.

method) for real crystals. This approach, as will be indicated below, also allows one to refine the meaning and the applicability limits of the correlations themselves, to give some numerical estimate of the parameters in the corresponding correlation relationships, and to indicate other, new similar correlations.

6.4.2 Band-to-Band Transitions and the Limited Nature of the Empirical Correlations

First, let us note that in the $A^N B^{8-N}$ crystals we are going to consider the value of E_g is equal or approximately equal to either the energy of the direct transition $\Gamma_{15}^v \rightarrow \Gamma_1^c$ or the energy of the indirect transition $\Gamma_{15}^v \rightarrow X_1^c$. It is not hard to see that in our EO LCAO method we have the following formula for the first case:

$$E_g = \left(\frac{1-\lambda^2}{1+\lambda^2}\right)\left\{\alpha_A^{(p)} - \alpha_B^{(p)} + \frac{2MZ}{R}\right\} - \frac{\mathscr{E}_{\text{promot}}(A) + \lambda^2 \mathscr{E}_{\text{promot}}(B)}{1+\lambda^2}$$
$$- \frac{8\lambda}{1+\lambda^2}\left\{\beta_{ss} + \frac{1}{3}\beta_{pp} - \frac{2}{3}\beta_\pi\right\} \tag{6.31}$$

For the second case the formula is

$$E_g = \left(\frac{1-\lambda^2}{1+\lambda^2}\right)\left\{\alpha_A^{(p)} - \alpha_B^{(p)} + \frac{2MZ}{R}\right\} - \frac{\mathscr{E}_{\text{promot}}(A)}{1+\lambda^2}$$
$$- \frac{8\lambda}{3(1+\lambda^2)}\{\beta_{pp} + \sqrt{3}\,\beta_{s_B p_A} - 2\beta_\pi\} \tag{6.32}$$

Thus the dependence of E_g on the parameters $\alpha_A^{(p)}, \ldots$ and $\beta_{ss}, \ldots$ is considerably different in both cases. It is logical to expect that a correlation between E_g and the different properties in any series of analogous crystals should be found if E_g in the entire series corresponds to transitions of one and the same nature, and it can break down in those cases when one type of transition becomes another.

The lack of adequately reliable experimental (in particular, thermochemical) data for the cubic modification of BN and also for BP, BAs, and AlP hampers the experimental proof of this prediction. Nevertheless, even with the available test data one can state in many cases that there is a breakdown in the simple correlation as one goes from crystals with an E_g of the type $\Gamma_{15}^v \rightarrow \Gamma_{15}^c$ to crystals with an E_g of the type $\Gamma_{15}^v \rightarrow X_1^c$. As is seen, for example, from Fig. 6.7, such a breakdown occurs for group IV elements as one goes from germanium [for which $E_g = \varepsilon(L_1^c) - \varepsilon(\Gamma_{25'}^v) = 0.78$ eV, which is very close to the value of $\varepsilon(\Gamma_{2'}^c) - \varepsilon(\Gamma_{25'}^v) = 0.8$ eV] to silicon [for which $E_g = \varepsilon(X_1^c) - \varepsilon(\Gamma_{25'}^v)$].

A similar effect is also observed for the $A^{III}B^V$ compounds. Here the value of E_g either is identical to the energy of the transition $\Gamma_{15}^v \rightarrow \Gamma_1^c$ (InSb, InAs, GaSb,

GaAs) or is quite close to it (InP, AlSb) up to AlSb. However, for GaP, for which $E_g = \varepsilon(X_1^c) - \varepsilon(\Gamma_{15}^v)$ and, probably, to an even greater degree for BP, this agreement does not exist, and Fig. 6.7 shows how the linear nature of the corresponding correlation relationship breaks down.

It is not hard to see that in both cases the linear nature of the correlation relationship is reestablished if E_g for Si and C is replaced by the transition $\Gamma_{25'}^v \to \Gamma_{2'}^c$, and for GaP and BP by the transition $\Gamma_{15}^v \to \Gamma_1^c$. Therefore, in speaking of correlations of the type (a)-(d) it is obviously generally better to search for them for transitions of one and the same type rather than for the forbidden band gap.

6.4.3 Forbidden Band Gap and the Heat of Atomization

Let us now examine how the EO LCAO method provides an explanation for the correlation (a). We will take the case when E_g is defined by the transition $\Gamma_{15}^v \to \Gamma_1^c$, as is the case for most of the $A^N B^{8-N}$ compounds.

By analogy with the group IV elements it is natural to assume that the resonance integrals β_{ss}, β_{pp}, and β_π in (6.31) vary for any "vertical" series of $A^N B^{8-N}$ crystals in approximately the same manner as the interaction integrals β of the hybrid tetrahedral sp^3 orbitals, which, in turn, vary as the thermochemical strengths Q_{at} of the bonds. Then, as follows from Eq. (6.31), E_g will be related to Q_{at} by the linear relationship

$$E_g \approx a + bQ_{at} \tag{6.33}$$

This was also established empirically in [207, 253, 254].

To show how one can find an estimate for the coefficients a and b in Eq. (6.33), let us consider the $A^{III}B^{V}$ compounds (since appropriate experimental data exist for them) and let us note that, according to Eq. (6.31), these coefficients have the form

$$a = \left(\frac{1-\lambda^2}{1+\lambda^2}\right)\left\{\alpha_A^{(p)} - \alpha_B^{(p)} + \frac{2MZ}{R}\right\} - \frac{\mathscr{E}_{\text{promot}}(A) + \lambda^2 \mathscr{E}_{\text{promot}}(B)}{1+\lambda^2} \tag{6.34}$$

$$b = -\frac{8\lambda}{1+\lambda^2} \cdot \frac{\beta_{ss} + (1/3)\,\beta_{pp} - (2/3)\,\beta_\pi}{Q_{at}} \tag{6.35}$$

Using the tentative effective charge value of $Z \sim 0.5$ as well as the Coulomb integrals $\alpha_A^{(p)}, \ldots$ from Table 3 for the $A^{III}B^{V}$ compounds, we obtain a value of $a = -7$ to -9 eV for the free term in Eq. (6.33). This agrees satisfactorily with the experimental value of $a_{exp} = -6.5$ eV (which can be found from the graph in Fig. 6.6).

A still better agreement with experimental data is obtained for the coefficient b in Eq. (6.33). In principle, to find this coefficient from Eq. (6.35) in the case of the $A^{III}B^{V}$ compounds it would be necessary to know the values of all the resonance integrals for these crystals. Here, however, one can proceed somewhat more simply by making use of data for the group IV elements already available to us.

As already repeatedly mentioned, one can assume that the resonance integrals for partially ionic $A^N B^{8-N}$ crystals are close to the corresponding integrals for the isoelectron covalent crystals, and as data from thermochemistry show [207] the same is true for their heats of atomization Q_{at}. Therefore for an approximate estimate one can take $\beta_{ss}, \ldots$ from the band structure of the A^{IV} elements (see Table 10) and Q_{at} from the corresponding thermochemical data [121]. In this situation (as we have seen in Sec. 5.3.2) the coefficient $2\lambda/(1+\lambda^2)$ in Eq. (6.35) can be assumed to be approximately equal to one or, more precisely, 0.95. It is not hard to see then that the proportionality constant b is practically independent of which of the group IV elements is used for the values of the resonance integrals and the heats of atomization and amounts to 0.2 (if E_g is measured in eV, as it usually is, and Q_{at} is in kcal/mol). Thus according to this estimate E_g varies by ~2 eV when Q_{at} increases by 10 kcal/mol. This agrees nearly exactly with the experimental estimate from the graph in Fig. 6.6.

6.4.4 Forbidden Band Gap and the Electronegativity Difference

Whereas the dependence of E_g on Q_{at} is determined by the third, "resonance" term in Eq. (6.31), for explaining the correlation between E_g and $\Delta X = X(\mathrm{B}) - X(\mathrm{A})$ the first term is of prime importance. This term is the Coulomb term in this formula and is proportional to the difference in the p levels of the ions A, B (with the Madelung potential of the lattice taken into consideration):

$$\tilde{\alpha}_{\mathrm{A}}^{(p)} - \tilde{\alpha}_{\mathrm{B}}^{(p)} = \left\{\alpha_{\mathrm{A}}^{(p)}(Z) + \frac{MZ}{R}\right\} - \left\{\alpha_{\mathrm{B}}^{(p)}(Z) - \frac{MZ}{R}\right\} \tag{6.36}$$

For atoms with p electrons and a small effective charge the presence of such a dependence is a trivial result of the relation (6.36). Actually, for $Z \approx 0$ the right side in Eq. (6.36) practically becomes the difference of the first ionization potentials $I_1^{\mathrm{B}} - I_1^{\mathrm{A}}$, which is approximately equal to twice the electronegativity difference $\Delta X = X(\mathrm{B}) - X(\mathrm{A})$:

$$\tilde{\alpha}_{\mathrm{A}}^{(p)} - \tilde{\alpha}_{\mathrm{B}}^{(p)} \approx \alpha_{\mathrm{A}}^{(p)} - \alpha_{\mathrm{B}}^{(p)} \approx I_1^{\mathrm{B}} - I_1^{\mathrm{A}} \approx 2\Delta X \tag{6.37}$$

since all the electronegativity scales are generally proportional to one another, and the Madelung electronegativity is defined as the average of the first ionization potential and the electron affinity*:

*For nearly all atoms, except the halogens, the affinity is about an order of magnitude less than value of the ionization potential.

$$X = \frac{1}{2}\{I_1 + F_1\} \approx \frac{1}{2} I_1 \tag{6.38}$$

In this case the forbidden band gap E_g should increase with an increase in ΔX although the regularity of this correlation will depend, of course, on the behavior of the promotion energies $\mathscr{E}_{\text{promot}}(\text{A})$ and $\mathscr{E}_{\text{promot}}(\text{B})$ entering into the second term on the right side of Eq. (6.31). This explanation is valid in the general case too for an effective charge $Z > 0$ although for $Z > 0$ the energies to remove an electron from the p levels of the ions A^{Z+}, B^{Z-} are no longer the same as the ionization potentials of the neutral atoms A^0, B^0. In this case, however, as seen from Fig. 6.10, consideration of the Madelung potential MZ/R changes the Coulomb integrals $\tilde{\alpha}_{\text{A}}^{(p)}(Z)$, $\tilde{\alpha}_{\text{B}}^{(p)}(Z)$ in such a manner that the proportionality between the difference $\tilde{\alpha}_{\text{A}}^{(p)} - \tilde{\alpha}_{\text{B}}^{(p)}$ and the quantity ΔX is retained although with a different proportionality constant*:

$$\tilde{\alpha}_{\text{A}}^{(p)} - \tilde{\alpha}_{\text{B}}^{(p)} \approx k \cdot \Delta X \quad (k \neq 2) \tag{6.39}$$

It is seen from this same figure that an approximate proportionality between the difference $\tilde{\alpha}_{\text{A}}^{(p)} - \tilde{\alpha}_{\text{B}}^{(p)}$ and the quantity ΔX also exists for the $\text{A}^{\text{II}}\text{B}^{\text{VI}}$ compounds, despite the fact that for the A^{II} atoms the value of the ionization

*In the calculation of $\tilde{\alpha}_{\text{A}}^{(p)} - \tilde{\alpha}_{\text{B}}^{(p)}$ the value $Z = 0.5$ is taken for the $\text{A}^{\text{III}}\text{B}^{\text{V}}$ compounds and the value $Z = 1$ for the $\text{A}^{\text{II}}\text{B}^{\text{VI}}$ compounds.

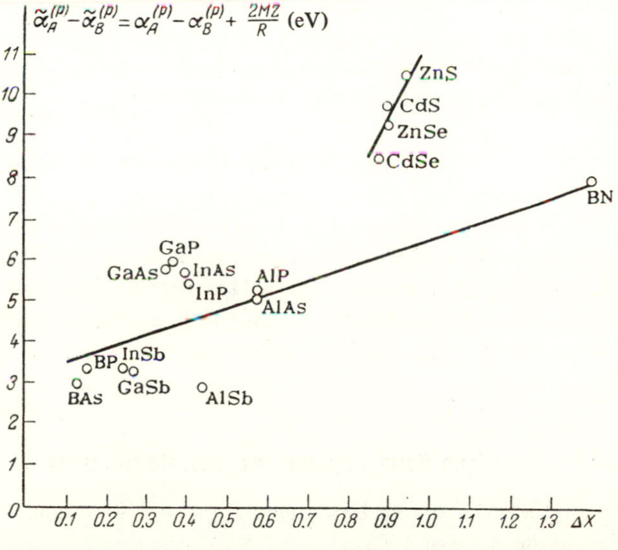

FIG. 6.10. Relationship between difference in Coulomb integrals $\tilde{\alpha}_{\text{A}}^{(p)} - \tilde{\alpha}_{\text{B}}^{(p)}$ and the electronegativity difference. The explanation for the electronegativity values is the same as in Fig. 6.8.

potential I_1^A, like the value of the electronegativity $X(A)$, corresponds to the removal of an electron from the s level and not the p level.

Let us point out again that there is another factor leading to an increase in the forbidden band gap E_g with an increase in ΔX: the increase in the ionicity of the bond. As follows from Eq. (6.31), an increase in the ionicity, i.e., a decrease in λ, leads to an increase in the coefficient $(1 - \lambda^2)/(1 + \lambda^2)$ in the first term on the right side of Eq. (6.31). The latter also contributes to an increase in E_g, despite some decrease in the "resonance" term in Eq. (6.31) (see Sec. 6.1.2 concerning this).

6.4.5 Forbidden Band Gap and the Pauling Binding Energy

Whereas the correlations (a) and (c) are independent, the relationship between E_g and the Pauling "binding energy" D follows from these correlations.

According to the well-known Pauling definition [31], when we speak of the energy $D(A–B)$ of a divalent ordinary and partially ionic bond A–B, we mean the sum

$$D\,(A-B) = D_{cov}\,(A-B) + D_{ion}\,(A-B) \tag{6.40}$$

Here, the "covalent" contribution D_{cov} is the geometric mean of the energies $D(A–A)$ and $D(B–B)$ of the corresponding ordinary covalent bonds A–A and B–B:

$$D_{cov}\,(A-B) = \sqrt{D\,(A-A) \cdot D\,(B-B)} \tag{6.41}$$

and the "ionic" contribution $D_{ion}(A–B)$ is proportional to the square of the electronegativity difference of the atoms A, B:

$$D_{ion}\,(A-B) = \text{const} \cdot (\Delta X)^2 \tag{6.42}$$

Thus like the right side of Eq. (6.31), the energy D of an ordinary Pauling bond is the sum of two terms, one of which is approximately proportional to Q_{at}, like the third term in Eq. (6.31), whereas the other varies similarly to ΔX, like the first term in (6.31). Thus, both quantities D and E_g increase with an increase in Q_{at} and ΔX. This explains the existence of the correlation between E_g and D.

6.4.6 Forbidden Band Gap and the Interatomic Distance

Finally, let us examine how the relation (6.31) explains the correlation between E_g and the value of $1/R$.

The "Coulomb" term on the right side of Eq. (6.31) contains the Madelung potential MZ/R. Therefore a larger E_g value should be observed for crystals with a smaller interatomic distance R, all other conditions being equal [constant heat of atomization, effective charges and Coulomb integrals $\alpha_A^{(p)}$, $\alpha_B^{(p)}$].

Such a correlation between E_g and $1/R$ is obviously retained in the general case too when the other quantities, entering into Eq. (6.31)–heats of atomization, effective charges, etc.–are not constant. It is not hard to see that small R values are characteristic of compounds of the light elements of the first periods of the periodic table. However, as experimental data show, these are exactly the compounds characterized by large values of the heats of atomization and effective charges and also the electronegativity difference ΔX, which also contributes to an increase in E_g.

6.4.7 Other Correlations between Band Structure and Properties

The correlations that have been considered up to now apply to the forbidden band gap E_g. Here we will consider how similar correlations can be found for other band structure parameters.

It is easy to see that the very possibility of explaining correlations (a)–(d) within the framework of band theory is essentially determined by two features of the semiempirical EO LCAO method. First, this method makes it possible to express E_g in analytic form and, secondly, in terms of the parameters that characterize both the individual atoms as well as the interaction among the valence bound atoms. As we have seen such an expression can also be obtained for other parameters of the band structure. This, accordingly, leads to new correlation relationships of a similar type.

We will consider three examples of such correlations. As was shown previously (see Sec. 4.2.1), the energy of the $\Gamma_{25'}^v \rightarrow \Gamma_{15}^c$ transition for covalent crystals is described by Eq. (4.14). This energy depends only on the resonance integrals β_{pp} and β_π and, consequently, is related to the heat of atomization Q_{at} by the relation

$$\varepsilon\,(\Gamma_{25'}^v \longrightarrow \Gamma_{15}^c) \approx \text{const} \cdot Q_{at} \tag{6.43}$$

This actually was already discussed in Sec. 4.1.1 and, as we have seen, matches the test data fairly well.

Another example relates to the correlation between the transition energy $\Gamma_{15}^v \rightarrow \Gamma_{15}^c$ in partially ionic crystals and the electronegativity difference ΔX. A comparison of Eq. (6.4) for the $\Gamma_{15}^v \rightarrow \Gamma_{15}^c$ transition with the analogous formula (6.31) shows that Eq. (6.4) contains a Coulomb term the same as in Eq. (6.31). Therefore it is natural to think that the band-to-band transition $\Gamma_{15}^v \rightarrow \Gamma_{15}^c$, like E_g, will be correlated with the difference $\Delta X = X(B) - X(A)$. Figure 6.8 confirms the presence of this correlation. In this situation it is worth considering the fact that for the $A^{III}B^V$ compounds it is much more pronounced than the indefinite correlation between E_g and ΔX.

The third example applies to the A^IB^{VII} compounds and illustrates well the advantage of the approach from the band theory side for looking for correlations of the type (a)-(d). It is seen from Fig. 6.8 that for the A^IB^{VII} compounds the correlation between E_g and ΔX, clearly pronounced, for example, for the $A^{II}B^{VI}$ compounds, is absent. This breakdown in the $E_g \longleftrightarrow \Delta X$ correlation in this case is related to the fact that the top of the valence band for the CuHal compounds (but not the AgHal) is formed by the $3d$ functions of the metal atoms, and not by the p functions of the halogen atoms. Therefore for an analogy with the $A^{II}B^{VI}$ and $A^{III}B^V$ compounds in the case of the A^IB^{VII} compounds it is necessary to replace the top of the valence band by the $\Gamma_{15}^v(p)$ level. Accordingly, for these compounds we should consider the energy of the $\Gamma_{15}^v(p) \to \Gamma_1^c$ transition, which is expressed by the same Eq. (6.31) for the A^IB^{VII} compounds, instead of the forbidden band gap E_g. As Fig. 6.8 shows, for the Cu and Ag halides the energy of the transition $\Gamma_{15}^v \to \Gamma_1^c$ actually correlates well with the value of ΔX, just as the energy of the transition $\Gamma_{15}^v \to \Gamma_{15}^c$.

6.5 BAND STRUCTURE AND THE CHARGES ON ATOMS

6.5.1 Band Structure as Method of Investigating Nature of the Chemical Bond

Until now we have been occupied with an investigation and calculation of the band structure of partially ionic crystals, based on a specified charge distribution. In this section, conversely, the analysis of experimental data on the band structure will be considered as a method of studying the nature of the chemical bond.

It is not hard to see that the experimental data on the band structure of the tetrahedral $A^{III}B^V$ and $A^{II}B^{VI}$ crystals yield evidence opposite to that of the ionic model of $(A^{III})^{3+}(B^V)^{3-}$ and $(A^{II})^{2+}(B^{VI})^{2-}$.

Actually, in this case, i.e., for $\lambda = 0$ (as already mentioned in Sec. 5.2.1), all branches of the dispersion law would assume the form $\varepsilon(\mathbf{k}) = \text{const}$, so that the width of all the filled bands would become zero, and the gap in the valence band and in the conduction band would be the same as the total width of these bands. In fact, as the experimental data show the dispersion law in the $A^N B^{8-N}$ crystals is quite far from the trivial form $\varepsilon(\mathbf{k}) = \text{const}$, and this makes it possible to use the structure of the bands for an approximate estimate of the values of the charges on the atoms.

For this purpose, in particular, it is convenient to use the following band structure parameters:

(1) width E_p of the p band X_5^v-Γ_{15}^v-L_3^v in the valence band;

(2) the energy of the $\Gamma_{15}^v \to \Gamma_{15}^c$ transition;

(3) the magnitude of the level splitting $\bar{\bar{X}}^v - \bar{X}^v$ $(X_3^v - X_1^v)$ in the valence band;

(4) the magnitude of the level splitting $\bar{\bar{X}}^c - \bar{X}^c$ $(X_3^c - X_1^c)$ in the conduction band.

As we will see, the values of the charges on the atoms found in this way generally fall within the same limits as the values given by other experimental methods* [205]. It must be emphasized, however, that our purpose is only to illustrate the possibility of such an estimate and that most of the charge values given below only are illustrative in nature primarily because of

(a) the absence of direct test data for many of the band structure parameters (parameters 1 and in particular 3);

(b) the incompleteness of the spectroscopic data for isolated atoms (this affects the estimate of the Coulomb integrals);

(c) the replacement—for the sake of simplicity—of the values of $\mathscr{E}_{\text{promot}}(\text{A})$, $\mathscr{E}_{\text{promot}}(\text{B})$ for the A^{Z+}, B^{Z-} ions by their values for neutral atoms;

(d) the simplifying assumptions adopted here for the estimate of the resonance integrals.

6.5.2 Width of the *p* Bands and the Effective Charge

The simplest method of estimating the effective charges is to compare the width ΔE_p of the p bands for the isoelectron analogs A^NB^{8-N} and A^{IV}. According to Eqs. (3.96) and (5.56), we obtain the formula

$$\Delta E_p\,(\text{A}^N\text{B}^{8-N}) : \Delta E_p\,(\text{IV}) = (^1/_4)\,\sqrt{(N-Z)\,(8-N+Z)} \times \frac{\beta_\pi\,(\text{A}^N\text{B}^{8-N})}{\beta_\pi\,(\text{IV})} \qquad (6.44)$$

for the ratio $\Delta E_p:\Delta E_p(\text{IV})$, the right side of which contains the function $f(Z) = 1/4\sqrt{(N-Z)(8-N+Z)}$, depending on the effective charge on the atoms, in the form of a coefficient. This formula can be used for an approximate determination of Z in those cases when the left side of Eq. (6.44) is known from experiment or, let us say, from sufficiently reliable EP and EC OPW data and when the ratio of the resonance integrals $\beta_\pi(\text{A}^N\text{B}^{8-N}):\beta_\pi(\text{IV})$ is known.

If it is again assumed that $\beta_\pi(\text{A}^N\text{B}^{8-N}) \approx \beta_\pi(\text{IV})$ for the isoelectron analogs, then the ratio $\Delta E_p = (\text{A}^N\text{B}^{8-N}) : \Delta E_p(\text{IV})$ will be determined only by the values of the functions $f(Z)$; which is shown in Fig. 6.11 for values of $N \approx 1, 2,$ and 3.

On the other hand according to the data from [280], this ratio is equal to 1.0–0.93 for $\text{A}^{\text{III}}\text{B}^{\text{V}}$ compounds and 0.8–0.75 for $\text{A}^{\text{II}}\text{B}^{\text{VI}}$ compounds.

*Let us note that in speaking of the effective charges it is always necessary to remember that the effective charge is not a rigorously defined physical quantity that only has to be measured. In reality this concept is introduced to describe different test data within the framework of the simplified model of nonoverlapping ions, and from an *a priori* point of view it could be that "single" values of the charge on the atoms do not exist at all. Each group of test data corresponds to its own "effective charge," varying, for example, from zero to the formal valence value. This result shows that the situation relative to the band structure is different and, thus, favors the very concept of an effective charge.

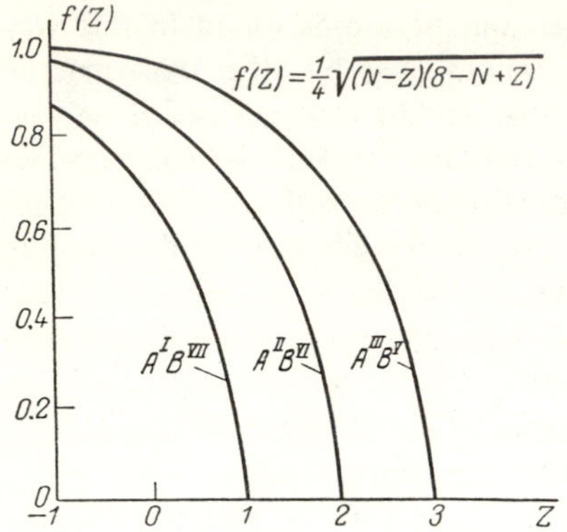

FIG. 6.11. Graphs of the functions $f(Z) = (1/4)\sqrt{(N-Z)(8-N+Z)}$ for $A^{III}B^{V}$, $A^{II}B^{VI}$, and $A^{I}B^{VII}$ compounds.

Accordingly, the graphs in Fig. 6.11 give values of $Z(A^{III}B^{V}) = 0\text{-}0.2$ and $Z(A^{II}B^{VI}) = 0.5\text{-}0.6$, which do not fall outside the limits of the "X-ray spectral" values of the charge on the atoms (see below, Sec. 6.5.4) although these values by themselves are too low. (Moreover, a high computation accuracy is not required here, in particular because there are no experimental data for the width of the p band for many $A^{N}B^{8-N}$ compounds.)

Even more acceptable results are obtained for the $A^{I}B^{VII}$ compounds. Thus for the Cu and Ag halides the EC OPW data [103] lead to values of $\Delta E_p(\text{CuCl}):\Delta E_p(\text{SiGe}) = 0.1$; $\Delta E_p(\text{CuBr}):\Delta E_p(\text{Ge}) = 0.23$; $\Delta E_p(\text{CuI}):\Delta E_p(\text{GeSn}) = 0.36$; and $\Delta E_p(\text{AgI}):\Delta E_p(\alpha\text{-Sn}) = 0.28$. This result corresponds to the following atom charges: $Z(\text{CuCl}) = 0.95$; $Z(\text{CuBr}) = 0.85$; $Z(\text{CuI}) = 0.65$; $Z(\text{AgI}) = 0.75$. These values are in completely satisfactory agreement with other experimental data [205] and also with the usual concepts on the nature of the variation of the ionicity as one goes from the chlorides to the iodides.

6.5.3 Comparison of Transitions between p Levels

A second method of estimating the effective charges is based on a comparison of the $\Gamma_{15}^{v} \to \Gamma_{15}^{c}$ transitions in partially ionic $A^{N}B^{8-N}$ crystals with the analogous $\Gamma_{24}^{v} \to \Gamma_{2'}^{c}$ transitions in the isoelectron crystals A^{IV} or $A^{IV}B^{IV}$.

If it is assumed, as before, that the resonance integrals for the isoelectron analogs are similar in value, then Eq. (6.4) is obtained for the transition $\Delta\varepsilon = \varepsilon(\Gamma_{15}^{v} \to \Gamma_{15}^{c})$. This relation is now rewritten in the form

$$\Delta\varepsilon - \frac{1}{4}\sqrt{(N-Z)(8-N+Z)}\,\Delta\varepsilon_{IV}$$

$$= \frac{1}{4}(4-N+Z)\left\{\alpha_A^{(p)}(0) - \alpha_B^{(p)}(0) + \left[\alpha_A^{(p)}(1) - \alpha_B^{(p)}(1) - \alpha_A^{(p)}(0) + \alpha_B^{(p)}(0) + \frac{2M}{R}\right]Z\right\} \tag{6.45}$$

which permits a direct determination of the effective charge from the experimental values of $\Delta\varepsilon$ and $\Delta\varepsilon_{IV}$.

To simplify the solution of Eq. (6.45) we assume that the values of the function $f(Z) = (1/4)\sqrt{(N-Z)(8-N+Z)}$ correspond to the charge $0 \lesssim Z \lesssim 1$. [We will see below that this assumption is not contradictory in the sense that the charges obtained with this assumption from Eq. (6.45) do not in fact fall outside the limits of this interval.] Then the coefficient $f(Z)$ for $\Delta\varepsilon_{IV}$ on the left side of Eq. (6.45) varies little over the entire region $0 < Z < 1$ and, as seen from Fig. 6.11, one can assume with an accuracy of ~10% that this coefficient $f(Z)$ is equal to one for $A^{III}B^{V}$ compounds (or equal to 0.95 with a ±5% accuracy). Similarly, one can assume that $f(Z)$ O 0.78 ± 0.10 for $A^{II}B^{VI}$ compounds.

In this case Eq. (6.45) becomes the quadratic equation

$$\frac{\alpha_A^{(p)}(0)-\alpha_B^{(p)}(0)}{4(\Delta\varepsilon - f\Delta\varepsilon_{IV})} + \frac{\alpha_A^{(p)}(1)-\alpha_B^{(p)}(1)-\alpha_A^{(p)}(0)+\alpha_B^{(p)}(0)+\frac{2M}{R}}{4(\Delta\varepsilon - f\cdot\Delta\varepsilon_{IV})}\cdot Z = \frac{1}{4-N+Z}$$

the graphical solution of which for a given family of crystals (i.e., for a constant N) is reduced to that of finding the intersection points of the family of straight lines with the standard hyperbola $y = 1/(4 - N + Z)$. Some examples of such a solution are given in Fig. 6.12, where the transition energies $\Delta\varepsilon$ and $\Delta\varepsilon_{IV}$ are taken from Herman's EC OPW data [103], and the Coulomb integrals $\alpha_A^{(p)}(0)$, etc., are taken from Table 3. As seen from Fig. 6.12 the described method also leads to reasonable values of the effective charges (although they also are only estimates).

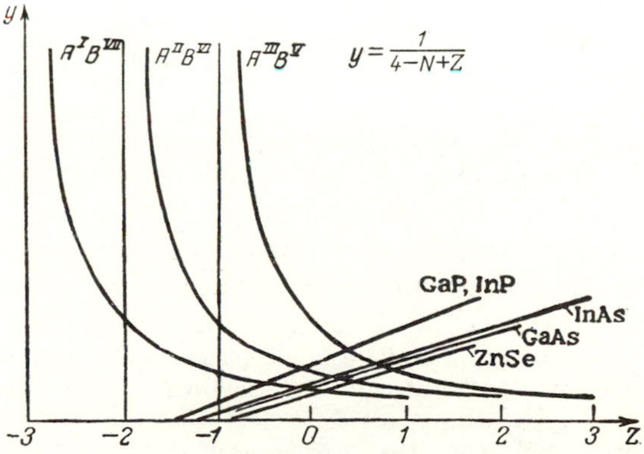

FIG. 6.12. Estimates of the effective charge from the energy of the $\Gamma_{15}^{v} \rightarrow \Gamma_{15}^{c}$ transition.

6.5.4 Effective Charges and Splitting of the X_1 Valence Levels

As is known, a characteristic feature of tetrahedral heteroatomic crystals, in contrast to homoatomic, is the removal of degeneracy for the doubly degenerate X_1^v and X_1^c levels, which are split into the X_1^v, X_3^v and X_1^c, X_3^c level. In the EO method the amount of splitting (5.53) serves as a measure of the difference in the interaction integrals for the first neighbor EOs and can be used to estimate the degree of ionicity of the corresponding A–B bonds.

For such an estimate it is necessary to switch to the EO LCAO approximation, making use of Eq. (5.53). We will first calculate the valence band. In this case we obtain from Eq. (5.53) the equation

$$\Delta\varepsilon + \frac{8\lambda}{\sqrt{3}\,(1+\lambda^2)}\{\beta_{sBpA} - \beta_{sApB}\} = \frac{1}{1+\lambda^2}\{\mathscr{E}_{\text{promot}}(B) - \lambda^2\mathscr{E}_{\text{promot}}(A)\} \tag{6.46}$$

where the "resonance" term is nearly an order of magnitude less than the Coulomb term or the amount of splitting. Ignoring this resonance term, we have the simpler formula

$$\Delta\varepsilon \approx \frac{\mathscr{E}_{\text{promot}}(B) - \lambda^2\mathscr{E}_{\text{promot}}(A)}{1+\lambda^2} \tag{6.47}$$

Hence

$$\lambda^2 \approx \frac{\mathscr{E}_{\text{promot}}(B) - \Delta\varepsilon}{\mathscr{E}_{\text{promot}}(A) + \Delta\varepsilon} \tag{6.48}$$

or

$$Z \approx N - 8\,\frac{\mathscr{E}_{\text{promot}}(B) - \Delta\varepsilon}{\mathscr{E}_{\text{promot}}(A) + \mathscr{E}_{\text{promot}}(B)} \tag{6.49}$$

so that, in principle, one can find the effective charge from spectroscopic data for isolated atoms A, B and from the experimental amount of splitting $\Delta\varepsilon = \varepsilon(X_3^v) - \varepsilon(X_1^v)$ although here too (even in the presence of precise values of the Coulomb integrals) the estimates have only a very tentative meaning, particularly in view of the absence of reliable data* on the value of $\Delta\varepsilon$ (theoretically such

*Direct experimental data on the amount of splitting $X_3^v - X_1^v$ have an accuracy of about 1 eV. The data of EC OPW and EP calculations [103] refer only to gallium compounds and, moreover, exhibit considerable scatter. For example, according to Herman's EC OPW data, the amount of splitting $X_3^v - X_1^v$ for GaAs and GaSb amounts to 5.2 and 4.6 eV, respectively, and according to the EP data in the same paper to 4.3 and 1.3 eV.

data can be obtained, for example, from X-ray spectra). Nevertheless, let us use the data from [280] for an illustration and assume, for the sake of simplicity, that $\mathscr{E}_{\text{promot}}$ does not depend on Z and is the same as the $\mathscr{E}_{\text{promot}}$ for the neutral atoms A^0, B^0 (see Table 3). Then we obtain from the simplified formula (6.49): $Z(\text{GaP}) \approx 0$; $Z(\text{GaAs}) \approx 0.4$; $Z(\text{InSb}) \approx 0$.

X-ray spectral estimates of the effective charges for $A^{III}B^{V}$ compounds vary from $Z(\text{InSb}) = 0$ [269, 270] to $Z(\text{AlN}) = 1.32$ [204]; that is, the estimates obtained from Eq. (6.49) necessarily lie within the same limits as the estimates given by X-ray spectroscopy although they differ from the exact values of the corresponding "X-ray spectral" charges [204, 205, 269-271].

Let us now compare the estimates from the simplified formulas (6.48), (6.49) with the estimate from the more exact formula (6.46). Let us note that the resonance term on the left side of Eq. (6.46) is always negative; consequently, taking it into consideration is equivalent to replacing the quantity $\Delta\varepsilon$ in the simplified formulas (6.48), (6.49) by the smaller quantity $\Delta\tilde{\varepsilon} = \Delta\varepsilon - \delta\varepsilon$, where $\delta\varepsilon = 8\lambda|[\sqrt{3}(1+\lambda^2)]^{-1} \cdot |\beta_{sBpA} - \beta_{sApB}|$. This decrease in the value of $\Delta\varepsilon$ can obviously only lead to a decrease in Z, so that when the refined formula (6.46) is used, the effective charge remains within the interval $0 < Z < 1$ as before. This fact makes it possible to replace the slowly varying coefficient $2\lambda/(1+\lambda^2)$ in the resonance term in Eq. (6.46) by one. Then the approximate formula

$$\delta\varepsilon = \frac{4}{\sqrt{3}} |\beta_{sBpA} - \beta_{sApB}| \tag{6.50}$$

is obtained for $\delta\varepsilon$, according to which we will have $\delta\varepsilon(\text{GaP}) \approx \delta\varepsilon(\text{GaAs}) \approx \delta\varepsilon(\text{GaSb}) \approx 0.4$ eV when the estimates (5.44), (5.48) are used. From this it is easy to see that the refined estimates for the effective charges $Z(\text{GaP})$, $Z(\text{GaAs})$, $Z(\text{InSb})$ will be quite close to the previous estimates.

6.5.5 Effective Charges and Splitting of the X_1 Levels in Conduction Band

The charge on atoms can be estimated most reliably from the amount of splitting X_3^c-X_1^c since the amount of splitting of the level X_1 in the conduction band is known from experiment [107, 146].

In this case the analog of Eq. (6.46) will be the relation

$$\Delta\varepsilon - \frac{8\lambda}{\sqrt{3}(1+\lambda^2)} \{\beta_{sBpA} - \beta_{sApB}\} = \frac{\mathscr{E}_{\text{promot}}(A) - \lambda^2 \mathscr{E}_{\text{promot}}(B)}{1+\lambda^2} \tag{6.51}$$

[It is derived analogously to Eq. (6.46), where the resonance term, however, is no longer small compared with the other two* and it must be taken into consideration.]

Assuming again that the coefficient $2\lambda/(1+\lambda^2) \approx 1$ (as in Sec. 6.5.3, this is a consistent assumption) and using the notations of the previous section, we obtain the estimate

$$Z \approx N - 8 \cdot \frac{\mathscr{E}_{\text{promot}}(\text{A}) - \Delta\tilde{\varepsilon}}{\mathscr{E}_{\text{promot}}(\text{A}) + \mathscr{E}_{\text{promot}}(\text{B})} \tag{6.52}$$

where, unlike in the valence band, $\Delta\tilde{\varepsilon} = \Delta\varepsilon + \delta\varepsilon$ and $\delta\varepsilon = [8\lambda/\sqrt{3}(1+\lambda)^2]$ $|\beta_{s\text{B}p\text{A}} - \beta_{s\text{A}p\text{B}}|$. Using the $\Delta\varepsilon$ value from optical spectra and again using the estimate (6.50) for $\delta\varepsilon$, we find that Z varies from 0 to 0.3 for the $\text{A}^{\text{III}}\text{B}^{\text{V}}$ compounds (AlSb, GaAs, GaSb, InAs, InSb) and $Z \approx$ 0.6–0.7 for the $\text{A}^{\text{II}}\text{B}^{\text{VI}}$ compounds (ZnSe, CdS, CdSe). This is in fairly good agreement with the data of other methods of estimating the effective charges [205].

6.5.6 Band Structure and the Sign of the Charge on Atoms

Finally, let us consider the question of the sign of the effective charges in the $\text{A}^{\text{III}}\text{B}^{\text{V}}$ compounds, which has not yet been discussed. The donor-acceptor nature of the bond in the A^NB^{8-N} crystals for $N = 1, 2, 3$ allows, in principle, a negative charge on the "electropositive" atom A^N, whereas experimental data are partially contradictory in nature. Thus on the basis of X-ray [273] and piezoelectric [274] data (opposite signs of the piezoelectric effect in $\text{A}^{\text{III}}\text{B}^{\text{V}}$ and $\text{A}^{\text{II}}\text{B}^{\text{VI}}$ compounds) many authors ascribe a negative sign to the charges on the A^{III} atoms.

On the other hand, many authors, also on the basis of X-ray data, reach the opposite conclusion [275-277], and the method of interpreting piezoelectric data has recently raised doubts [278].

At the same time the Szigeti method [279], often used to estimate charges, cannot give the sign of the charge generally, and probably the very same thing should be said about studies of the spin orbital splitting of levels in the valence band, which is not very sensitive to the sign of the charge and is compatible both with the structure $(\text{A}^{\text{III}})^{Z+}(\text{B}^{\text{V}})^{Z-}$ and $(\text{A}^{\text{III}})^{Z-}(\text{B}^{\text{V}})^{Z+}$ (although the compatibility is somewhat worse, of course).

Nevertheless, based on the data as a whole (and especially bearing in mind the results of X-ray spectral studies [204, 205, 269-272] it is more probable that

*Since $\mathscr{E}_{\text{promot}}(\text{B}) > \mathscr{E}_{\text{promot}}(\text{A})$ and $\lambda^2 < 1$, then the subtrahend and the diminuend in the difference $\mathscr{E}_{\text{promot}}(\text{A}) - \lambda^2\mathscr{E}_{\text{promot}}(\text{B})$ compensate each other to a considerable extent. This compensation does not occur in the valence band where the difference $\mathscr{E}_{\text{promot}}(\text{B}) - \lambda^2\mathscr{E}_{\text{promot}}(\text{A})$ is involved. On the other hand, as experimental data show, the amount of splitting X_3^c–X_1^c is also small for the conduction band.

TABLE 22. Dependence of the Energy of the $\Gamma_{15}^{v} \rightarrow \Gamma_{15}^{c}$ Transition on the Sign of the Charge on the Atoms for $A^{N}B^{8-N}$ Compounds

$\Delta\varepsilon - \Delta\varepsilon_{IV}$	Crystals										
	AlSb	GaP	GaAs	GaSb	InP	InAs	InSb	ZnS	ZnSe	CdS	CdSe
Calculation											
$Z < 0$	0.1—0.2	0.2—0.4	0.1—0.2	0.1—0.2	0.3—0.6	0.1—0.2	0.1—0.2	1.5	1	2	1.5
$Z > 0$	0.9	1.5	1.6	1.1	1.8	1.7	1	6.7	5.8	6.3	5.5
Experiment	—	1.7	1.7	0.7	1.9	1.9	1	—	—	—	—
	1.4	(1.9)	(1.3)	(1.1)	(1.4)	(1.6)	(0.7)	6.1	4.7	7.1	5.6

Note: The values (—0.3)-(—0.5) and (—1) are taken as the Z values in the first row of the table for $A^{III}B^{V}$ and $A^{II}B^{VI}$ compounds, respectively. In the second row values of +0.5 and +1 are taken for Z in the first and second case, respectively. The values of $\varepsilon(\Gamma_{15}^{v} \rightarrow \Gamma_{15}^{c}) - \varepsilon(\Gamma_{15'}^{v} \rightarrow \Gamma_{15}^{c})$, taken from experiment ([146], in parentheses) or from EP data [107] and EC OPW data [103], are listed in third and fourth rows. All energies are in eV.

the charge on the A^{III} atoms is positive, and as we will see, band structure studies will, in all probability, support this charge sign assignment.

As shown above, all the methods listed on page 222 for estimating the charge on the basis of experimental data on the band structure give a positive sign for the charge on the A^{III} atoms. At the same time it is not hard to see that the assumption of the opposite charge sign on the atoms would lead to a contradiction of experiments. In fact, let us assume that the effective charges in the $A^{III}B^{V}$ compounds are equal in absolute magnitude to about half the charge of an electron, but there is a negative charge on the A^{III} atoms. Then, as seen from the graph in Fig. 6.11, the value of the function $f(Z) = (1/4)\sqrt{(N-Z)(8-N+Z)}$ would amount to ~0.99 for the $A^{III}B^{V}$ compounds; consequently, the p band X_5^v-Γ_{15}^v-L_3^v should contract by only 1% as one goes from the group IV elements to the $A^{III}B^{V}$ compounds. This, of course, represents a poorer match with test data (a contraction of almost 7%) than the previous value of 10%, obtained for positive charges on the A^{III} atoms.

Even more significant in this respect is a comparison of the energies of the transition $\Delta\varepsilon = \varepsilon(\Gamma_{15}^v \to \Gamma_{15}^c)$, calculated for $Z > 0$ and $Z < 0$.

The values of the differences $\Delta\varepsilon - \Delta\varepsilon_{IV}$, where $\Delta\varepsilon_{IV} = \varepsilon(\Gamma_{25'}^v \to \Gamma_{15}^c)$ calculated for $Z > 0$ and $Z < 0$, are compared in Table 22 with the experimental data. It is seen from this table that a "plus" sign on the A^{III} atoms leads to fairly good agreement with experiment, whereas the opposite assumption leads to values of difference $\Delta\varepsilon - \Delta\varepsilon_{IV}$ that are smaller than the experimental values by about an order of magnitude.

Conclusion

We have examined a number of the problems in the theory of the chemical bond in crystals, which we shall now summarize. Apparently, the main conclusion that can be drawn from the entire discussion is the following: There is a much greater parallelism between molecular quantum chemistry and the band structure theory of crystals (or even between the chemistry and physics of solids) than is usually assumed to exist. Thus in all probability the same factors determine both the unique character of the band structure of diamond as well as the features of the chemical bond in hydrocarbons. In a similar manner the phenomenon of the "Sedgwick inert pair," well known to all chemists, leads to the regular evolution of the band structure in the vertical series of $A^N B^{8-N}$ semiconductors and, in particular, is associated with the localization of the absolute minimum of the conduction band of the germanium crystal at the point L.

The existence of such a close relationship between these two fields of science leads, in turn, to two important consequences. First, the vast arsenal of data from inorganic and organic chemistry can be successfully applied to explain (and predict) the band structure of crystals; second, the data on the electron structure of crystals can shed light on the unanswered questions of inorganic and organic chemistry.* Another consequence is that the methods, developed in molecular quantum chemistry for studying the electron structure of molecules, can be used with the same success for investigating the electron structure of crystals. It is possible that in many cases this can be done with even greater success since in some respects the theory of "infinite" crystals is simpler than the theory of small molecules. (For example, in the Hartree-Fock scheme the difference in the one-electron energies for a filled and for an empty orbital is the same as the "true" energy of the one-electron excitation of the system.)

In conclusion let us say a few words about questions that were not answered in this book and, in particular, about some promising future investigations. The dynamics of charge carriers–electrons and holes–and above all their

*See [283] for an example of the use of data on the electron structure of a solid in the theory of molecules. The question of the interpretation of the photoelectron spectra of the alkanes is examined in these papers (within the framework of the equivalent orbital method). It is also shown how the use of the band structure data of diamond makes it possible to correct the previously adopted scale of Brailsford–Ford parameters [128] and to obtain a complete set of the parameters, including the interaction of the C–C and C–H bonds out to fourth-neighbor bonds.

fundamental dynamic characteristics–effective masses–are closely related to the band structure. The use of effective masses occupies an important place in solid state theory. However, it is far from the usual problems of quantum chemistry. In addition, one can assume *a priori* that the approximate nature of the quantum chemical methods, aggravated by the use of parameters taken from atomic and molecular data, severely hinders the study of effective masses since the latter imposes more stringent requirements on the accuracy of the calculations than the study of the band structure [the effective masses are determined by the second derivatives of the function $\varepsilon(\mathbf{k})$ with respect to the components of the vector $\mathbf{k}$]. Even more interesting is the fact that the effective masses, as it turns out, can be determined from atomic and molecular parameters in a number of cases, and the explicit analytic form of the corresponding expressions for the effective masses makes it possible to examine their dependence on the properties and on the character of the chemical bond [284].

We have limited our consideration to the one-electron, and therefore the "non-self-consistent," version of the theory. Although, as already mentioned, the ignoring (in explicit form) of interelectron interaction does not affect the interpretation of the one-electron transitions, a reformulation of the theory in the language of any of the versions of the self-consistent field method could lead to new results since the use of such a method leads to a change in the parameter scheme. In addition, this formulation of the theory would make it possible to investigate both the "collective" (charge distribution, binding energy) as well as the one-electron properties of $A^N B^{8-N}$ crystals in a more proper manner.

In striving for intelligibility and simplicity of the presentation, in this book we have not touched questions related to interelectron correlation. In this regard we direct the reader's attention to the work of Tolpygo and Lyapin [285]. Also omitted, unfortunately, is the so-called Phillips-van Vectan "dielectric" theory of the chemical bond (see, for example, [286]) that has become very popular recently. This theory, in terms of its initial concepts, is so far removed from traditional quantum chemical methods that any detailed consideration of it would lead to a considerable increase in the volume of this book and would make it less intelligible to the chemist.

References

1. Landau, L. D. and Ye. M. Lifshits: "Kvantovaya mekhanika (Quantum Mechanics)," Fizmatgiz Press, Moscow, 1963.
2. Blokhintsev, D. I.: "Osnovy Kvantovoy mekhaniki (Fundamentals of Quantum Mechanics)," 4th ed., Vysshaya Shkola Press, Moscow, 1963.
3. Gombash, P.: "The Many-Particle Problem in Quantum Mechanics," translated from German, 2nd ed., Foreign Literature Press, Moscow, 1953.
4. Bethe, H.: "Quantum Mechanics," (translated from English), ed. V. L. Bonch-Bruyevich, Mir Press, Moscow, 1965.
5. Hartree, D.: "Atomic Structure Calculations," (translated from English), ed. V. A. Fok, Foreign Literature Press, Moscow, 1960.
6. Fock, V.: *Z. Phys.*, vol. 61, pp. 126–148, 1930.
7. Roothaan, C.: *Rev. Mod. Phys.*, vol. 23, no. 2, pp. 69–89, 1951.
8. Slater, J. C.: *Phys. Rev.*, vol. 81, no. 3, pp. 385–390, 1951.
9. Herman, F. and S. Skillman: "Atomic Structure Calculations," Prentice-Hall, Englewood Cliffs, New Jersey, 1963.
10. Brattsev, V. F.: "Tablitsy atomnykh volnovykh funktsiy (Tables of Atomic Wave Functions)," ed. M. G. Veselov, Nauka Press, Moscow–Leningrad, 1966.
11. Clementi, E.: *IBM J. Res. Dev. Suppl., Tables of Atomic Functions,* vol. 9, pp. 2, 1965.
12. Richardson, J. W. et al.: *J. Chem. Phys.*, vol. 36, no. 4, pp. 1057–1061, 1962; vol. 38, no. 5, pp. 796–801, 1963.
13. Moore, C. E.: "Atomic Energy Levels," vols. 1, 2, 3, National Bureau of Standards, Washington, D.C., 1949, 1952, 1958.
14. Slater, J. C.: *Phys. Rev.*, vol. 98, no. 4, pp. 1039–1045, 1955.
15. Dewar, M.: "Theory of Molecular Orbitals in Organic Chemistry," (translated from English), ed. M. Ye. Dyatkina, Mir Press, Moscow, 1972.
16. Morrel, J., S. Kettle and T. Tedder: "Valence Theory," (translated from English), ed. M. G. Veselov, Mir Press, Moscow, 1968.
17. Slater, J. C.: "Electronic Structure of Molecules," (translated from English), ed. D. A. Bochvar, Mir Press, Moscow, 1965.
18. Bersuker, I. B.: "Stroyeniye i svoystva koordinatsionnykh soyedineniy (The Structure and Properties of Coordination Compounds)," Khimiya Press, Moscow, 1971.
19. Johnson, K. H.: *J. Chem. Phys.*, vol. 45, no. 8, pp. 3085–3095, 1966; *Int. J. Q. Chem. Sympos.*, pp. 361–367, 1967. Johnson, K. H. and F. C. Smith, *Phys. Rev. Lett.*, vol. 24, pp. 139–142, 1970; Vojtik, J. and M. Tomasek: *Chem. Phys. Lett.*, vol. 6, pp. 189–191, 1970; Fricker, H. S. and P. W.

Anderson: *J. Chem. Phys.*, vol. 55, no. 10, pp. 5028–5035, 1971; Fricker, H. S.: *J. Chem. Phys.*, vol. 55, no. 10, pp. 5034–5042, 1971; Johnson, K. H. and F. C. Smith: *Phys. Rev. B.*, vol. 5, no. 3, pp. 831–844, 1972; Slater, J. C. and K. H. Johnson: *Phys. Rev. B.*, vol. 5, no. 3, pp. 844–853, 1972.

20. Pariser, R. and R. G. Parr: *J. Chem. Phys.*, vol. 21, no. 3, pp. 466–471, 1953; no. 4, pp. 767–776, 1953; Pople, J. A.: *Trans. Faraday Soc.*, vol. 49, no. 12, pp. 1375–1385, 1953.
21. Pople, J. A. et al.: *J. Chem. Phys.*, vol. 43, no. 10, part 2, pp. S129–S135, S136–S151, 1965.
22. Veselov, M. G. and M. M. Mestechkin: *Litov. Fiz. Sb.*, vol. 3, no. 1–2, pp. 269–276, 1963; Fischer-Hjalmars, J.: *J. Chem. Phys.*, vol. 42, no. 6, pp. 1962–1972, 1965; *Theor. Chim. Acta,* vol. 4, pp. 332, 1966; Fischer-Hjalmars, J.: In "Modern Quantum Chemistry," (translated from English), ed. A. M. Brodskiy and V. V. Tolmachev, vol. 1, Mir Press, Moscow, 1968.
23. Shustorovich, Ye. M.: "Novoye v uchenii o valentnosti (Recent Progress in Valence Studies)," Znaniye Press, Moscow, 1968.
24. Heine, V.: "Group Theory in Quantum Mechanics," (translated from English), ed. V. Ya Feynberg, Foreign Literature Press, Moscow, 1963.
25. Hochstrasser, R.: "Molecular Aspects of Symmetry," (translated from English), ed. M. Ye. Dyatkina, Mir Press, Moscow, 1968.
26. Dewar, M.: "Hyperconjugation," (translated from English), ed. M. Ye. Dyatkina, Mir Press, Moscow, 1965.
27. Rudenberg, K.: In "Modern Quantum Chemistry," (translated from English), ed. A. M. Brodskiy and V. V. Tolmachev, vol. 1, Mir Press, Moscow, 1968.
28. Coulson, C. A.: *Trans. Faraday Soc.*, vol. 38, part 9, pp. 433–444, 1942.
29. Lennard-Jones, J. E.: *Proc. R. Soc. A*, vol. 198, pp. 1–13, 14–26, 1949; Lennard-Jones, J. E. and J. A. Pople: *Proc. R. Soc. A*, vol. 202, pp. 166–181, 1950; vol. 210, pp. 190–206, 1951; Hall, G. G.: *Proc. R. Soc. A*, vol. 202, pp. 336–344, 1950; Hall, G. G. and J. E. Lennard-Jones: *Proc. R. Soc.*, vol. 202, pp. 155–165, 1951; vol. 205, pp. 357–374.
30. Gel'fand, I. M.: "Lektsii po lineynoy algebre (Lectures on Linear Algebra)," 4th ed., Nauka Press, Moscow, 1971.
31. Pauling, L.: "Nature of the Chemical Bond," (translated from English), ed. Ya. K. Syrkin, Goskhimizdat Press, Moscow–Leningrad, 1948.
32. Shmid, L. A.: *Phys. Rev.*, vol. 92, no. 6, pp. 1373–1379, 1953.
33. O-ohata, K.: *J. Phys. Soc. Jap.*, vol. 15, pp. 1258–1263, 1960; no. 8, pp. 1449–1455.
34. Goto, F.: *J. Phys. Soc. Jap.*, vol. 21, no. 5, pp. 895–906, 1966.
35. "Poluprovodnikovyye veshchestva. Voprosy khimicheskoy svyazi (Semiconductor Substances. Chemical Bond Problems)" (collection of papers, translated from English and German), ed. V. P. Zhuze, Foreign Literature Press, Moscow, 1960.
36. Syushe, J. P.: "Physical Chemistry of Semiconductors," (translated from English), 2d ed., ed. N. A. Goryunova, Metallurgiya Press, Moscow, 1969.
37. Ziman, J.: "Principles of Solid State Theory," (translated from English), ed. V. L. Bonch-Bruyevich, Mir Press, Moscow, 1966.

38. Kittel, C.: "Quantum Theory of Solids," (translated from English), Nauka Press, Moscow, 1967.
39. Tsidil'kovskiy, I. M.: "Elektrony i dyrki v poluprovodnikakh (Electrons and Holes in Semiconductors)," Nauka Press, Moscow, 1972.
40. Collwine, J.: "Energy Band Structure Theory," (translated from English), ed. S. V. Vonskovskiy, Mir Press, Moscow, 1969.
41. Jones, G.: "Brillouin Zone Theory and Electron States in Crystals," (translated from English), ed. V. L. Bonch-Bruyevich, Mir Press, Moscow, 1968.
42. Slater, J.: "Quantum Theory of Molecules and Solids," vol. 2, McGraw-Hill, New York, 1965.
43. Levin, A. A.: "Kvantovaya khimiya kovalentnykh kristallov (Quantum Chemistry of Covalent Crystals)," Znaniye Press, Moscow, 1970.
44. Knox, R. and A. Gold: "Symmetry in Solids," (translated from English), ed. V. L. Bonch-Bruyevich, Nauka Press, Moscow, 1970.
45. Streitwolf, G.: "Group Theory in Solid State Physics," (translated from German), ed. S. V. Vonkovskiy, Mir Press, Moscow, 1971.
46. Sokolov, A. V. and V. P. Shirokovskiy: *Usp. Fiz. Nauk,* vol. 71, no. 3, pp. 485–514, 1960.
47. Kittel, C.: "Introduction to Solid State Physics," (translated from English), 2d ed., Fizmatgiz Press, Moscow, 1962.
48. Shubnikov, A. V., Ye. Ye. Flint and G. B. Bokiy: "Osnovy kristallografii (Fundamentals of Crystallography)," ed. A. V. Shubnikov, Press of USSR Academy of Sciences, Moscow–Leningrad, 1940; Bokiy, G. B.: "Kristallokhimiya (Crystal Chemistry)," 3rd ed., Nauka Press, Moscow, 1971.
49. Hall, G. G.: *Proc. Phys. Soc.,* vol. 69, pp. 1124–1132, 1956.
50. Herman, F., I. B. Ortenburger and J. P. Van Dyke: *Int. L. Quant. Chem.,* Symp. no. 3, pp. 827–846, 1970.
51. Ziman, J. M.: *Solid State Phys.,* vol. 26, pp. 1–103, 1971.
52. Painter, G. S., D. E. Ellis and A. R. Lubinsky: *Phys. Rev. B,* vol. 4, no. 10, pp. 3610–3622, 1971.
53. Messmer, R. P.: *Chem. Phys. Lett.,* vol. 11, no. 5, pp. 589–592, 1971.
54. Pugh, D.: *Mol. Phys.,* vol. 20, no. 5, pp. 835–848, 1971.
55. Chaney, R. C., C. C. Lin and E. E. Lafon: *Phys. Rev. B,* vol. 3, no. 2, pp. 459–472, 1971.
56. Painter, G. S. and D. E. Ellis: *Phys. Rev. B,* vol. 1, no. 12, pp. 4747–4752, 1971.
57. Chaney, R. C., E. E. Lafon and C. C. Lin: *Phys. Rev. B,* vol. 4, no. 8 pp. 2734–2741, 1971; Hayns, M. R.: *J. Phys. C,* vol. 5, no. 1, pp. 15–20, 1972; Drost, D. M. and J. L. Fry: *Phys. Rev. B,* vol. 5, no. 2, pp. 684–697, 1972; Honig, J. M., W. E. Wahnsiedler and J. O. Dimmok: *J. Solid State Chem.,* vol. 5, no. 2, pp. 452–461, 1972; Soules, T. F. et al.: *Phys. Rev. B,* vol. 6, no. 4, pp. 1519–1532, 1972.
58. Slater, J. C.: *Phys. Rev.,* vol. 51, no. 10, pp. 846–851, 1937.
59. Saffren, M. M. and J. C. Slater: *Phys. Rev.,* vol. 92, no. 5, pp. 1126–1128, 1953.
60. Slater, J. C.: *Adv. Quant. Chem.,* vol. 1, pp. 35–58, 1964.

61. Dimmok, J. O.: *Solid State Phys.*, vol. 26, pp. 103–274, 1971.
62. Matteheiss, L. F., J. H. Wood and A. C. Switendick: *Methods Comput. Phys.*, vol. 8, pp. 64–146, 1968.
63. Bhatnagar, S.: *Phys. Rev.*, vol. 183, no. 3, pp. 657–663, 1969.
64. Scope, P. M.: *Phys. Rev.*, vol. 139, no. 3A, pp. 934–940, 1965.
65. Switendick, A. C.: *J. Appl. Phys.*, vol. 37, no. 3, pp. 1022–1023, 1966; Snow, E. C. and J. T. Waber: *Phys. Rev.*, vol. 157, no. 3, pp. 570–578, 1967; Snow, E. C.: *Phys. Rev.*, vol. 158, no. 3, pp. 158, 683–688, 1967; De Cicco, P. D. and A. Kitz: *Phys. Rev.*, vol. 163, no. 2, pp. 486–491, 1967; Snow, E. C.: *Phys. Rev.*, vol. 171, no. 3, pp. 785–789, 1968; 172, pp. 708–711, 1968.
66. Ruge, W. E.: *Phys. Rev.*, vol. 181, no. 3, pp. 1024–1032, 1969.
67. Keown, R.: *Phys. Rev.*, vol. 150, no. 2, pp. 568–573, 1966.
68. Rössler, U. and M. Lietz: *Phys. Stat. Sol.*, vol. 17, no. 2, pp. 597–604, 1966.
69. Maschke, K. and U. Rössler: *Phys. Stat. Sol.*, vol. 28, no. 2, pp. 577–581, 1968.
70. Korringa, J.: *Physica*, vol. 13, no. 6–7, pp. 393–400, 1947.
71. Kohn, W. and N. Rostoker: *Phys. Rev.*, vol. 94, no. 5, pp. 1111–1120, 1954.
72. Segall, B. and F. S. Ham: *Methods Comput. Phys.*, vol. 8, pp. 251–294, 1968.
73. Segall, B.: *Phys. Rev.*, vol. 105, no. 2, pp. 108–115.
74. Eckelt, P., O. Madelung and J. Treusch: *Phys. Rev. Lett.*, vol. 18, no. 16, pp. 656–658, 1967.
75. Eckelt, P.: *Phys. Stat. Sol.*, vol. 23, no. 1, pp. 307–312, 1967.
76. Treusch, J., P. Eckelt and O. Madelung: "II–VI Semiconducting Compounds, Proceedings of an International Conference," ed. D. G. Thomas, Benjamin, New York–Amsterdam, 1967.
77. Madelung, O. and J. Treusch: In "Trudy IX Mezhdunarodnoy Konferentsii po fizike poluprovodnikov (Proceedings of IX International Conference on Semiconductor Physics)," vol. 1, Nauka Press, Moscow, 1969.
78. Rössler, U.: *Phys. Rev.*, vol. 184, no. 3, pp. 733–738, 1969.
79. Overhof, H. and U. Rössler: *Phys. Stat. Sol.*, vol. 37, pp. 691–698, 1970.
80. Onodera, Y. and M. Okazaki: *J. Phys. Soc. Jap.*, vol. 21, no. 4, p. 816, 1966.
81. Liberman, D., J. T. Waber and D. I. Gromer: *Phys. Rev.*, vol. 137, no. 1A, pp. 27–34, 1965.
82. Herman, F.: In "Proceedings of an International Conference on Physics of Semiconductors, Paris," Dunod, Paris, 1964.
83. Herman, F. and R. L. Kortum: In "Quantum Theory of Atoms, Molecules and the Solid State," Academic Press, New York–London, 1966.
84. Herman, F. and R. L. Kortum: *J. Phys. Soc. Jap. Suppl.*, vol. 21, pp. 7–14, 1966.
85. Herman, F., R. L. Kortum and C. D. Kuglin: *Int. J. Quant. Chem., 1st Symp.*, pp. 533–566, 1967.
86. Euwema, R. N., T. C. Collins et al.: *Phys. Rev.*, vol. 162, no. 3, pp. 710–715, 1967.

87. Collins, T. C., R. N. Euwema and J. S. De Witt: In "II–VI Semiconducting Compounds, Proceedings of an International Conference," ed. D. G. Thomas, Benjamin, New York–Amsterdam, 1967.
88. Stuckel, D. J. and R. N. Euwema: *Phys. Rev.,* vol. 179, no. 3, pp. 740–751, 1969.
89. Stuckel, D. J. and R. N. Euwema: *Phys. Rev.,* vol. 186, no. 3, pp. 754–757, 1969.
90. Stuckel, D. J., R. N. Euwema et al.: *Phys. Rev. B,* vol. 1, no. 2, pp. 779–790, 1970.
91. Collins, T. C., D. J. Stuckel and R. N. Euwema: *Phys. Rev. B,* vol. 1, no. 2, pp. 724–730, 1970.
92. Stuckel, D. J. and R. N. Euwema: *Phys. Rev. B,* vol. 1, no. 4, pp. 1635–1643, 1970.
93. Stuckel, D. J.: *Phys. Rev. B,* vol. 1, no. 8, pp. 3458–3463, 1970.
94. Stuckel, D. J., R. N. Euwema and T. C. Collins: *Int. J. Quant. Chem., III Symp.,* pp. 789–796, 1970.
95. Stuckel, D. J.: *Phys. Rev. B,* vol. 1, no. 12, pp. 4791–4797, 1970.
96. Collins, T. S. et al.: *Int. J. Quant. Chem., IV Symp.,* no. 4, pp. 77–85, 1971.
97. Collins, T. S. et al.: *Int. J. Quant. Chem., V Symp.,* no. 5, pp. 451–458, 1971.
98. Rezer, B. I. and V. P. Shirokovskiy: *Fiz. Met. Metalloved.,* vol. 32, no. 5, pp. 934–946, 1971.
99. Kunz, B.: *Phys. Rev.,* vol. 180, no. 3, pp. 934–941, 1969; *Phys. Rev. B,* vol. 3, no. 2, pp. 491–497, 1971.
100. Herman, F. and S. Skillman: In "Proceedings of an International Conference on Semiconductor Physics, Prague, 1960," Publishing House of the Czechoslovak Academy of Sciences, Prague, 1961.
101. Kohn, W. and L. J. Sham: *Phys. Rev.,* vol. 140, no. 4A, pp. 1133–1138, 1965; Liberman, D. A.: *Phys. Rev.,* vol. 171, no. 1, pp. 1–3, 1968.
102. Page, L. J. and E. H. Hygh: *Phys. Rev. B,* vol. 1, no. 8, pp. 3472–3479, 1970.
103. Herman, F. et al.: *Methods Comput. Phys.,* vol. 8, pp. 193–250, 1968.
104. Heine, W., M. Cohen and D. Ware: "Pseudopotential Theory," (translated from English), ed. V. L. Bonch-Bruyevich, Mir Press, Moscow, 1973.
105. Brust, D.: *Methods Comput. Phys.,* vol. 8, pp. 33–63, 1968.
106. Saslow, W., T. K. Bergstresser and M. L. Cohen: *Phys. Rev. Lett.,* vol. 16, no. 9, pp. 350–356, 1966; Saslow, W.: *Phys. Rev. Lett.,* vol. 21, p. 715, 1968.
107. Cohen, M. L. and T. K. Bergstresser: *Phys. Rev.,* vol. 141, no. 2, pp. 789–796, 1966.
108. Van Haeringen, W. and H. G. Junginger: *Solid State Commun.,* vol. 7, no. 16, pp. 1135–1137, 1969.
109. Junginger, H. G. and W. Van Haeringen: *Phys. Stat. Sol.,* vol. 37, no. 2, pp. 709–719, 1970.
110. Dresselhaus, G. and M. S. Dresselhaus: *Phys. Rev.,* vol. 160, no. 3, pp. 649–679, 1967; "Trudy IV Mezhdunarodnoy konferentsii po fizike poluprovodnikov (Transactions of IX International Conference of Semi-

conductor Physics)," vol. 1, Nauka Press, Moscow, 1968; Nauka Press, Leningrad, 1969.

111. Poplavnoy, A. S. et al.: *Izv. Vuzov, Ser. Fiz.*, no. 6, p. 95, 1970; Poplavnoy, A. S. and Yu. I. Polygalov: *Neorg. Mater.*, vol. 7, no. 10, pp. 1706–1710, 1711–1714, 1971; Goryunova, N. A. and A. S. Poplavnoi: *Phys. Stat. Sol.*, vol. 39, no. 1, pp. 9–17, 1970.
112. Wannier, G. H.: *Phys. Rev.*, vol. 52, no. 3, pp. 191–197, 1937.
113. Hall, G. G.: *Proc. R. Soc. A*, vol. 205, pp. 541–552, 1951.
114. Hall, G. G.: *Philos. Mag.*, vol. 43, no. 338, pp. 338–343, 1952.
115. Hall. G. G.: *Philos. Mag.*, vol. 3, no. 29, pp. 429–439, 1958.
116. Koster, G. F.: *Phys. Rev.*, vol. 89, no. 1, pp. 67–77, 1953.
117. Coulson, C. A., L. B. Redei and D. Stocker: *Proc. R. Soc. A*, vol. 270, pp. 357–372, 1962.
118. Levin, A. A.: *Z. Mod. Khim.*, vol. 11, no. 3, pp. 520–526, 1970.
119. Buberman, G. S.: *Usp. Fiz. Nauk*, vol. 103, pp. 675–704, 1971.
120. Wyckoff, R. W. C.: "Crystal Structures," 2d ed., vol. 1, New York, 1963.
121. Benson, S.: *J. Chem. Educ.*, vol. 42, no. 9, pp. 502–518, 1965.
122. Wells, A. F.: "Structural Inorganic Chemistry," 3rd ed., Oxford, 1962.
123. Knox, B. E. and H. B. Palmer: *Chem. Rev.*, vol. 61, no. 3, pp. 247–255, 1961.
124. Roberts, J. and M. Cassario: "Fundamentals of Organic Chemistry," (translated from English), ed. A. N. Nesmeyanov, vol. 1, Mir Press, Moscow, 1968.
125. Gunn, S. R. et al.: *J. Phys. Chem.*, vol. 65, no. 5, pp. 779–783, 1961.
126. Cotton, F. and J. Wilkinson: "Modern Inorganic Chemistry," (translated from English), ed. K. V. Astakhov, vol. 2, Mir Press, Moscow, 1969.
127. Levin, A. A.: *Dokl. Akad. Nauk SSSR*, vol. 181, no. 5, pp. 1168–1196, 1968.
128. Brailsford, D. F. and B. Ford: *Molec. Phys.*, vol. 18, pp. 621–630, 1970.
129. Schmidt, W. and B. T. Wilkins: *Angew. Chem.*, vol. 84, pp. 168–170, 1972.
130. Levin, A. A., Ya. K. Syrkin and M. Ye. Dyatkina: *Fiz. Tekh. Poluprov.*, vol. 1, no. 5, pp. 687–695, 1967.
131. Levin, A. A., N. M. Enden and A. A. Dyatkina: *Z. Mod. Khim.*, vol. 10, no. 6, pp. 1091–1097, 1969.
132. Wolfsberg, W. and L. Helmholz: *J. Chem. Phys.*, vol. 20, no. 5, pp. 837–843, 1952.
133. Bash, H., A. Viste and H. B. Gray: *J. Chem. Phys.*, vol. 44, no. 1, pp. 10–19, 1966.
134. Newton, M. D. et al.: *Proc. Natl. Acad. Sci. U.S.A.*, vol. 53, pp. 1089–1091, 1965.
135. Streitweiser, E.: "Theory of Molecular Orbits for Organic Chemists," (translated from English), ed. M. Ye. Dyatkina, Mir Press, Moscow, 1965.
136. Pople, J. A. and D. P. Santry: *Molec. Phys.*, vol. 7, pp. 269–286, 1963–1964.
137. Mestechkin, M. M.: "Coll.: Voprosy kvantovoy khimii (Problems of Quantum Chemistry)," Leningrad State University Press, Leningrad, 1963.

138. Sokolov, N. D.: *Usp. Khim.,* vol. 36, no. 12, pp. 2195–2208, 1967.
139. Gray, H.: "Electrons and the Chemical Bond," (translated from English), ed. M. Ye. Dyatkina, Mir Press, Moscow, 1967.
140. Musgrave, M. J. P. and J. A. Pople: *Proc. R. Soc. A,* vol. 268, pp. 474–484, 1962; McMurry, H. L. et al.: *J. Phys. Chem. Solids,* vol. 28, no. 12, pp. 2359–2368, 1967.
141. Shorygin, P. P. et al.: *Teor. Eksp. Khim.,* vol. 2, pp. 190–195, 1966.
142. Stocker, D.: *Proc. R. Soc. A,* vol. 270, pp. 397–410, 1962.
143. Doggett, G.: *J. Phys. Chem. Solids,* vol. 27, no. 1, pp. 99–110, 1966.
144. Levin, A. A. et al.: *Z. Mod. Khim.,* vol. 7, no. 6, pp. 907–908, 1966.
145. Roberts, R. A. and W. C. Walker: *Phys. Rev.,* vol. 161, no. 3, pp. 730–735, 1967.
146. Phillips, J.: "Optical Spectra of Solids," (translated from English), ed. V. P. Zhuze, Mir Press, Moscow, 1968.
147. Tauts, Ya.: "Optical Properties of Semiconductors," (translated from English), ed. V. P. Zhuze, Mir Press, Moscow, 1967.
148. Clark, C. D. et al.: *Proc. R. Soc. A,* vol. 277, pp. 312–329, 1964.
149. Macfarlane, G. G.: *Phys. Rev.,* vol. 111, no. 5, pp. 1245–1254, 1958.
150. Grobman, W. D. and D. E. Eastman: *Phys. Rev. Lett.,* vol. 29, no. 22, pp. 1508–1512, 1972.
151. Ley, L.: *Phys. Rev. Lett.,* vol. 29, no. 16, pp. 1088–1092, 1972.
152. Wiech, G.: *Z. Phys.,* vol. 207, pp. 428–445, 1967.
153. Tomboulian, D. H. and D. E. Bedo: *Phys. Rev.,* vol. 104, no. 3, pp. 590–597, 1956.
154. Holliday, J. E.: In "Röntgenspektren und Chemische Bindung (X-Ray Spectra and the Chemical Bond)," Leipzig Physical Chemical Institute of Karl Marx University, Leipzig, 1966; Holliday, J.: "The Chemical Bond in Crystals," (translated from German), ed. N. N. Sirota, Nauka i Tekhnika Press, Minsk, 1969.
155. Kleinman, L. and J. C. Phillips: *Phys. Rev.,* vol. 116, no. 4, pp. 880–884, 1959.
156. Chalklin, F. C.: *Proc. R. Soc.,* vol. 194, pp. 42–62, 1948.
157. Thomas, J. M. et al.: *Trans. Faraday Soc.,* vol. 67, pp. 1875–1886, 1971.
158. Gora, T. et al.: *Phys. Rev. B,* vol. 5, no. 6, pp. 2309–2314, 1972.
159. Levin, A. A. et al.: *Z. Mod. Khim.,* vol. 8, no. 3, pp. 561–562, 1967.
160. Sagawa, T.: *J. Phys. Soc. Jap.,* vol. 21, pp. 49–52, 1966.
161. Nefedov, V. I. et al.: *Z. Mod. Khim.,* vol. 12, no. 6, pp. 893–898, 1971.
162. Bassani, F. and M. Yoshimine: *Phys. Rev.,* vol. 130, no. 1, pp. 20–33, 1963.
163. Levin, A. A. and A. A. Dyatkina: *Z. Mod. Khim.,* vol. 9, no. 4, pp. 680–685, 1968.
164. Gubanov, A. I. and A. A. Nran'yan: *Fiz. Tverd. Tela,* vol. 1, no. 7, pp. 1044–1052, 1959.
165. Nran'yan, A. A.: *Fiz. Tverd. Tela,* vol. 2, no. 3, pp. 474–481, 1960; no. 7, pp. 1650–1655, 1960.
166. Cohen, N. Y. et al.: *Proc. Phys. Soc.,* vol. 82, pp. 65–73, 1963.
167. Pugh, D.: *J. Phys. C,* vol. 3, no. 1, pp. 47–49, 1970.

168. Levin, A. A., Ya. K. Syrkin and M. Ye. Dyatkina: *Z. Mod. Khim.*, vol. 8, no. 2, pp. 317–320, 1967.
169. Fritz, G. and J. Grabe: *Z. Anorg. Allgem. Chem.*, vol. 311, pp. 325–330, 1961.
170. Benkeser, R. A., R. F. Grossman, and G. M. Stenton: *J. Am. Chem. Soc.*, vol. 84, no. 24, pp. 4723–4726, 4727–4730, 1962.
171. Urry, G.: Abstr. 133rd Mtg. Am. Chem. Soc., San Francisco, 1968.
172. Aylett, B. J.: In "Advances in Inorganic Chemistry and Radiochemistry," Academic Press, New York, 1968.
173. Syrkin, Ya. K.: *Usp. Khim.*, vol. 31, no. 4, pp. 397–416, 1962.
174. Hinze, J. and H. H. Jaffe: *J. Am. Chem. Soc.*, vol. 84, pp. 540–546, 1962.
175. Saravia, L. R. and D. Brust: *Phys. Rev.*, vol. 171, no. 3, pp. 916–924, 1968.
176. Levin, A. A. and A. A. Dyatkina: *Z. Mod. Khim.*, vol. 9, no. 5, pp. 863–866, 1968.
177. Mulliken, R. S.: *J. Am. Chem. Soc.*, vol. 72, no. 10, pp. 4493–4503, 1950.
178. Coulson, C.: "Valence," (translated from English), ed. N. D. Sokolov, Mir Press, Moscow, 1965.
179. Vilkov, L. V., N. G. Rambidi and V. P. Spiridonov: *Z. Mod. Khim.*, vol. 8, no. 5, pp. 786–812, 1967.
180. Lorquet, A. J.: *Theor. Chim. Acta*, vol. 5, no. 3, pp. 192–207, 1966.
181. Levin, A. A.: *Fiz. Tekh. Poluprov.*, vol. 3, no. 12, pp. 1864–1865, 1969.
182. Saravia, L. R. and D. Brust: *Phys. Rev.*, vol. 170, no. 3, pp. 683–686, 1968.
183. Levin, A. A. and A. A. Dyatkina: *Z. Mod. Khim.*, vol. 10, no. 5, pp. 949–950, 1969.
184. Levin, A. A. and A. A. Dyatkina: *Z. Mod. Khim.*, vol. 10, no. 4, p. 759, 1969.
185. Slater, J. C. and G. F. Koster: *Phys. Rev.*, vol. 94, no. 6, pp. 1498–1524, 1954.
186. Levin, A. A.: *Dokl. Akad. Nauk SSSR*, vol. 183, pp. 148–149, 1968.
187. Levin, A. A. and A. A. Dyatkina: *Z. Mod. Khim.*, vol. 10, no. 3, pp. 563–564, 1969.
188. Levin A. A.: *Dokl. Akad. Nauk. SSSR*, vol. 189, pp. 808–809, 1969.
189. Ohno, K., Y. Mizuno, and M. Mizushima: *J. Chem. Phys.*, vol. 28, no. 4, pp. 691–699, 1958.
190. Jackel, G. S. and W. Gordy: *Phys. Rev.*, vol. 176, no. 2, pp. 443–452, 1968.
191. Leman, G. and J. Friedel: *J. Appl. Phys. Suppl.*, vol. 33, no. 1, pp. 281–285, 1962.
192. Levin, A. A.: *Z. Mod. Khim.*, vol. 12, no. 3, pp. 550–551, 1971.
193. Brust, D. and L. Liu: *Phys. Rev.*, vol. 154, no. 3, pp. 647–653, 1967.
194. Drickammer, H.: In "High-Pressure Physics," (translated from English), ed. L. F. Vereshchagin, Foreign Literature Press, Moscow, 1963.
195. Neuringer, L. J.: *Phys. Rev.*, vol. 113, no. 6, pp. 1495–1503, 1959.
196. Cardona, M. and W. Paul: *J. Phys. Chem. Solids*, vol. 17, pp. 138–142, 1960.

197. Slikhouse, T. E. and H. G. Drickammer: *J. Phys. Chem. Solids,* vol. 7, no. 213, pp. 210–213, 1958.
198. Paul, W. and G. L. Pearson: *Phys. Rev.,* vol. 98, no. 6, pp. 1755–1757, 1955.
199. Keyes, R. W.: In "Semiconductors and Semimetals," vol. 4, Academic Press, New York–London, 1968.
200. Edwards, A. L. and H. G. Drickammer: *Phys. Rev.,* vol. 122, no. 4, pp. 1149–1157, 1961.
201. Madelung, O.: "The Physics of Semiconductor Compounds of Group III and V Elements," (translated from English), ed. B. I. Boltaks, Mir Press, Moscow, 1967; Hilsum, K. and A. Rose-Eans: "Type $A^{III}B^{IV}$ Semiconductors," (translated from English), ed. N. P. Sazhin and G. V. Zakhvatkin, Foreign Literature Press, Moscow, 1963.
202. Parthe, E.: "Cristallochimie des structures tetraedriques (Crystal Chemistry of Tetrahedral Structures)," 2d ed. Gordon-Breach, Paris–London–New York, 1972.
203. Levin, A. A., N. M. Enden and A. A. Dyatkina: *Z. Mod. Khim.,* vol. 10, no. 6, pp. 1091–1097, 1969.
204. Barinskiy, R. L. and V. I. Nefedov: "Rentgenospektral'noye opredeleniye zaryada atomov v molekulakh (X-Ray Spectral Determination of the Charge of the Atoms in Molecules)," Nauka Press, Moscow, 1966.
205. Levin, A. A., Ya. K. Syrkin and M. Ye. Dyatkina: *Usp. Khim.,* vol. 38, no. 2, pp. 193–221, 1969; Levin, A. A.: *Z. Mendeleyeva,* vol. 17, p. 308, 1972.
206. Moffitt, W.: *Annu. Rep. Progr. Phys.,* vol. 17, pp. 173–200, 1954.
207. Sirota, N. N.: In "Semiconductors and Semimetals," vol. 4, Academic Press, New York–London, 1968.
208. Boron–Nitrogen Chemistry, "Advances in Chemistry," no. 42, American Chemical Society, Washington, 1964.
209. Mulliken, R., C. Ricke et al.: *J. Chem. Phys.,* vol. 17, no. 12, pp. 1248–1267, 1949.
210. Herman, F.: *J. Electron.,* vol. 1, no. 2, pp. 103–114, 1955.
211. Hilsum, C.: In "Semiconductors and Semimetals," vol. 1, Academic Press, New York–London, 1966.
212. Levin, A. A.: *Z. Mod. Khim.,* vol. 11, no. 6, pp. 1101–1108, 1970.
213. Wiech, G.: *Z. Phys.,* vol. 216, pp. 472–487, 1968.
214. Hund, F. and B. Mrowka: *Ber. Verhand. Sächs. Acad. Wiss. Math.-Phys. k1,* vol. 87, p. 325, 1935.
215. Fomichev, V. A. and M. A. Rumsh: *J. Phys. Chem. Solids,* vol. 29, no. 6, pp. 1015–1024, 1968.
216. Spicer, W. E. and R. K. Iden: In "Trudy IX Mezhdunarodnoy konferentsii po fizike poluprovodnikov (Transactions of IX International Conference on Semiconductor Physics)," vol. I, Nauka Press, Moscow, 1969.
217. Stuckel, D. J.: *Phys. Rev.,* vol. 1, no. 12, pp. 4790–4797, 1970.
218. Nemoshkalenko, V. V. and V. G. Aleshin: *Fiz. Tverd. Tela,* vol. 12, no. 1, pp. 59–62, 1970.
219. Aleshin, V. G. and V. V. Nemoshkalenko: "Zonnaya struktura i rentgenovskiye emissionnyye spektry kristallov (The Band Structure and

the X-Ray Emission Spectra of Crystals)," preprint IMF, Press of the USSR Academy of Sciences, Kiev, 1970.
220. Wiff, D. R. and R. Keown: *J. Chem. Phys.*, vol. 47, no. 9, pp. 3113–3119, 1967.
221. Levin, A. A.: *Phys. Stat. Sol.*, vol. 36, no. 2, pp. 511–513, 1969.
222. Minden, H. T.: *Appl. Phys. Lett.*, vol. 17, pp. 358–360, 1970.
223. Hemstreet, L. A. and C. Y. Fong: *Phys. Rev. B*, vol. 6, no. 4, pp. 1464–1480, 1972.
224. Cardona, M., K. L. Shakllee and F. H. Pollak: *Phys. Rev.*, vol. 154, no. 4, pp. 696–720, 1967.
225. Wang, C. C., M. Cardona and A. G. Fisher: *RCA Rev.*, vol. 25, p. 159, 1964.
226. Archer, R. J., R. Y. Koyama et al.: *Phys. Rev. Lett.*, vol. 12, no. 9, pp. 538–540, 1964.
227. Monemar, B.: *Solid State Commun.*, vol. 8, no. 16, pp. 1295–1298, 1970.
228. Vorob'yev, V. G., Z. S. Medvedeva and V. V. Sobolev: *Neorg. Mater.*, vol.3, p. 1079, 1967.
229. Kleinman, L. and J. C. Phillips: *Phys. Rev.*, vol. 117, no. 2, pp. 460–464, 1960.
230. Aleshin, V. G., V. P. Smirnov and B. V. Gantsevich: *Fiz. Tverd. Tela*, vol. 10, no. 9, pp. 2884–2885, 1968.
231. Aleshin, V. G. and V. P. Smirnov: *Fiz. Tverd. Tela*, vol. 11, no. 7, pp. 1920–1927, 1969.
232. Poplavnoy, A. S.: *Fiz. Tverd. Tela*, vol. 8, no. 7, pp. 2238–2240, 1966.
233. Levin, A. A. and A. A. Dyatkina: *Z. Mod. Khim.*, vol. 11, no. 4, pp. 740–745, 1970.
234. Rumsh, M. A. et al.: *Vestn. LGU, Ser. Fiz. Khim.*, no. 16, issue 3, pp. 49–64, 1968.
235. Philipp, H. R. and E. A. Taft: *Phys. Rev.*, vol. 127, no. 1, pp. 159–161, 1962.
236. Phillips, J. C.: *J. Chem. Phys.*, vol. 48, no. 12, pp. 5740–5741, 1968.
237. Baer, A. D.: *Bull. Am. Phys. Soc.*, vol. 13, p. 478, 1968.
238. Fomichev, B. A., I. I. Zhukova and I. K. Polushina: *J. Phys. Chem. Solids*, vol. 29, no. 6, pp. 1025–1032, 1968.
239. Archer, R. J. et al.: *Phys. Rev. Lett.*, vol. 12, no. 19, pp. 538–540, 1964.
240. Stearns, R.: *J. Appl. Phys.*, vol. 36, no. 1, pp. 330–331, 1965.
241. Ku, S. M.: *J. Electrochem. Soc.*, vol. 113, no. 8, p. 813, 1966.
242. Lorenz, M. R. and A. Onton: In "Proceedings, 10th International Conference on Physics and Semiconductors," Cambridge, Mass., 1970, p. 444.
243. Lorenz, M. R., R. Chicotka and G. Pettit: *Solid State Commun.*, vol. 8, no. 9, pp. 693–697, 1970.
244. Merz, J. L. and R. T. Lynch: *J. Appl. Phys.*, vol. 39, no. 5, pp. 1988–1993, 1968.
245. Onton, A.: In "Proceedings, 10th International Conference on Physics and Semiconductors," Cambridge, Mass., 1970, p. 107.
246. Yim, W. M.: *J. Appl. Phys.*, vol. 42, p. 2854, 1971.
247. Monemar, B.: *Solid State Commun.*, vol. 8, no. 24, p. 2171, 1970.

248. Herman, F. et al.: *Phys. Rev. Lett.*, vol. 11, no. 12, pp. 541–545.
249. Stuckel, D. J. and R. N. Euwema: *Phys. Rev.*, vol. 188, no. 3, pp. 1193–1196, 1969.
250. Levin, A. A.: *Z. Mod. Khim.*, vol. 11, no. 5, pp. 953–954, 1970.
251. Cardona, M.: *Phys. Rev.*, vol. 129, no. 1, pp. 69–78, 1963.
252. Levin, A. A.: *Fiz. Tekh. Poluprov.*, vol. 5, no. 8, pp. 1481–1487, 1971.
253. Sirota, N. N.: In "Fizika i fiziko-khimicheskiy analiz (Physics and Physico-Chemical Analysis)," no. 1, Press of the Moscow Institute of Non-Ferrous Metals and Gold, Moscow, 1957; Sirota, N. N.: In "Khimicheskaya svyaz' v poluprovodnikakh i termodinamika (The Chemical Bond in Semiconductors and Thermodynamics)," Nauka i Tekhnika, Minsk, 1966.
254. Ormont, B. F.: *Z. Neorg. Khim.*, vol. 5, no. 2, pp. 255–263, 1960.
255. Manca, P.: *J. Phys. Chem. Solids,* vol. 20, no. 314, pp. 268–273, 1961.
256. Mooser, E. and W. B. Pearson: *J. Electron.*, vol. 1, no. 6, pp. 629–645, 1955.
257. Welker, H.: *Z. Naturforsch.*, vol. 7a, no. 11, pp. 744–749, 1952.
258. Goodman, C. H. L.: *Phys. Chem. Solids,* vol. 6, no. 4, pp. 305–314, 1958.
259. Goodman, C. H. L.: *J. Electron.*, vol. 1, no. 1, pp. 115–121, 1955.
260. Gordy, W. and D. Thomas: *J. Chem. Phys.*, vol. 24, no. 2, pp. 439–444, 1956.
261. Haissinsky, M.: *J. Phys. Radium,* vol. 7, no. 1, pp. 7–11, 1946.
262. Allred, A. L.: *J. Inorg. Nucl. Chem.*, vol. 17, no. 314, pp. 215–221, 1961.
263. Heywang, W. and B. Seraphin: *Z. Naturforsch.*, vol. 11a, no. 6, pp. 425–429, 1956.
264. Seraphin, B.: *Z. Naturforsch.*, vol. 9a, no. 5, pp. 450–456, 1956.
265. Adavi, I.: *Phys. Rev.*, vol. 105, no. 3, pp. 789–792, 1956.
266. Hooge, F. N.: *Z. Phys. Chem.*, vol. 24, no. 314, pp. 275–282, 1960.
267. Gubanov, A. I.: *Z. Tekh. Fiz.*, vol. 26, pp. 2170–2178, 1956.
268. Raychaudhuri, A.: *Proc. Natl. Inst. Sci. India,* vol. 25A, pp. 201–205, 1959.
269. Domashevskaya, E. P. and Ya. A. Uray: In "Röntgenspektren und Chemisches Bindung (X-Ray Spectra and the Chemical Bond)," Physical Chemical Institute of Karl Marx University, Leipzig, 1966.
270. Uray, Ya. V. and E. P. Domashevskaya: In "Khimicheskaya svyaz' v poluprovodnikakh i termodinamika (The Chemical Bond in Semiconductors and Thermodynamics)," Nauka i Tekhnika Press, Minsk, 1969; Domashevskaya, E. P., Ya. A. Uray and O. Ya Gukov: In "Khimicheskaya svyaz' v kristallakh (The Chemical Bond in Crystals)," Nauka i Tekhnika Press, Minsk, 1969.
271. Leongardt, G. et al.: In "Khimicheskaya svyaz' v polumetallakh (The Chemical Bond in Semimetals)," Nauka i Tekhnika Press, Minsk, 1972.
272. Gusatinskiy, A. N. and S. A Nemnonov: In "Khimicheskaya svyaz' v poluprovodnikakh i termodinamika (The Chemical Bond in Semiconductors and Thermodynamics)," Nauka i Tekhnika Press, Minsk, 1966; Sharay, V. G. and O. I. Pashkovskiy: In "Khimicheskaya svyaz' v kristallakh poluprovodnikov i polumetallov (The Chemical Bond in Semiconductor and Semimetal Crystals)," Nauka i Tekhnika Press, Minsk, 1973.

273. Attard, A. E. and L. V. Asaroff: *J. Appl. Phys.*, vol. 34, no. 4, pp. 774–776, 1963; Attard, A. E.: *Br. J. Appl. Phys.*, vol. 1, no. 3, pp. 390–391, 1968; Attard, A. E. et al.: *J. Phys. C*, vol. 2, no. 5, pp. 816–823, 1969.
274. Zerbst, M. and H. Boroffka: *Z. Naturforsch.*, vol. 18a, no. 5, pp. 642–645, 1963; Hambleten, K. G.: *Phys. Lett.*, vol. 16, pp. 241–242, 1965.
275. Sirota, N. N.: In "Khimicheskaya svyaz' v poluprovodnikakh i tverdykh telakh (The Chemical Bond in Semiconductors and Solids)," Nauka i Tekhnika Press, Minsk, 1965; Sirota, N. N., A. U. Sheleg and Ye. M. Gololobov: In "Khimicheskaya svyaz' v kristallakh (The Chemical Bond in Crystals)," Nauka i Tekhnika Press, Minsk, 1969; Sirota, N. N., Ye. M. Gololobov and A. U. Sheleg: In "Khimicheskaya svyaz' v poluprovodnikakh (The Chemical Bond in Semiconductors)," Nauk i Tekhnika Press, Minsk, 1969.
276. Uno, R. and A. Ishigaki: In "Khimicheskaya svyaz' v poluprovodnikakh (The Chemical Bond in Semiconductors)," Nauki i Tekhnika Press, Minsk, 1969.
277. Kröger, F. A.: *Phys. Lett.*, vol. 15, no. 3, p. 218, 1965.
278. Arlt, G. and P. Quadflieg: *Phys. Stat. Sol.*, vol. 25, no. 1, pp. 323–330, 1968.
279. Szigeti, B.: *Trans. Faraday Soc.*, vol. 45, pp. 155–166, 1949; Szigeti, B.: *Proc. R. Soc. A*, vol. 204, pp. 51–62, 1951.
280. Chelikowsky, J., D. J. Chadi and M. L. Cohen: *Phys. Rev. B*, vol. 8, no. 6, pp. 2786–2794, 1973.
281. Shulze, K. R., I. Topol and E. Hess: *Phys. Stat. Sol. (b)*, vol. 55, no. 1, pp. K75–K77, 1973.
282. Hess, E. et al.: *Phys. Stat. Sol. (b)*, vol. 55, no. 1, pp. 187–192, 1973.
283. Dyachkov, P. N. and A. A. Levin: *Theor. Chim. Acta*, in press.
284. Levin, A. A. and A. A. Dyatkina: *Z. Mod. Khim.*, vol. 15, in press.
285. Lyapin, V. G. and K. B. Tolpygo: In "Khimicheskaya svyaz' v kristallakh (The Chemical Bond in Crystals)," Nauka i Tekhnika Press, Minsk, 1969; Lyapin, V. G. et al.: In "Khimicheskaya svyaz' v poluprovodnikakh i polumetallov (The Chemical Bond in Semiconductors and Semimetals)," Nauka i Tekhnika Press, Minsk, 1972.
286. Phillips, J. C.: "Covalent Bonding in Crystals, Molecules and Polymers," Univ. of Chicago Press, Chicago–London, 1969; Phillips, J. C.: *Rev. Mod. Phys.*, vol. 42, no. 3, pp. 317–356, 1970.

Index